Edward Albert Bowser

An Elementary Treatise on the Differential and Integral Calculus

Edward Albert Bowser

An Elementary Treatise on the Differential and Integral Calculus

ISBN/EAN: 9783337811839

Printed in Europe, USA, Canada, Australia, Japan

Cover: Foto ©berggeist007 / pixelio.de

More available books at **www.hansebooks.com**

BOWSER'S MATHEMATICS.

ACADEMIC ALGEBRA. With numerous Examples.

COLLEGE ALGEBRA. With numerous Examples.

PLANE AND SOLID GEOMETRY. With numerous Exercises.

AN ELEMENTARY TREATISE ON ANALYTIC GEOMETRY, embracing Plane Geometry, and an Introduction to Geometry of Three Dimensions.

AN ELEMENTARY TREATISE ON THE DIFFERENTIAL AND INTEGRAL CALCULUS. With numerous Examples.

AN ELEMENTARY TREATISE ON ANALYTIC MECHANICS. With numerous Examples.

AN ELEMENTARY TREATISE ON HYDROMECHANICS. With numerous Examples.

AN

ELEMENTARY TREATISE

ON THE

DIFFERENTIAL AND INTEGRAL

CALCULUS.

WITH NUMEROUS EXAMPLES.

BY

EDWARD A. BOWSER, LL.D.,

PROFESSOR OF MATHEMATICS AND ENGINEERING IN RUTGERS COLLEGE

THIRTEENTH EDITION.

New York:

D. VAN NOSTRAND COMPANY,

23 MURRAY STREET & 27 WARREN STREET.

1892.

PREFACE.

THE present work on the Differential and Integral Calculus is designed as a text-book for colleges and scientific schools. The aim has been to exhibit the subject in as concise and simple a manner as was consistent with rigor of demonstration, to make it as attractive to the beginner as the nature of the Calculus would permit, and to arrange the successive portions of the subject in the order best suited for the student.

I have adopted the method of *infinitesimals*, having learned from experience that the fundamental principles of the subject are made more intelligible to beginners by the method of infinitesimals than by that of *limits*, while in the practical applications of the Calculus the investigations are carried on entirely by the method of infinitesimals. At the same time, a thorough knowledge of the subject requires that the student should become acquainted with both methods; and for this reason, Chapter III is devoted exclusively to the method of limits. In this chapter, all the fundamental rules for differentiating algebraic and transcendental functions are obtained by the method of limits, so that the student may compare the two methods. This chapter may be omitted without interfering with the continuity of the work, but the omission of at *least* the first part of the chapter is not recommended.

To familiarize the student with the principles of the subject, and to *fix* the principles in his mind, a large number of examples is given at the ends of the chapters. These examples have been carefully selected with the view of illustrating the most important points of the subject. The greater part of them will present no serious difficulty to the student, while a few may require some analytical skill.

In preparing this book, I have availed myself pretty freely of the writings of the best American and English and French authors. Many volumes have been consulted whose titles are not mentioned, as credit could not be given in every case, and probably I am indebted to these volumes for more than I am aware of. The chief sources upon which I have drawn are indicated by the references in the body of the work, and need not be here repeated. For examples, I have drawn upon the treatises of Gregory. Price, Todhunter, Williamson, Young, Hall, Rice and Johnson, Ray, and Olney, while quite a number has been taken from the works of De Morgan, Lacroix, Serret, Courtenay, Loomis, Church, Byerly, Docharty, Strong, Smyth, and the Mathematical Visitor; and I would hereby acknowledge my indebtedness to all the above-named works, both American and foreign, for many valuable hints, as well as for examples. A few examples have been prepared specially for this work.

I have again to express my thanks to Mr. R. W. Prentiss, Fellow in Mathematics at the Johns Hopkins University, for reading the MS and for valuable suggestions.

E. A. B.

RUTGERS COLLEGE,
NEW BRUNSWICK, N. J., *June,* 1880.

TABLE OF CONTENTS.

PART I.

DIFFERENTIAL CALCULUS.

CHAPTER I.

FIRST PRINCIPLES.

CHAPTER II.

DIFFERENTIATION OF ALGEBRAIC AND TRANSCENDENTAL FUNCTIONS.

TRIGONOMETRIC FUNCTIONS.

CIRCULAR FUNCTIONS.

CHAPTER III.

LIMITS — DERIVED FUNCTIONS.

CHAPTER IV.

SUCCESSIVE DIFFERENTIALS AND DERIVATIVES.

CHAPTER V.

DEVELOPMENT OF FUNCTIONS.

CHAPTER VI.

EVALUATION OF INDETERMINATE FORMS.

CHAPTER VII.

FUNCTIONS OF TWO OR MORE VARIABLES. — CHANGE OF THE INDEPENDENT VARIABLE.

CHAPTER VIII.

MAXIMA AND MINIMA OF FUNCTIONS OF A SINGLE VARIABLE.

CHAPTER IX.

TANGENTS, NORMALS, AND ASYMPTOTES.

CHAPTER X.

DIRECTION OF CURVATURE—SINGULAR POINTS—TRACING OF CURVES.

CHAPTER XI.

RADIUS OF CURVATURE, EVOLUTES AND INVOLUTES— ENVELOPES.

———◆◆◆———

PART II.

INTEGRAL CALCULUS.

CHAPTER I.

ELEMENTARY FORMS OF INTEGRATION.

CHAPTER II.

INTEGRATION OF RATIONAL FRACTIONS.

CHAPTER III.

INTEGRATION OF IRRATIONAL FUNCTIONS BY RATIONALIZATION.

CHAPTER IV.

INTEGRATION BY SUCCESSIVE REDUCTIONS.

—————

CHAPTER V.

INTEGRATION BY SERIES—SUCCESSIVE INTEGRATION—INTEGRATION OF FUNCTIONS OF TWO VARIABLES. DEFINITE INTEGRALS.

CHAPTER VI.

LENGTHS OF CURVES.

CHAPTER VII.

AREAS OF PLANE CURVES.

CHAPTER VIII.

AREAS OF CURVED SURFACES.

CHAPTER IX.

VOLUMES OF SOLIDS.

PART I.

DIFFERENTIAL CALCULUS.

CHAPTER I.

FIRST PRINCIPLES.

1. Constants and Variables.—In the Calculus, as in Analytic Geometry, there are two kinds of quantities used, *constants* and *variables*.

A *constant quantity*, or simply a *constant*, is one whose value does not change in the same discussion, and is represented by one of the leading letters of the alphabet.

A *variable quantity*, or simply a *variable*, is one which admits of an infinite number of values within certain limits that are determined by the nature of the problem, and is represented by one of the final letters of the alphabet.

For example, in the equation of the parabola,

$$y^2 = 2px,$$

x and y are *variables*, as they represent the co-ordinates of any point of the parabola, and so may have an indefinite number of different values. $2p$ is a *constant*, as it represents the latus rectum of the parabola, and so has but one fixed value. Any given *number* is constant.

2. Independent and Dependent Variables. — An *independent* variable is one to which any arbitrary value may be assigned at pleasure. A *dependent* variable is one whose value varies in consequence of the variation of the independent variable or variables with which it is connected.

Thus, in the equation of the circle

$$x^2 + y^2 = r^2,$$

if we assign to x any arbitrary value, and find the corresponding value of y, we make x the independent variable, and y the dependent variable. If we were to assign to y any arbitrary value, and find the corresponding value of x, we would make y the independent variable and x the dependent variable.

Frequently, when we are considering two or more variables, it is in our power to make whichever we please the independent variable. But, having once chosen the independent variable, we are not at liberty to change it throughout our operations, *unless* we make the corresponding transformations which such a change would require.

3. Functions.—One quantity is called a *function* of another, when it is so connected with it that no change can take place in the latter without producing a corresponding change in the former.

For example, the sine, cosine, tangent, etc., of an angle are said to be functions of the angle, as they depend upon the angle for their value. Also, the area of a square is a function of its side; the volume of a sphere is a function of its radius. In like manner, any algebraic expression in x, as

$$x^3 - 2bx^2 + bx + c,$$

is a function of x. Also, we may have a function of two or more variables: a rectangle is a function of its two sides; a parallelopiped is a function of its three edges ; the expression $\tan (ax + by)$ is a function of two variables, x and y; $x^2 + y^2 + z^2$ is a function of three variables, $x, y,$ and z ; etc.

When we wish to write that one quantity is a function of

one or more others, and wish, at the same time, to indicate several forms of functions in the same discussion, we use such symbols as the following:

$$y = f(r) ; \qquad y = F(x) ; \qquad y = \phi(x) ; \qquad y = f'(x) ;$$

$$y = f(x, z) ; \qquad \phi(x, y) = 0 ; \qquad f(x, y, z) = 0 ;$$

which are read: "y equals the f function of x; y equals the large F function of x; y equals the ϕ function of x; y equals the f prime function of x; y equals the f function of x and z; the ϕ function of x and y equals zero; the f function of x, y, and z equals zero;" or sometimes "$y = f$ of x, $y = F$ of x," etc. If we do not care to state precisely the *form* of the function, we may read the above, "$y =$ a function of x; $y =$ a function of x and z; a function of x and y $= 0$; a function of x, y, and $z = 0$."

For example, in the equation

$$y = ax^2 + bx + c,$$

y is a function of x, and may be expressed, $y = f(x)$.

Also, the equation

$$ax^2 + bxy + cy^2 = 0$$

may be expressed, $f(x, y) = 0$.

In like manner, the equations

$$y = ax^3 + bx^2z + cz^3,$$

and $\qquad\qquad\qquad y = ax^2 + bxz + dz^2,$

may be expressed, $y = f(x, z)$ and $y = \phi(x, z)$.

Every function of a single variable may be represented geometrically by the ordinate of a curve of which the variable is the corresponding abscissa. For if y be any function of x, and we assign any value to x and find the corresponding value of y, these two values may be regarded as the co-ordinates of a point which may be constructed. In the same way, any number of values may be assigned to x, and the corresponding values of y found, and a series of points constructed. These points make up a curve of which the variable ordinate is y and the corresponding abscissa is x.

In like manner it may be shown that a function of two variables may be represented geometrically by the ordinate of a *surface* of which the variables are the corresponding abscissas.

4. Algebraic and Transcendental Functions.—An *algebraic function* is one in which the only operations indicated are addition, subtraction, multiplication, division, involution, and evolution; as,

$$(a + bx^2)^n \; ; \qquad (x^2 - bxy)^{\frac{1}{2}} \; ; \qquad \sqrt{\dfrac{x^2 - a^2}{(x^2 + a^2)^{\frac{5}{2}}}} \; ; \quad \text{etc.}$$

Transcendental functions are those which involve other operations, and are subdivided into *trigonometric, circular, logarithmic,* and *exponential.*

A *trigonometric* function is one which involves sines, tangents, cosines, etc., as variables. For example,

$$y = \sin x; \quad y = \tan^2 x; \quad y = \cos x \sec x; \quad \text{etc.}$$

A *circular* function is one in which the concept is a variable arc, as $\sin^{-1} x$,* $\cos^{-1} x$, $\sec^{-1} y$, $\cot^{-1} x$, etc., read, "*the arc whose sine is x, the arc whose cosine is x,*" etc. It is the inverse of the trigonometric function ; thus, from the trigonometric function, $y = \sin x$, we obtain the circular function, $x = \sin^{-1} y$. In the first function we *think* of the *right line*, the *sine*, the arc being given to tell us *which* sine ; in the second we *think* of the *arc*, the sine being given to tell us *which* arc. The circular functions are often called *inverse trigonometric functions.*

---

* This notation was suggested by the use of the negative exponents in algebra. If we have $y = ax$, we also have $x = a^{-1}y$, where y is a function of x, and x is the corresponding inverse function of y. It may be worth while to caution the beginner against the error of supposing that $\sin^{-1} y$ is equivalent to $\dfrac{1}{\sin y}$; while it is true that a^{-1} is equivalent to $\dfrac{1}{a}$.

A *logarithmic* function is one which involves logarithms of the variables; as,

$$y = \log x; \qquad y = \log \sqrt{a - x};$$

$$y^2 = 3 \log \sqrt{\frac{a^2 - x^2}{x^2 + x^2}}; \quad \text{etc.}$$

An *exponential* function is one in which the variable enters as an exponent; as,

$$y = a^x; \qquad y = x^z; \qquad u = x^{x^y}; \quad \text{etc.}$$

5. Increasing and Decreasing Functions. —An *increasing* function is one that increases when its variable increases, and decreases when its variable decreases.

For example, in the equations

$$y = ax^3, \quad y = \log x, \quad y = \sqrt{a^2 + x^2}, \quad y = a^x,$$

y is an increasing function of x.

A *decreasing* function is one that decreases when its variable increases, and increases when its variable decreases. Thus, in the equations

$$y = \frac{1}{x}, \quad y = (a - x)^3, \quad y = \log \frac{1}{x}, \quad x^2 + y^2 = r^2,$$

y is a decreasing function of x. In the expression,

$$y = (a - x)^2,$$

y is a decreasing function for all values of $x < a$, but increasing for all values $> a$. In the expression

$$y = \sin x,$$

y is an increasing function for all values of x between $0°$ and $90°$, decreasing for all values of x between $90°$ and $270°$, and increasing for all values of x between $270°$ and $360°$.

6. Explicit and Implicit Functions.—An *explicit* function is one whose value is directly expressed in terms of the variable and constants.

For example, in the equations

$$y = (a - x)^2, \quad y = \sqrt{a^2 - x^2}, \quad y = 2ax^3 - 3x^{\frac{1}{2}},$$

y is an explicit function of x.

An *implicit* function is one whose value is not directly expressed in terms of the variables and constants.

For example, in the equations

$$y^3 - 3axy + x^3 = 4, \qquad x^2 - 3xy + 2y = 16,$$

y is an implicit function of x, or x is an implicit function of y. If we solve either equation with respect to y, we shall have y as an explicit function of x; also, if we solve for x, we shall have x as an explicit function of y.

7. Continuous Functions.—A function of x is said to be a *continuous* function of x, between the limits a and b, when, for every value of x between these limits, the corresponding value of the function is finite, and when an infinitely small change in the value of x produces only an infinitely small change in the value of the function. If these conditions are not fulfilled, the function is *discontinuous*.

For example, both conditions are fulfilled in the equations

$$y = ax + b, \qquad y = \sin x,$$

in which, as x changes, the value of the function also changes, but changes gradually as x changes gradually, and there is no abrupt passage from one value to another; if x receives a very small change, the corresponding change in the function of x is also very small.

The expression $\sqrt{r^2 - x^2}$ is a continuous function of x for all values of x between $+ r$ and $- r$, while $\sqrt{x^2 - r^2}$ is discontinuous between the same limits.

2. Infinites and Infinitesimals.—An *infinite quantity,* or an *infinite,* is a quantity which is greater than any assignable quantity.

An *infinitesimal* is a quantity which is less than any assignable quantity.

An infinite is not the largest possible quantity, nor is an infinitesimal the smallest; there would, in this case, be but *one* infinite or infinitesimal. Infinites may differ from each other and from a quantity which transcends every assignable quantity, that is, from *absolute* infinity. So may infinitesimals differ from each other and from *absolute* zero.

The terms *infinite* and *infinitesimal* are not applicable to quantities *in themselves considered,* but only in their *relation* to each other, or to a common standard. A magnitude which is infinitely great in comparison with a *finite* magnitude is said to be *infinitely great.* Also, a magnitude which is infinitely small in comparison with a finite magnitude is said to be *infinitely small.* Thus, the diameter of the earth is very great in comparison with the length of *one inch,* but very small in comparison with the distance of the *earth from the pole star;* and it would accordingly be represented by a very large or a very small number, according to which of these distances is assumed as the unit of comparison.

The symbols ∞ and 0 are used to represent an infinite and an infinitesimal respectively, the relation of which is

$$\infty = \frac{1}{0} \quad \text{and} \quad 0 = \frac{1}{\infty}.$$

The cipher 0 is an abbreviation to denote an indefinitely small quantity, or an infinitesimal—that is, a quantity which is less than any assignable quantity—and does not mean *absolute* zero; neither does ∞ express absolute infinity.

If a represents a finite quantity, and x an infinite, then $\frac{a}{x}$ is an infinitesimal. If x is an infinitesimal and a is finite, $\frac{a}{x}$ is infinite; that is, the reciprocal of an infinite is infinitesimal, and the reciprocal of an infinitesimal is infinite.

A number is *infinitely* great in comparison with another,

when *no number can be found sufficiently large to express the ratio between them.* Thus. x is infinitely great in relation to a, when no number can be found large enough to express the quotient $\frac{x}{a}$. Also, a is infinitely small in relation to x when no number can be found small enough to express the quotient $\frac{a}{x}$; x and $\frac{a}{x}$ represent an infinite and an infinitesimal.

One million in comparison with *one millionth* is a very large number, but not *infinitely* large, since the ratio of the first to the second can be expressed in figures: it is *one trillion :* though a *very* large number, it is *finite.* So, also, one millionth in comparison with one million is a very small number, but not *infinitely* small, since a number can be found small enough to express the ratio of the first to the second: it is *one trillionth,* and therefore finite.

9. Orders of Infinites and Infinitesimals.—But even though $\frac{x}{a}$ is greater than any quantity to which we can assign a value, we may suppose another quantity as large in relation to x as x is in relation to a; for, whatever the magnitude of x, we may have the proportion

$$a : x :: x : \frac{x^2}{a},$$

in which $\frac{x^2}{a}$ is as large in relation to x as x is in relation to a, for $\frac{x^2}{a}$ will contain x as many times as x will contain a; hence, $\frac{x^2}{a}$ may be regarded as an infinite of the *second order,* $\frac{x}{a}$ being an infinite of the *first order.*

Also, even though $\frac{a}{x}$ is less than any quantity to which we can assign a value, we may suppose another quantity as small in relation to a as a is in relation to x; for we may have the proportion,

$$x : a \ :: \ a : \frac{a^2}{x},$$

in which $\frac{a^2}{x}$ is as small in relation to a as a is in relation to

c, for $\frac{a^2}{x}$ is contained as many times in a as a is contained

in x; hence, $\frac{a^2}{x}$ may be regarded as an infinitesimal of the

second order, $\frac{a}{x}$ being an infinitesimal of the *first order.*

We may, again, suppose quantities infinitely greater and infinitely less than these just named ; and so on indefinitely. Thus, in the series

$$ax^3, \quad ax^2, \quad ax, \quad a, \quad \frac{a}{x}, \quad \frac{a}{x^2}, \quad \frac{a}{x^3}, \quad \text{etc.,}$$

if we suppose *a finite* and *x infinite*, it is clear that any term is infinitely small with respect to the one that immediately precedes it, and infinitely large with respect to the one that immediately follows it; that is, ax^3, ax^2, ax are *infinites* of the *third, second,* and *first orders,* respectively; $\frac{a}{x}$, $\frac{a}{x^2}$, $\frac{a}{x^3}$ are *infinitesimals* of the *first, second,* and *third orders,* respectively, while a is finite.

If two quantities, as x and y, are infinitesimals of the first order, their product is an infinitesimal of the second order ; for we have the proportion,

$$1 : x \ :: \ y : xy.$$

Hence, if x is infinitely small in relation to 1, xy is infinitely small in relation to y; that is, it is an infinitesimal of the *second order* when x and y are infinitesimals of the *first order.*

Likewise, the product of two *infinites* of the first order is an infinite of the second order.

The product of an infinite and an infinitesimal of the same order is a *finite quantity.* The product of an infinite and an infinitesimal of different orders is an infinite or an infinitesimal, according as the order of the infinite is higher or lower than that of the infinitesimal, and the order of the product is the sum of the orders of the factors.

For example, in the expressions

$$ax^3 \times \frac{a}{x^3} = a^2, \qquad ax^2 \times \frac{a}{x} = a^2x, \qquad ax \times \frac{a}{x^3} = \frac{a^2}{x^2},$$

the first product is *finite;* the second is an *infinite* of the *first order;* the third is an *infinitesimal* of the *second order.*

Though two quantities are each infinitely small, they may have any ratio whatever.

Thus, if a and b are finite and x is infinite, the two quantities $\frac{a}{x}$ and $\frac{b}{x}$ are infinitesimals; but their *ratio* is $\frac{a}{b}$, which is *finite.* Indeed, two very small quantities may have a much larger ratio than two very large quantities, for the value of a ratio depends on the *relative,* and not on the *absolute* magnitude of the *terms* of the ratio. The ratio of the fraction *one-millionth* to *one-ten-millionth* is *ten,* while the ratio of *one million* to *ten million* is *one-tenth.* The latter numbers are respectively a million times a million, and ten million times ten million, times as great as the first, and yet the ratio of the last two is only one-hundredth as great as the ratio of the first two.

Assume the series

$$\frac{1}{10^6}, \quad \left(\frac{1}{10^6}\right)^2, \quad \left(\frac{1}{10^6}\right)^3, \quad \left(\frac{1}{10^6}\right)^4, \quad \left(\frac{1}{10^6}\right)^5, \quad \left(\frac{1}{10^6}\right)^6, \text{ etc}$$

in which the first fraction is one-millionth, the second one-millionth of the first, and so on. Now suppose the first fraction is one-millionth of an *inch* in length, which may be regarded as a very small quantity of the *first order;* the second, being one-millionth of the first, must be regarded as a small quantity of the *second order,* and so on. Now, if we continue this series indefinitely, it is *clear* that we can make the terms become *as small as we please* without ever reaching *absolute* zero. It is also clear that, however small the terms of this series become, the ratio of any term to the one that immediately follows it is one million.

10. Geometric Illustration of Infinitesimals.—The following geometric results will help to illustrate the theory of infinitesimals.

Let A and B be two points on the circumference of a circle : draw the diameter AE, and draw EB produced to meet the tangent AD at D. Then, as the triangles EAB and ADB are similar, we have,

$$\frac{BE}{AE} = \frac{AB}{AD}, \qquad (1)$$

and $\qquad \dfrac{AB}{AE} = \dfrac{BD}{AD}. \qquad (2)$

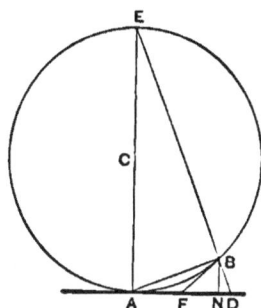

Fig. I.

Now suppose the point B to approach the point A till it becomes infinitely near to it, then BE becomes ultimately equal to AE; but, from (1), when

$$BE = AE,$$

we have $\qquad AB = AD.$

Also, $\dfrac{AB}{AE}$ becomes infinitely small, that is, AB becomes an *infinitely small quantity in comparison with* AE. Hence, from (2), BD becomes infinitely small in comparison with AD or AB ; that is, *when AB is an infinitesimal of the first order, BD is an infinitesimal of the second order.*

Since DE − AE < BD, it follows that, *when one side of a right-angled triangle is regarded as an infinitely small quantity of the first order, the difference between the hypothenuse and the remaining side is an infinitely small quantity of the second order.*

Draw BN perpendicular to AD; then, since AB > AN, we have,

$$AD - AB < AD - AN < DN ;$$

therefore, $\qquad \dfrac{AD - AB}{BD} < \dfrac{DN}{BD} < \dfrac{AD}{DE}.$

But AD is infinitely small in comparison with DE, therefore AD − AB is infinitely small in comparison with BD;

but BD is *an infinitesimal of the second order* (see above), hence AD — AB is *an infinitesimal of the third order.*

In like manner it may be shown that BD — BN is an infinitesimal of the *fourth* order, and so on. ['The student who wishes further illustration is referred to Williamson's Dif. Cal., p. 35, from which this was taken.]

11. Axioms.—From the nature of an infinite quantity, a finite quantity can have no value when added to it, and must therefore be dropped.

An infinitesimal can have no value when added to a finite quantity, and must therefore be dropped.

If an infinite or an infinitesimal be multiplied or divided by a finite quantity, its order is not changed.

If an expression involves the sum or difference of infinites of different orders, its value is equal to the infinite of the highest order, and all the others can have no value when added to it, and must be dropped.

If an expression involves the sum or difference of infinitesimals of different orders, its value is equal to the infinitesimal of the lowest order, and all the others can have no value when added to it, and must be dropped.

These axioms are *self-evident*, and, therefore, *axioms* in the strict sense. For example, suppose we were to compare the mass of the sun with that of the earth : the latter weighs about six sextillion tons, the former weighs about 355000 times as much. If a weight of one grain were added to or subtracted from either, it would not affect the ratio *appreciably ;* and yet the grain, compared with either, is *finite*—it can be expressed in figures, though on the verge of an infinitesimal. If we divide this grain into a great many equal parts—a sextillion, for instance—and add one of these parts to the sun or the earth, the error of the ratio will be still less ; hence, when the subdivision is continued *indefinitely,* it is evident that we may obtain a fraction *less than any assignable quantity, however small,* which, when added to the sun or the earth, will affect the above ratio by a quantity less than any to which we can assign a value.

By reason of the terms that may be omitted, in virtue of the principles contained in these axioms, the equations formed in the solution of a problem will be greatly simplified. It may be remarked that in the *method of limits*,* when exclusively adopted, it is usual to *retain* infinitely small quantities of higher orders until the *end* of the calculation, and then to neglect them on proceeding to the limit; while, in the *infinitesimal* method, such quantities are neglected from the *beginning*, from the knowledge that they cannot affect the *final result*, as they necessarily disappear in the limit. The advantage derived from neglecting these quantities will be evident when it is remembered how much the difficulty in the solution of a problem is increased when it is necessary to introduce into its equations the second, third, and in general the higher powers of the quantities to be considered.

EXAMPLES.

1. Find the value of the fraction $\dfrac{3x + a}{5x + b}$, if x is infinite, and a and b finite.

Since a and b are finite, they have no value in comparison with x, and must therefore be dropped, giving us $\dfrac{3x}{5x} = \frac{3}{5}$ as the required value of the fraction.

2. Find the value of the fraction $\dfrac{2x - a}{3x + b}$, if x is infinitesimal, and a and b finite.

Since x is an infinitesimal, it has no value in comparison with a and b, and must therefore be dropped, giving us $-\dfrac{a}{b}$ for the required value of the fraction.

3. Find the value of $\dfrac{8x^2 + 2x}{2x^2 + x}$, when x is infinite; also when x is infinitesimal.

Ans. When x is infinite, 4; when infinitesimal, 2.

4. Find the value of $\dfrac{ax^3 + bx^2 + cx + e}{mx^3 + nx^2 + px + q}$, when x is infinite; and when infinitesimal.

Ans. When x is infinite, $\dfrac{a}{m}$; when infinitesimal, $\dfrac{e}{q}$.

* For a discussion of limits, see Chapter **III.**

5. Find the value of $\dfrac{ax^6 + 3x^2 + 2}{bx^4 - 4x + 1}$, when x is infinite; and when infinitesimal.

Ans. When x is infinite, ∞; when infinitesimal, 2.

6. Find the value of $\dfrac{4x^4 + 3x^2 + 2x - 1}{2x^5 + 4x^2 + 2x}$, when x is infinite; and when infinitesimal.

Ans. When x is infinite, 0; when infinitesimal, ∞.

7. Find the value of $\dfrac{5x^2 - 7m^2x}{4x^3 - mx}$, when x is infinite; and when infinitesimal.

Ans. When x is infinite, 0; when infinitesimal, $7m$.

8. Find the value of $\dfrac{x^5}{2a - x^2}$, when x is infinite; and when infinitesimal.

Ans. When x is infinite, ∞; when infinitesimal, 0.

9. Find the value of $\dfrac{7x - 2y}{4x + 3y}$, when x and y are infinitesimals.

Ans. We do not know, since the *relation* between x and y is unknown.

CHAPTER II

DIFFERENTIATION OF ALGEBRAIC AND TRANSCENDENTAL FUNCTIONS.

12. Increments and Differentials.—If any variable, as x, be supposed to receive any change, such change is called an *increment ;* this increment of x is usually denoted by the notation Δx, read "difference x," or "delta x," where Δ is taken as an abbreviation of the word *difference.* If the variable is *increasing,* the increment is $+$; but if it is *decreasing,* the increment is $-$.

When the *increment,* or *difference,* is supposed *infinitely small,* or an *infinitesimal,* it is called a *differential,* and is represented by dx, read "differential x," where d is taken as an abbreviation of the word *differential,* or infinitely small difference. The symbols Δ and d, when prefixed to a variable or function, have not the effect of multiplication ; that is, dx is not d times x, and Δx is not Δ times x, but their power is that of an operation performed on the quantity to which they are prefixed.

If u be a function of x, and x becomes $x + \Delta x$, the corresponding value of u is represented by $u + \Delta u$; that is, the increment of u corresponding to a *finite* increment of x is denoted by Δu, read "difference u."

If x becomes $x + dx$, the corresponding value of u is represented by $u + du$; that is, the infinitely small increment of u caused by an infinitely small increment in x, on which u depends, is denoted by du, read "differential u." Hence, dx is the infinitesimal increment of x, or the infinitesimal quantity by which x is increased ; and du is the corresponding infinitesimal increment of u.

The differential du or dx is $+$ or $-$ according as the variable is increasing or decreasing, *i.e.*, the first value is always to be taken from the second.

13. Consecutive Values. — *Consecutive values* of a function or variable are values which differ from each other by *less than any assignable quantity.*

Consecutive points are points nearer to each other than any assignable distance.

Thus, if two points were one-millionth of an inch apart, they might be considered *practically* as consecutive points; and yet we might have a million points between them, the distance between any two of which would be a millionth of a millionth of an inch; and so we might have a million points between any two of these last points, and so on; that is, however close two points might be to each other, we could still suppose *any number* of points between them.

A *differential* has been defined as an *infinitely small increment,* or an *infinitesimal;* it may also be defined as the *difference between two consecutive values* of a variable or function. The difference is always found by taking the first value from the second.

In the *Differential Calculus,* we investigate the relations between the infinitesimal increments of variables from given relations between finite values of those variables.

The operation of finding the differential of a function or a variable is called *differentiation.*

14. Differentiation of the Algebraic Sum of a Number of Functions.

Let
$$u = v + y - z, \qquad (1)$$
in which u, v, y, z, are functions of x.*

---

* We might also, in a similar manner, find the differential of a function of *several variables;* but we prefer to reserve the inquiry into the differentials of functions of *several variables* for a later chapter, and confine ourselves at present to functions of a *single variable.*

Give to x the infinitesimal increment dx, and let du, dv, dy, dz, be the corresponding infinitesimal increments of u, v, y, z, due to the increment which x takes. Then (1) becomes

$$u + du = v + dv + y + dy - (z + dz). \qquad (2)$$

Subtracting (1) from (2), we have

$$du = dv + dy - dz, \qquad (3)$$

which is *the differential required.*

Therefore, *the differential of the algebraic sum of any number of functions is found by taking the algebraic sum of their differentials.*

15. To Differentiate

$$y = ax \pm b. \qquad (1)$$

Give to x the infinitesimal increment dx, and let dy be the corresponding infinitesimal increment of y due to the ncrement which x takes. Then (1) becomes

$$y + dy = a(x + dx) \pm b. \qquad (2)$$

Subtracting (1) from (2), we get

$$dy = adx, \qquad (3)$$

which is *the required differential.*

Hence, *the differential of the product of a constant by a variable is equal to the constant multiplied by the differential of the variable: also, if a constant be connected with a variable by the sign + or −, it disappears in differentiation.*

This may also be proved geometrically as follows:

Let AB (Fig. 2) be the line whose equation is $y = ax + b$, and let (x, y) be any point P on this line. Give OM $(= x)$

the infinitesimal increment MM' $(= dx)$, then the corresponding increment of MP $(= y)$ will be CP' $(= dy)$. Now in the triangle CPP' we have

$$CP' = CP \tan CPP'; *$$

or letting $a = \tan CPP'$, and substituting for CP' and CP their values dy and dx, we have,

$$dy = adx.$$

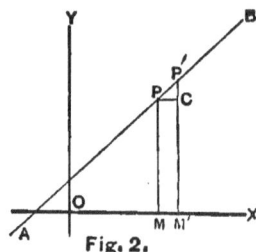

Fig. 2.

It is evident that the constant b will disappear in differentiation, from the very nature of constants, which do not admit of increase, and therefore can take no increment.

16. Differentiation of the Product of two Functions.

Let $$u = yz, \tag{1}$$

where y and z are both functions of x. Give x the infinitesimal increment dx, and let du, dy, dz be the corresponding increments of u, y, and z, due to the increment which x takes. Then (1) becomes

$$\begin{aligned} u + du &= (y + dy)(z + dz) \\ &= yz + zdy + ydz + dz\,dy. \end{aligned} \tag{2}$$

Subtracting (1) from (2), and omitting $dz\,dy$, since it is an infinitesimal of the second order, and added to others of the first order (Art. 11), we have

$$du = zdy + ydz, \tag{3}$$

which is *the required differential.*

Hence, *the differential of the product of two functions is equal to the first into the differential of the second, plus the second into the differential of the first.*

---

* In the Calculus as in the Analytic Geometry, the radius is always regarded as 1, unless otherwise mentioned.

This may also be proved geometrically as follows:

Let z and y represent the lines AB and BC respectively; then will u represent the area of the rectangle ABCD. Give AB and BC the infinitesimal increments Ba $(= dz)$ and Cc $(= dy)$ respectively. Then the rectangle ABCD will be increased by the rectangles BaCh, DCbc, and Chcd, the values of which are ydz, zdy, and $dzdy$ respectively; therefore

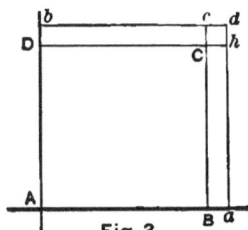

$$du = ydz + zdy + dz\,dy.$$

But $dz\,dy$ being an infinitesimal of the second order and connected with others of the first order, must be dropped (Art. 11); if this were not done, infinitesimals would not be what they are (Art. 8); the very fact of dropping the term $dz\,dy$ *implies* that its value, as compared with that of $ydz + zdy$ is *infinitely small.*

The statement that $ydz + zdy + dz\,dy$ is *rigorously* equal to $ydz + zdy$ is not true, and yet by taking dz and dy sufficiently small, the error may be made as small as we please.

Or, we may introduce the idea of motion, and consider that dz and dy represent the *rate* at which AB and BC are increasing at the *instant* they are equal to z and y respectively. The rate at which the rectangle ABCD is enlarging at this *instant* depends upon the length of BC and the rate at which it is moving to the right + the length of DC and the rate at which it is moving upward. If we let dz represent the rate at which BC is moving to the right, and dy the rate at which DC is moving upward at the *instant* that AB $= z$ and BC $= y$, we shall have $du = zdy + ydz$ as the rate at which the rectangle ABCD is enlarging at this *instant*. (See Price's Calculus, vol. i, p. 41.)

17. Differentiation of the Product of any Number of Functions.

Let
$$u = vyz. \tag{1}$$

Then giving to x the infinitesimal increment dx, and letting du, dv, dy, dz be the corresponding increments of u, v, y, z, (1) becomes

$$u + du = (v + dv)(y + dy)(z + dz). \tag{2}$$

Subtracting (1) from (2), and omitting infinitesimals of higher orders than the first, we have

$$du = yz\, dv + vz\, dy + vy\, dz, \tag{3}$$

and so on for any number of functions.

Hence, *the differential of the product of any number of functions is equal to the sum of the products of the differential of each into the product of all the others.*

Cor.—Dividing (3) by (1), we have

$$\frac{du}{u} = \frac{dv}{v} + \frac{dy}{y} + \frac{dz}{z}. \tag{4}$$

That is, *if the differential of each function be divided by the function itself, the sum of the quotients will be equal to the differential of the product of the functions divided by the product.*

18. Differentiation of a Fraction.

Let
$$u = \frac{x}{y},$$

then
$$uy = x; \tag{1}$$

therefore, by Art. 16, we have

$$u\,dy + y\,du = dx.$$

Substituting for u its value, we have

$$\frac{x}{y} dy + y du = dx.$$

Solving for du, we get

$$du = \frac{y dx - x dy}{y^2},$$

which is the required differential.

Hence, *the differential of a fraction is equal to the denominator into the differential of the numerator, minus the numerator into the differential of the denominator, divided by the square of the denominator.*

Cor. 1.—If the numerator be constant, the first term in the differential vanishes, and we have

$$du = \left(-\frac{x dy}{y^2} \right).$$

Hence, *the differential of a fraction with a constant numerator is equal to minus the numerator into the differential of the denominator divided by the square of the denominator.*

Cor. 2.—If the denominator be constant, the second term vanishes, and we have

$$du = \frac{dx}{y},$$

which is the same result we would get by applying the rule of Art. 15.

19. Differentiation of any Power of a Single Variable.

Let $$y = x^n.$$

1st. *When n is a positive integer.*

Regarding x^n as the product x, x, x, etc., of n equal factors, each equal to x, and applying the rule for differentiating a product (Art. 17), we get

$$dy = x^{n-1} dx + x^{n-1} dx + x^{n-1} dx + \text{etc., to } n \text{ terms.}$$

$$\therefore \quad dy = nx^{n-1} dx. \qquad (1)$$

2d. *When* **n** *is a positive fraction.*

Let $\qquad\qquad y = x^{\frac{m}{n}};$

then $\qquad\qquad y^n = x^m.$

Differentiating this as just shown, we have,

$$ny^{n-1} dy = mx^{m-1} dx.$$

Therefore, $\qquad dy = \dfrac{m}{n} \dfrac{x^{m-1}}{y^{n-1}} dx$

$$= \frac{m}{n} \frac{x^{m-1} y}{y^n} dx$$

$$= \frac{m}{n} \frac{x^{m-1} x^{\frac{m}{n}}}{x^m} dx \quad (\text{since } y^n = x^m).$$

$$\therefore \quad dy = \frac{m}{n} x^{\frac{m}{n} - 1} dx.$$

3d. *When* **n** *is a negative exponent, integral or fractional.*

Let $\qquad\qquad y = x^{-n};$

then $\qquad\qquad y = \dfrac{1}{x^n}.$

Differentiating by Art. 18, Cor. 1, we have

$$dy = - \frac{nx^{n-1} dx}{x^{2n}} = - nx^{-n-1} dx. \qquad (3)$$

Combining the results in (1), (2), and (3), we have the following rule: *The differential of any constant power of a variable is equal to the product of the exponent, the variable with its exponent diminished by unity, and the differential of the variable.*

Cor.—If $n = \frac{1}{2}$, we have from (1),

$$dy = \frac{1}{2}x^{\frac{1}{2}-1} dx = \frac{1}{2}x^{-\frac{1}{2}} dx = \frac{dx}{2\sqrt{x}}.$$

Hence, *the differential of the square root of a varia-ble is equal to the differential of the variable divided by twice the square root of the variable.*

EXAMPLES.

1. Differentiate $y = 9 + 2x + x^3 + x^2y^3 - x^7 + 2axy$.

By Art. 14, we differentiate each term separately, and take the algebraic sum. By Art. 15, the constant 9 disappears in the differentiation ; and the differential of $2x$ is the constant 2, multiplied by the differential of the variable x, giving $2dx$. By Art. 19, the differential of x^2 is $2x\, dx$. The term x^2y^3 is the product of two functions ; therefore, Art. 16, its differential is $x^2\, d\,(y^3) + y^3\, d\,(x^2)$, which, Art. 19, gives $3x^2y^2\, dy + 2y^3x\, dx$. In like manner proceed with the other terms, giving the proper rule in each case. The answer is

$$dy = 2dx + 3x^2\, dx + 3x^2y^2\, dy + 2xy^3\, dx - 7x^6\, dx \\ + 2ax\, dy + 2ay\, dx.$$

2. $\quad u = ax^3y^2.$ $\qquad\qquad du = 3ax^2y^2\, dx + 2ax^3y\, dy.$

3. $\quad u = 2ax - 3x^2 - abx^4 - 7.$

$$du = (2a - 6x - 4abx^3)\, dx.$$

4. $\quad u = x^2y^{\frac{5}{2}}.$ $\qquad\qquad du = \frac{5}{2}x^2y^{\frac{3}{2}}\, dy + 2xy^{\frac{5}{2}}\, dx.$

5. $\quad u = 2bz^{-2} + 3ax^{\frac{5}{3}}z^{\frac{1}{2}}.$

$$du = 5ax^{\frac{2}{3}}z^{\frac{1}{2}}\, dx + \frac{3ax^{\frac{5}{3}}dz}{2\sqrt{z}} - \frac{4bdz}{z^3}.$$

6. $\quad u = x^{\frac{1}{2}}y^{\frac{1}{2}}.$ $\qquad\qquad du = \frac{xdy + ydx}{2x^{\frac{1}{2}}y^{\frac{1}{2}}}.$

7. $u = ax^{\frac{5}{3}} + \dfrac{a}{x^2} - bx^{\frac{1}{2}}$.

$$du = \left(\tfrac{5}{3}ax^{\frac{2}{3}} - \frac{2a}{x^3} - \frac{b}{2\sqrt{x}} \right) dx.$$

8. $y^2 = 2px$, to find the value of dy.

$$dy = \frac{p}{y}\, dx.$$

9. $a^2y^2 + b^2x^2 = a^2b^2$, to find the value of dy.

$$dy = -\frac{b^2x}{a^2y}\, dx.$$

10. $x^2 + y^2 = r^2$, to find the value of dy.

$$dy = -\frac{x}{y}\, dx.$$

11. $u = \dfrac{a}{b - 2y^2}.$ $\qquad\qquad du = \dfrac{4ay\, dy}{(b - 2y^2)^2}.$

12. $u = (a + bx + cx^2)^5$.

Regarding the quantity within the parenthesis as a varia·ble, we have, by Art. 19,

$$du = 5\,(a + bx + cx^2)^4\, d\,(a + bx + cx^2)$$
$$= 5\,(a + bx + cx^2)^4\,(b + 2cx)\, dx.$$

13. $u = \dfrac{2x^2 - 3}{4x + x^2}.$

By Art. 18,

$$du = \frac{(4x + x^2)\, d\,(2x^2 - 3) - (2x^2 - 3)\, d\,(4x + x^2)}{(4x + x^2)^2}$$
$$= \frac{(4x + x^2)\,4x\, dx - (2x^2 - 3)\,(4 + 2x)\, dx}{(4x + x^2)^2}$$
$$= \frac{(8x^2 + 6x + 12)\, dx}{(4x + x^2)^2}.$$

14. $u = \dfrac{2x^4}{a^2 - x^2}.$ $\qquad\qquad du = \dfrac{8a^2x^3 - 4x^5}{(a^2 - x^2)^2}\, dx.$

15. $y = \dfrac{1 + x}{1 + x^2}.$ $dy = \dfrac{(1 - 2x - x^2)\, dx}{(1 + x^2)^2}.$

16. $y = \dfrac{1}{x^n}.$ $dy = -\dfrac{n}{x^{n+1}}\, dx.$

17. $y = (ax^2 - x^3)^4.$
$$dy = 4\,(ax^2 - x^3)^3\,(2ax - 3x^2)\, dx.$$

18. $y = (a + bx^2)^{\frac{5}{3}}.$ $dy = \tfrac{10}{3}\,(a + bx^2)^{\frac{2}{3}}\, bx\, dx.$

19. $y = \dfrac{a}{(b^2 + x^2)^3}.$ $dy = -\dfrac{6ax}{(b^2 + x^2)^4}\, dx.$

20. $y = \sqrt{x^3 - a^3}$ (Art. 19, Cor.).
$$dy = \dfrac{d\,(x^3 - a^3)}{2\sqrt{x^3 - a^3}} = \dfrac{3x^2\, dx}{2\sqrt{x^3 - a^3}}.$$

21. $y = \sqrt{2ax - x^2}.$ $dy = \dfrac{(a - x)\, dx}{\sqrt{2ax - x^2}}.$

22. $y = \dfrac{1}{\sqrt{1 - x^2}}.$ $dy = \dfrac{x\, dx}{(1 - x^2)^{\frac{3}{2}}}.$

23. $y = \sqrt{ax} + \sqrt{c^2 x^3}.$ $dy = \dfrac{a^{\frac{1}{2}} + 3cx}{2\sqrt{x}}\, dx.$

24. $y = \sqrt{ax^2 + bx + c}.$ $dy = \dfrac{(2ax + b)\, dx}{2\sqrt{ax^2 + bx + c}}.$

25. $y = (x^3 + a)\,(3x^2 + b)$ (Art. 16).
$$dy = (x^3 + a)\, d\,(3x^2 + b) + (3x^2 + b)\, d\,(x^3 + a)$$
$$= (15x^4 + 3bx^2 + 6ax)\, dx.$$

26. $y = (1 + 2x^2)\,(1 + 4x^3).$
$$dy = 4x\,(1 + 3x + 10x^3)\, dx.$$

27. $y = (a - x)\sqrt{a + x}.$ $dy = -\dfrac{(a + 3x)\, dx}{2\sqrt{a + x}}.$

28. $y = (a + x)\sqrt{a - x}.$ $dy = \dfrac{(a - 3x)\, dx}{2\sqrt{a - x}}.$

1. In the parabola $y^2 = 4x$, which is increasing the most rapidly at $x = 3$, the abscissa or the ordinate ? How does the relative rate of change vary as we recede from the vertex ?

Differentiating $y^2 = 4x$, we get $dy = \dfrac{2}{y} dx$, which shows that if we give to x the infinitely small increment dx, the corresponding increment of y is $\dfrac{2}{y}$ times as great; that is, the ordinate changes $\dfrac{2}{y}$ times as fast as the abscissa. At $x = 3$, we have $y = \sqrt{12}$. Hence, at this point,

$$dy = \frac{2}{\sqrt{12}} dx = \frac{1}{\sqrt{3}} dx;$$

that is, the ordinate is increasing a little over one-half as fast as the abscissa at $x = 3$.

At $x = 1$, $y = 2$, and $dy = dx$; that is, x and y are increasing equally; in general, at the focus the abscissa and ordinate of a parabola are increasing equally. At $x = 4$, $y = 4$, and $dy = \frac{1}{2}dx$; that is, y is increasing $\frac{1}{2}$ as fast as x. At $x = 9$, $y = 6$, and $dy = \frac{1}{3}dx$; that is, y is increasing $\frac{1}{3}$ as fast as x. At $x = 36$, $y = 12$, and $dy = \frac{1}{6}dx$; that is, y is increasing $\frac{1}{6}$ as fast as x, and so on. We see from the equation $dy = \dfrac{2}{y} dx$, as well as from the *figure of the parabola*, that the larger x becomes, and therefore y, the less rapidly y increases, while x continues to increase uniformly.

2. If the side of an equilateral triangle is increasing uniformly at the rate of $\frac{1}{2}$ an inch per second, at what rate is its altitude increasing ? Is the relative rate of increase of the side and altitude constant or variable ?

Let $x = $ a side of the triangle and $y = $ its altitude. Then $y^2 = \frac{3}{4}x^2$, and $dy = \frac{\sqrt{3}}{2} dx$, which shows that when x takes the infinitely small increment dx, the corresponding increment of y is $\frac{\sqrt{3}}{2}$ times as great; that is, the altitude y always changes $\frac{\sqrt{3}}{2}$ times as fast as the side x. When x is increasing at the rate of $\frac{1}{2}$ an inch per second, y's increasing $\frac{\sqrt{3}}{2}$ times $\frac{1}{2}$, or $\frac{\sqrt{3}}{4}$ inches per second.

3. A boy is running on a horizontal plane directly towards the foot of a tower 60 feet in height. How much faster is he nearing the foot than the top of the tower? How far is he from the foot of the tower when he is approaching it twice as fast as he is approaching the top? When he is 100 feet from the foot of the tower, how much faster is he approaching it than the top?

Let $x = $ the boy's distance from the foot of the tower, and $y = $ his distance from the top. Then we have

$$y^2 = x^2 + \overline{60}^2.$$

$$\therefore \quad dx = \frac{y}{x} dy ;$$

that is, the boy is nearing the foot $\frac{y}{x}$ times as fast as he is the top.

2d. When he is approaching the foot of the tower twice as fast as he is the top, we have $dx = 2dy$, which in

$$dx = \frac{y}{x} dy$$

gives us $y = 2x$, and this in $y^2 = x^2 + \overline{60}^2$ gives us

$$3x^2 = \overline{60}^2, \qquad \text{or} \qquad x = \frac{60}{\sqrt{3}} = 34.64.$$

3d. When he is 100 feet from the foot,

$$y = \sqrt{100^2 + 60^2} = 116.62,$$

and $\dfrac{y}{x} = \dfrac{116.62}{100}$, which in $dx = \dfrac{y}{x} dy$ gives

$$dx = 1.1662 \, dy \, ;$$

that is, he is approaching the foot of the tower 1.1662 times as fast as he is the top.

4. In the parabola $y^2 = 12x$, find the point at which the ordinate and abscissa are increasing equally; also the point at which the ordinate is increasing half as fast as the abscissa. *Ans.* The point (3, 6); and the point (12, 12).

5. If the side of an equilateral triangle is increasing uniformly at the rate of 2 inches per second, at what rate is the altitude increasing. *Ans.* $\sqrt{3}$ inches per second.

6. If the side of an equilateral triangle is increasing uniformly at the rate of 5 inches per second, at what rate is the area increasing when the side is 10 feet?
Ans. $\frac{25}{12}\sqrt{3}$ sq. ft. per second.

7. A vessel is sailing northwest at the uniform rate of 10 miles per hour; at what rate is she making north latitude? *Ans.* 7.07 + miles per hour.

8. A boy is running on a horizontal plane directly toward the foot of a tower, at the rate of 5 miles per hour; at what rate is he approaching the top of the tower when he is 60 feet from the foot, the tower being 80 feet high?
Ans. 3 miles per hour.

LOGARITHMIC AND EXPONENTIAL FUNC-TIONS.

20. To Differentiate $y = \log x$.—We have

$$y + dy = \log (x + dx) = \log \left[x \left(1 + \frac{dx}{x} \right) \right]$$

$$= \log x + \log \left(1 + \frac{dx}{x} \right).$$

Subtracting, we have,

$$dy = \log \left(1 + \frac{dx}{x} \right) = m \left(\frac{dx}{x} - \frac{\overline{dx}^2}{2x^2} + \text{etc.} \right)$$

(from Algebra, where m is the modulus of the system).

$$\therefore \quad dy = d (\log x) = m \frac{dx}{x} \ (\text{Art. 11}).$$

This result may also be obtained as follows:

Let $\qquad\qquad y = ax.$ $\qquad\qquad$ (1)

$$\therefore \ \log y = \log a + \log x. \qquad\qquad (2)$$

By Art. 15, $\qquad dy = a \, dx, \qquad\qquad$ (3)

and $\qquad d (\log y) = d (\log x).$ $\qquad\qquad$ (4)

Dividing (4) by (3), we get,

$$\frac{d (\log y)}{dy} = \frac{d (\log x)}{a \, dx}$$

$$= \frac{d (\log x)}{\frac{y}{x} \, dx}, \text{ from (1)};$$

or, $\qquad\qquad \dfrac{d (\log y)}{d (\log x)} = \dfrac{\dfrac{dy}{y}}{\dfrac{dx}{x}}.$

Multiply both terms of the second fraction by the arbitrary factor m, and we have

$$\frac{d\,(\log y)}{d\,(\log x)} = \frac{\dfrac{m\,dy}{y}}{\dfrac{m\,dx}{x}}. \tag{5}$$

We may suppose m to have such a value as to make

$$d\,(\log y) = m\frac{dy}{y}; \tag{6}$$

therefore, $\quad\quad d\,(\log x) = m\dfrac{dx}{x} \tag{7}$

Similarly, let $\quad\quad y = bz.$

$$\therefore\ \ \log y = \log b + \log z. \tag{8}$$

Differentiating, $\quad\quad dy = b\,dz,$

and $\quad\quad d\,(\log y) = d\,(\log z).$

Dividing and substituting,

$$\frac{d\,(\log y)}{d\,(\log z)} = \frac{\dfrac{dy}{y}}{\dfrac{dz}{z}}.$$

But $\quad\quad d\,(\log y) = m\dfrac{dy}{y}$

$$\therefore\ \ d\,(\log z) = m\frac{dz}{z}. \tag{9}$$

In the same way we may show that the differential of the logarithm of any other quantity is equal to m times the differential of the quantity divided by the quantity, and hence the factor m is a constant, *provided* that the logarithms be taken in each case *in the same system ;* of course, if the logarithms in (8) be taken in a different system from those in (2), the numerical values of $\log y$ in the two equations are different, and therefore the m in (7) is different from the m in (9). Since m is a constant in the same system and different for different systems, it varies with the *base* of the system, as the only other quantities involved in logarithms are *the number* and *its logarithm.* That is, m is a function of the base; its value will be computed hereafter. (See Rice and Johnson's Calculus, p. 39; also, Olney's Calculus, p. 25.)

Hence, *the differential of the logarithm of a quantity is equal to the modulus of the system into the differential of the quantity divided by the quantity.*

Cor.—If the logarithm be taken in the Naperian * system, the modulus is unity, and we have

$$dy = \frac{dx}{x}.$$

Hence, *the differential of the logarithm of a quantity in the Naperian system is equal to the differential of the quantity divided by the quantity.*

21. To Differentiate $y = a^x$.

Passing to logarithms, we have,

$$\log y = x \log a.$$

Differentiating, we have

$$m \frac{dy}{y} = dx \log a;$$

or

$$dy = \frac{y\, dx \log a}{m}.$$

$$\therefore \quad dy = d(a^x) = \frac{a^x}{m} \log a\, dx.$$

---

* So called from the name of the inventor of logarithms; also sometimes called *natural* logarithms, from being those which occur first in the investigation of a method of calculating logarithms. They are sometimes called *hyperbolic* logarithms, from having been originally derived from the hyperbola.

Hence, *the differential of an exponential function with a constant base is equal to the function into the logarithm of the base into the differential of the exponent, divided by the modulus.*

Cor. 1.—If we take *Naperian* logarithms, we have

$$dy = d\,(a^x) = a^x \log a\, dx \text{ (since } m = 1).$$

Cor. 2.—If $a = e$, the Naperian base, then

$$\log a = \log e = 1,$$

and therefore $\qquad dy = d\,(e^x) = e^x\, dx.$

Sch.—In analytical investigations, the logarithms used are almost exclusively Naperian, the base of which system is represented by the letter e. Since the form of the differential is the simplest in the Naperian system, we shall in all cases understand our logarithms to be Naperian, unless otherwise stated.

22. To Differentiate $u = y^x$.

Passing to logarithms, we have,

$$\log u = x \log y.$$

Differentiating, we have,

$$\frac{du}{u} = \log y\, dx + x\frac{dy}{y};$$

or, $\qquad du = u \log y\, dx + ux\frac{dy}{y};$

$$\therefore\quad du = y^x \log y\, dx + xy^{x-1}\, dy.$$

Hence, *to differentiate an exponential function with a variable base, differentiate first as though the base were constant and the exponent variable, and second as though the base were variable and the exponent constant, and take the sum of the results.*

EXAMPLES.

1. $y = x \log x.$ $\qquad\qquad$ $dy = (\log x + 1)\, dx.$

2. $y = \log x^2.$ $\qquad\qquad$ $dy = \dfrac{2\, dx}{x}.$

3. $y = a^{\log x}.$ $\qquad\qquad$ $dy = \dfrac{a^{\log x} \log a \; dx}{x}.$

4. $y = a^{e^z}.$ $\qquad\qquad$ $dy = a^{e^z} e^x \log a \; dx.$

5. $y = x^{x^x}.$ $\quad dy = x^{x^x}\left[\log x \,(1 + \log x) + \dfrac{1}{x}\right] x^x \, dx.$

6. $y = \log \sqrt{1 - x^2}.$ $\qquad\qquad$ $dy = -\dfrac{x\, dx}{1 - x^2}.$

7. $y = \log\left(x + \sqrt{1 + x^2}\right).$ $\qquad$ $dy = \dfrac{dx}{\sqrt{1 + x^2}}.$

8. $y = \log\left(\dfrac{a + x}{a - x}\right) = \log\,(a + x) - \log\,(a - x).$

$$dy = \dfrac{2a\, dx}{a^2 - x^2}.$$

9. $y = \log\sqrt{\dfrac{1 + x}{1 - x}} = \tfrac{1}{2}\log\,(1 + x) - \tfrac{1}{2}\log\,(1 - x).$

$$dy = \dfrac{dx}{1 - x^2}.$$

10. $y = \log\,(\log x).$ $\qquad\qquad$ $dy = \dfrac{dx}{x \log x}.$

11. $y = \log^2 x.$ $\qquad\qquad$ $dy = 2 \log x \dfrac{dx}{x}.$

12. $y = x^x.$ $\qquad\qquad$ $dy = x^x \,(\log x + 1)\, dx.$

13. $y = \log \dfrac{\sqrt{x^2 + 1} - x}{\sqrt{x^2 + 1} + x}.$

Multiplying both terms by the numerator to rationalize the denominator, we get

$$y = \log \left[\sqrt{x^2 + 1} - x \right]^2.$$

$$\therefore \quad dy = - \frac{2dx}{\sqrt{x^2 + 1}}.$$

14. $y = e^x (x - 1).$ $dy = e^x x \, dx.$

15. $y = e^x (x^2 - 2x + 2).$ $dy = e^x x^2 \, dx.$

16. $y = \dfrac{e^x - 1}{e^x + 1}.$ $dy = \dfrac{2e^x}{(e^x + 1)^2} \, dx.$

17. $y = e^x \log x.$ $dy = e^x \left(\log x + \dfrac{1}{x} \right) dx.$

18. $y = e^{\log \sqrt{a^2 + x^2}}.$

     Then $\log y = \log \sqrt{a^2 + x^2}.$

$$\therefore \quad y = \sqrt{a^2 + x^2}.$$

$$dy = \frac{xdx}{\sqrt{a^2 + x^2}}.$$

19. $y = \dfrac{e^x}{1 + x}.$ $dy = \dfrac{xe^x \, dx}{(1 + x)^2}.$

20. $y = \log \dfrac{x}{\sqrt{x^2 + 1} + x}.$ $dy = \dfrac{dx}{x} - \dfrac{dx}{\sqrt{x^2 + 1}}.$

21. $y = \log \left(x + a + \sqrt{2ax + x^2} \right).$ $dy = \dfrac{dx}{\sqrt{2ax + x^2}}.$

22. $y = x^{a\sqrt{-1}}.$ $dy = a\sqrt{-1} \, x^{a\sqrt{-1} - 1} \, dx.$

23. $y = \dfrac{a^x}{x^x}.$ $dy = \left(\dfrac{a}{x} \right)^x \left(\log \dfrac{a}{x} - 1 \right) dx.$

23. Logarithmic Differentiation.—When the function to be differentiated consists of products and quotients, the differentiation is often performed with greater facility by first passing to logarithms. This process is called logarithmic differentiation.

EXAMPLES.

1. $u = x (a^2 + x^2) \sqrt{a^2 - x^2}$.

Passing to logarithms, we have

$$\log u = \log x + \log (a^2 + x^2) + \tfrac{1}{2} \log (a^2 - x^2).$$

$$\therefore \quad \frac{du}{u} = \frac{dx}{x} + \frac{2x\, dx}{a^2 + x^2} - \frac{x\, dx}{a^2 - x^2}.$$

$$\therefore \quad du = \left[(a^2 + x^2)\sqrt{a^2 - x^2} + 2x^2\sqrt{a^2 - x^2} - \frac{x^2 (a^2 + x^2)}{\sqrt{a^2 - x^2}} \right] dx$$

or,

$$du = \frac{a^4 + a^2 x^2 - 4x^4}{\sqrt{a^2 - x^2}}\, dx.$$

2. $u = \dfrac{1 + x^2}{1 - x^2}$. Passing to logarithms, we have

$$\log u = \log (1 + x^2) - \log (1 - x^2).$$

$$\therefore \quad \frac{du}{u} = \frac{2x\, dx}{1 + x^2} + \frac{2x\, dx}{1 - x^2} = \frac{4x\, dx}{(1 + x^2)(1 - x^2)}.$$

$$\therefore \quad du = \frac{4x\, dx}{1 - x^4}.$$

3. $u = (a^x + 1)^2$. $\qquad du = 2a^x (a^x + 1) \log a\, dx$.

4. $u = \dfrac{a^x - 1}{a^x + 1}$. $\qquad\qquad du = \dfrac{2a^x \log a\, dx}{(a^x + 1)^2}$.

5. $u = \dfrac{\sqrt{1 + x}}{\sqrt{1 - x}}$. $\qquad\qquad du = \dfrac{dx}{(1 - x)\sqrt{1 - x^2}}$.

ILLUSTRATIVE EXAMPLES.

1. Which increases the more rapidly, a number or its logarithm? How much more rapidly is the number 4238 increasing than its common logarithm, supposing the two to be increasing uniformly? While the number increases by 1, how much will its logarithm increase, supposing the

latter to increase uniformly (which it does not) while the number increases uniformly.

Let $x =$ the number, and y its logarithm; then we have

$$y = \log x;$$

$$\therefore \ dy = \frac{m}{x} dx,$$

which shows that if we give to the number (x) the infinitely small increment (dx), the corresponding increment of y is $\frac{m}{x}$ times as great; that is, the logarithm (y) is increasing $\frac{m}{x}$ times as fast as the number. Hence, the increase in the common logarithm of a number is $>$, $=$, $<$ the increase of the number, according as the number (x) $<$, $=$, $>$ the modulus (m).

When $x = 4238$, we have

$$dy = \frac{m}{4238} dx = \frac{.43429448}{4238} dx;$$

hence, $\qquad dx = \frac{4238}{.43429448} dy = $ about $9758\, dy;$

that is, the increment of the logarithm is $\frac{.43429448}{4238}$ part of the increment of the number, and the number is increasing about 9758 times as fast as its logarithm.

While the number increases by 1, its logarithm will increase (supposing it to increase uniformly with the number) $\frac{.43429448}{4238}$ times $1 = .00010247$; that is, the logarithm of 4239 would be .00010247 larger than the logarithm of 4238, if it were increasing uniformly, while the number increased from 4238 to 4239.

REMARK.—While a number is increasing *uniformly*, its logarithm is increasing *more and more slowly*; this is evident from the equation $dy = \frac{m}{x} dx$, which shows that if the number receives a very small increment, its logarithm receives a very small increment; **but on giving** to the number a second very small increment equal to the first, the corresponding increment of the logarithm is a little *less* than the first, and so on; and yet the supposition that the relative rate of change of a number and its logarithm is *constant* for comparatively small changes in the number is sufficiently accurate for practical purposes, and is the assumption made in using the *tabular difference* in the tables of logarithms.

2. The common logarithm of 327 is 2.514548. What is the logarithm of 327.12, supposing the relative rate of change of the number and its logarithm to continue uniformly the same from 327 to 327.12 that it is at 327?

Ans. 2.514707.

3. Find what should be the tabular difference in the table of logarithms for numbers between 4825 and 4826; in other words, find the increment of the logarithm while the number increases from 4825 to 4826. *Ans.* .0000900.

4. Find what should be the tabular difference in the table of logarithms for numbers between 9651 and 9652.

Ans. .0000450.

5. Find what should be the tabular difference in the table of logarithms for numbers between 7235 and 7236.

Ans. .0000601.

TRIGONOMETRIC FUNCTIONS.

24. To Differentiate $y = \sin x.$ (1)

Give to x the infinitely small increment dx, and let dy represent the corresponding increment of y; then we have

$$y + dy = \sin (x + dx)$$
$$= \sin x \cos dx + \cos x \sin dx. \qquad (2)$$

Because the arc dx is infinitely small, its sine is equal to the arc itself and its cosine equals 1; therefore (2) may be written

$$y + dy = \sin x + \cos x \, dx. \qquad (3)$$

Subtracting (1) from (3), we have

$$dy = \cos x\, dx. \tag{4}$$

Hence, *the differential of the sine of an arc is equal to the cosine of the arc into the differential of the arc*

25. To Differentiate $y = \cos x$.

Give to x the infinitely small increment dx, and we have

$$y + dy = \cos (x + dx)$$
$$= \cos x \cos dx - \sin x \sin dx$$
$$= \cos x - \sin x\, dx \quad \text{(Art. 24)}.$$
$$\therefore \quad dy = - \sin x\, dx.$$

Otherwise thus:

We have $\qquad y = \cos x = \sin (90° - x).$

Differentiating by Art. 24, we have

$$dy = \cos (90° - x)\, d (90° - x)$$
$$= \sin x\, d (90° - x).$$
$$\therefore \quad dy = - \sin x\, dx.$$

Hence, *the differential of the cosine of an arc is negative and equal to the sine of the arc into the differential of the arc.* (The negative sign shows that the cosine decreases as the arc increases.)

26. To Differentiate $y = \tan x$.

We have $\qquad y = \tan x = \dfrac{\sin x}{\cos x}$

Differentiating by Arts. 18, 24, and 25, we have

$$dy = \frac{\cos x\, d \sin x - \sin x\, d \cos x}{\cos^2 x}$$
$$= \frac{\cos^2 x + \sin^2 x}{\cos^2 x}\, dx = \frac{dx}{\cos^2 x}$$
$$= \sec^2 x\, dx. \qquad \therefore \quad dy = \sec^2 x\, dx.$$

Otherwise thus:

Give to x the infinitesimal increment dx, and we have

$$y + dy = \tan (x + dx)$$

$$\therefore \ dy = \tan (x + dx) - \tan x$$

$$= \frac{\tan x + \tan dx}{1 - \tan x \tan dx} - \tan x$$

$$= \frac{\tan x + dx}{1 - \tan x \, dx} - \tan x \ (\text{since } \tan dx = dx)$$

$$= \frac{dx + \tan^2 x \, dx}{1 - \tan x \, dx} = \sec^2 x \, dx$$

(since $\tan x \, dx$, being an infinitesimal, may be dropped from the denominator).

$$\therefore \ dy = \sec^2 x \, dx.$$

Hence, *the differential of the tangent of an arc is equal to the square of the secant of the arc into the differential of the arc.*

27. To Differentiate $y = \cot x.$

We have $\quad y = \cot x = \tan (90° - x)$.

$$\therefore \ dy = \sec^2 (90° - x) \ d (90° - x)$$

$$\therefore \ dy = - \operatorname{cosec}^2 x \, dx.$$

The minus sign shows that the cotangent decreases as the arc increases.

Hence, *the differential of the cotangent of an arc is negative, and equal to the square of the cosecant of the arc into the differential of the arc.*

28. To Differentiate $y = \sec x.$

We have $\quad y = \sec x = \dfrac{1}{\cos x}.$

$$\therefore \quad dy = -\frac{d\cos x}{\cos^2 x} = \frac{\sin x \, dx}{\cos^2 x} = \sec x \tan x \, dx.$$

Hence, *the differential of the secant of an arc is equal to the secant of the same arc, into the tangent of the arc, into the differential of the arc.*

29. To Differentiate $y = \operatorname{cosec} x.$

We have $y = \operatorname{cosec} x = \sec(90° - x).$

$$\therefore \quad dy = d\sec(90° - x)$$
$$= \sec(90° - x)\tan(90° - x)\, d(90° - x)$$
$$= -\operatorname{cosec} x \cot x \, dx.$$

Hence, *the differential of the cosecant of an arc is negative, and equal to the cosecant of the arc, into the cotangent of the arc, into the differential of the arc.*

30. To Differentiate $y = \operatorname{vers} x.$

We have $y = \operatorname{vers} x = 1 - \cos x.$

$$\therefore \quad dy = d(1 - \cos x) = \sin x \, dx.$$

Hence, *the differential of the versed-sine of an arc is equal to the sine of the arc into the differential of the arc.*

31. To Differentiate $y = \operatorname{covers} x.$

We have $y = \operatorname{covers} x = \operatorname{vers}(90° - x).$

$$\therefore \quad dy = d\operatorname{vers}(90° - x) = \sin(90° - x)\, d(90° - x)$$
$$= -\cos x \, dx.$$

Hence, *the differential of the coversed-sine of an arc is negative, and equal to the cosine of the arc into the differential of the arc.*

32. Geometric Demonstration.—The results arrived at in the preceding Articles admit also of easy demonstration by geometric construction.

Let P and Q be two consecutive points * in the arc of a circle described with radius $= 1$. Let $x = $ arc AP ; then

$$dx = \text{arc PQ.}$$

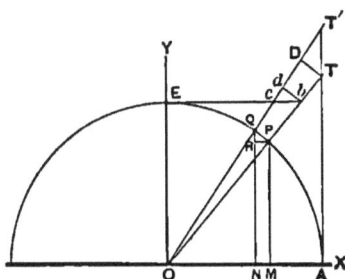

Fig. 4.

From the figure we have,

PM $= \sin x$;   NQ $= \sin (x + dx)$;

$$\therefore \quad \text{QR} = d \sin x.$$

OM $= \cos x$;   ON $= \cos (x + dx)$;

$\therefore$ NM $= - d \cos x$ (minus because decreasing).

AT $= \tan x$;   AT$'$ $= \tan (x + dx)$;

$$\therefore \quad \text{TT}' = d \tan x.$$

OT $= \sec x$;   OT$'$ $= \sec (x + dx)$;

$$\therefore \quad \text{DT}' = d \sec x.$$

Now, since RP and QP are perpendicular respectively to MP and OP, and since DT and TT$'$ are also perpendicular to OT and OA respectively, the two infinitely small triangles PQR and DTT$'$ are similar to MOP. Hence we have the following equations:

$$d \sin x = \text{RQ} = \text{QP} \cos \text{PQR}$$
$$= \cos x \, dx.$$
$$\therefore \quad d \sin x = \cos x \, dx.$$

---

* All that is meant here is that P and Q are to be *reasoned* upon as though they were consecutive points ; of course, strictly speaking. consecutive points can never be represented *geometrically*, since their distance apart is less than any assignable distance. When we say that P and Q are consecutive points, we may regard the distance PQ in the figure as representing the infinitesimal distance between two consecutive points, *highly magnified.*

$$d \cos x = - \mathrm{PR} = - \mathrm{PQ} \sin \mathrm{PQR}$$
$$\dot{=} - \sin x \, dx.$$

$$\therefore \quad d \cos x = - \sin x \, dx.$$

$$\boldsymbol{d} \tan x = \mathrm{TT'} = \mathrm{DT} \sec \mathrm{DTT'} = \mathrm{DT} \sec x$$
$$= \mathrm{OT \cdot QP} \sec x \text{ (since } \mathrm{DT} = \mathrm{OT \cdot QP})$$
$$= \sec^2 x \, dx.$$

$$\therefore \quad d \tan x = \sec^2 x \, dx.$$

$$d \sec x = \mathrm{DT'} = \mathrm{DT} \tan \mathrm{DTT'}$$
$$= \mathrm{OT \cdot QP} \tan x = \sec x \, \boldsymbol{\tan x \, dx.}$$

$$\therefore \quad d \sec x = \sec x \tan x \, dx.$$

Also, $$cb = - d \, (\cot x),$$

and $$cd = - d \, (\operatorname{cosec} x).$$

But the triangle cbd is similar to the triangle OPM, since cb and db are respectively perpendicular to MP and OP. Hence we have

$$d \cot x = - cb = - db \operatorname{cosec} dcb$$
$$= - b\mathrm{O \cdot QP} \operatorname{cosec} x = - \operatorname{cosec}^2 x \, \boldsymbol{dx.}$$

$$\therefore \quad d \cot x = - \operatorname{cosec}^2 x \, dx.$$

$$\boldsymbol{d} \operatorname{cosec} x = - cd = - db \cot dcb$$
$$= - \mathrm{O}b \cdot \mathrm{QP} \cot x$$
$$= - \operatorname{cosec} x \cot x \, dx.$$

$$\therefore \quad d \operatorname{cosec} x = - \operatorname{cosec} x \cot x \, dx.$$

From the figure we see that the differential of the *versed-*sine is the same numerically as that of the cosine, but with a contrary sign, *i. e.*, as the versed-sine increases the cosine decreases ; also the differential of the coversed-sine has the same value numerically as that of the sine.

E X A M P L E S.

1. $y = \sin mx.$ By Art. 24 we have,

$$dy = \cos mx \, d(mx) = m \cos mx \, dx.$$

2. $y = \sin(x^2).$

$$dy = \cos(x^2) \, d(x^2) = 2x \cos(x^2) \, dx.$$

3. $y = \sin^m x.$

$$dy = m \sin^{m-1} x \, d(\sin x) = m \sin^{m-1} x \cos x \, dx.$$

4. $y = \cos^3 x.$

$$dy = 3 \cos^2 x \, d(\cos x) = -3 \cos^2 x \sin x \, dx$$
$$= 3(\sin^3 x - \sin x) \, dx.$$

5. $y = \sin 2x \cos x.$

$$dy = \sin 2x \, d \cos x + \cos x \, d \sin 2x$$
$$= -\sin 2x \sin x \, dx + 2 \cos 2x \cos x \, dx.$$

6. $y = \cot^2(x^3).$ $dy = -6x^2 \cot x^3 \operatorname{cosec}^2 x^3 \, dx.$

7. $y = \sin^3 x \cos x.$ $dy = \sin^2 x \, (3 - 4 \sin^2 x) \, dx.$

8. $y = 3 \sin^4 x.$ $dy = 12 \sin^3 x \cos x \, dx.$

9. $y = \sec^2 5x.$ $dy = 10 \sec^2 5x \tan 5x \, dx.$

10. $y = \log \sin x.$

$$dy = \frac{d(\sin x)}{\sin x} \text{ (Art. 20)} = \frac{\cos x}{\sin x} dx = \cot x \, dx.$$

11. $y = \log(\sin^2 x) = 2 \log \sin x.$ $dy = 2 \cot x \, dx.$

12. $y = \log \cos x.$ $dy = -\tan x \, dx.$

13. $y = \log \tan x.$ $dy = \dfrac{d \tan x}{\tan x} = \dfrac{2dx}{\sin 2x}.$

14. $y = \log \cot x.$ $dy = -\dfrac{2dx}{\sin 2x}.$

15. $y = \log \sec x.$ $dy = \tan x \, dx.$

16. $y = \log \operatorname{cosec} x.$ $dy = -\cot x \, dx.$

17. $y = \log \sqrt{\dfrac{1 + \cos x}{1 - \cos x}}$

$= \tfrac{1}{2} \log (1 + \cos x) - \tfrac{1}{2} \log (1 - \cos x).$

$$dy = - \frac{dx}{\sin x}.$$

18. $y = e^x \cos x.$ $dy = e^x \, d \cos x + \cos x \, de^x$

$= - e^x \sin x \, dx + e^x \cos x \, dx$

$= e^x (\cos x - \sin x) \, dx.$

19. $y = x \sin x + \cos x.$ $dy = x \cos x \, dx.$

20. $y = x e^{\cos x}.$ $dy = e^{\cos x} (1 - x \sin x) \, dx.$

21. $y = e^{\cos x} \sin x.$ $dy = e^{\cos x} (\cos x - \sin^2 x) \, dx.$

22. $y = \log \sqrt{\sin x} + \log \sqrt{\cos x}$

$= \tfrac{1}{2} \log \sin x + \tfrac{1}{2} \log \cos x.$

$\therefore \;\; dy = \tfrac{1}{2} (\cot x - \tan x) \, dx = \cot 2x \, dx.$

23. $y = \log (\cos x + \sqrt{-1} \sin x).$ $dy = \sqrt{-1} \, dx.$

24. $y = \log \sqrt{\dfrac{1 + \sin x}{1 - \sin x}}.$ $dy = \dfrac{dx}{\cos x}.$

25. $y = \log \tan (45° + \tfrac{1}{2}x).$ $dy = \dfrac{dx}{\cos x}.$

26. $y = \sin (\log x).$ $dy = \dfrac{1}{x} \cos (\log x) \, dx.$

ILLUSTRATIVE EXAMPLES.

1. Which increases faster, the arc or its tangent? When is this difference least, and when greatest? What is the value of the arc when the tangent is increasing twice as fast as the arc, and when increasing four times as fast as the arc?

From $y = \tan x$, we get $dy = \sec^2 x \, dx$, which shows that if we give to the arc (x) the infinitesimal increment dx, the corresponding increment of the tangent (y) is $\sec^2 x$ times as great; that is, the tangent (y) is increasing secant square times as fast as the arc, and hence is *generally* increasing more rapidly than the arc. When $x = 0$, $\sec x = 1$; therefore, at this point, the tangent and the arc are increasing at the same rate. When $x = 90°$, the secant is infinite; therefore, at this point, the tangent is increasing infinitely faster than the arc.

When the tangent is increasing twice as fast as the arc, we have $dy = 2dx$, or $\sec^2 x = 2$, which gives $x = 45°$; hence at $45°$ the tangent is increasing twice as fast as the arc.

When the tangent is increasing four times as fast as the arc, we have $dy = 4dx$, or $\sec^2 x = 4$, which gives $x = 60°$; hence at $60°$ the tangent is increasing four times as fast as the arc.

2. Assuming that the relative rate of increase of the sine, as compared with the arc, remains constantly the same as at $60°$, how much does the sine increase when the arc increases from $60°$ to $60°$ $20'$.

Let $x =$ the arc and y its sine; then we have $y = \sin x$, $\therefore dy = \cos x\, dx$, which shows that the increment of the sine is cosine times the increment of the arc. Now the arc of $20' = \dfrac{3.14159}{180 \times 3} = .0058177 = dx$; therefore,

$$dy = \cos 60°\, dx = \tfrac{1}{2} \times .0058176 = .0029088,$$

which is the increase of the sine on the above supposition, and is a little greater than the increase as found from a table of natural sines, as it should be. since we have supposed the sine to increase *uniformly* while the arc was increasing uniformly from $60°$ to $60°$ $20'$, whereas the sine is increasing *more* and *more* slowly while the arc is increasing uniformly. This is evident from the equation $dy = \cos x\, dx$, and also from geometric considerations.

3. The natural cosine of $5° 31'$ is .995368. Assuming that the relative rate of change of the cosine and the arc remains the same as at this point, while the arc increases to $5° 32'$, what is the cosine of $5° 32'$?   *Ans.* .995340.

4. The logarithmic sine of $13° 49'$ is 9.3780633. Assuming that the relative rate of change of the logarithmic sine and the arc remains the same as at this point, while the arc increases to $13° 49' 10''$, what is the logarithmic sine of $13° 49' 10''$?   *Ans.* 9.3781489.

5. The log cot $58° 21' = 9.789863$. On the same supposition as above, what is the decrease of this logarithm for 1 second increase of arc.   *Ans.* .00000471.

6. A wheel is revolving in a vertical plane about a fixed centre. At what rate, as compared with its angular velocity, is a point in its circumference ascending, when it is $60°$ above the horizontal plane through the centre of motion.

Ans. Half as fast.

CIRCULAR FUNCTIONS.

33. To Differentiate $y = \sin^{-1} x$.*

We have $\qquad x = \sin y$;

therefore, $dx = \cos y \, dy = \sqrt{(1 - \sin^2 y)} \, dy$

$$= \sqrt{1 - x^2} \, dy.$$

$$\therefore \quad dy = \frac{dx}{\sqrt{1 - x^2}} = d\,(\sin^{-1} x).$$

34. To Differentiate $y = \cos^{-1} x$.

We have, $\qquad x = \cos y$;

therefore, $dx = -\sin y \, dy = -\sqrt{1 - \cos^2 y} \, dy$

$$= -\sqrt{1 - x^2} \, dy.$$

$$\therefore \quad dy = -\frac{dx}{\sqrt{1 - x^2}} = d\,(\cos^{-1} x).$$

---

* This notation, as already explained, means $y =$ the arc whose sine is x.

35. To Differentiate $y = \tan^{-1} x$.

We have
$$x = \tan y \, ;$$

therefore,
$$dx = \sec^2 y \, dy = (1 + \tan^2 y) \, dy$$
$$= (1 + x^2) \, dy.$$

$$\therefore \ dy = \frac{dx}{1 + x^2} = d \, (\tan^{-1} x).$$

36. To Differentiate $y = \cot^{-1} x$.

We have
$$x = \cot y \, ;$$

therefore,
$$dx = - \cosec^2 y \, dy = - (1 + \cot^2 y) \, dy$$
$$= - (1 + x^2) \, dy.$$

$$\therefore \ dy = - \frac{dx}{1 + x^2} = d \, (\cot^{-1} x).$$

37. To Differentiate $y = \sec^{-1} x$.

We have
$$x = \sec y \, ;$$

therefore,
$$dx = \sec y \, \tan y \, dy = \sec y \, \sqrt{\sec^2 y - 1} \, dy$$
$$= x\sqrt{x^2 - 1} \, dy.$$

$$\therefore \ dy = \frac{dx}{x\sqrt{x^2 - 1}} = d \, (\sec^{-1} x).$$

38. To Differentiate $y = \cosec^{-1} x$.

We have
$$x = \cosec y \, ;$$

therefore,
$$dx = - \cosec y \, \cot y \, dy$$
$$= - \cosec y \, \sqrt{\cosec^2 y - 1} \, dy$$
$$= - x\sqrt{x^2 - 1} \, dy.$$

$$\therefore \ dy = - \frac{dx}{x\sqrt{x^2 - 1}} = d \, (\cosec^{-1} x).$$

39. To Differentiate $y = \text{vers}^{-1} x$.

We have $\qquad\qquad x = \text{vers } y\,;$

therefore, $\quad dx = \sin y\, dy = \sqrt{1 - \cos^2 y}\, dy$

$$= \sqrt{1 - (1 - \text{vers } y)^2}\, dy$$

$$= \sqrt{2 \text{ vers } y - \text{vers}^2 y}\, dy$$

$$= \sqrt{2x - x^2}\, dy.$$

$$\therefore\quad dy = \frac{dx}{\sqrt{2x - x^2}} = d\,(\text{vers}^{-1} x)$$

40. To Differentiate $y = \text{covers}^{-1} x$.

We have $\qquad\qquad x = \text{covers } y\,;$

therefore, $\quad dx = -\cos y\, dy = -\sqrt{1 - \sin^2 y}\, dy$

$$= -\sqrt{1 - (1 - \text{covers } y)^2}\, dy$$

$$= -\sqrt{2 \text{ covers } y - \text{covers}^2 y}\, dy$$

$$= -\sqrt{2x - x^2}\, dy.$$

$$\therefore\quad dy = -\frac{dx}{\sqrt{2x - x^2}} = d\,(\text{covers}^{-1} x).$$

EXAMPLES.

1. Differentiate $y = \sin^{-1} \dfrac{x}{a}$.

We have, by Art. 33,

$$dy = \frac{d\,\dfrac{x}{a}}{\sqrt{1 - \dfrac{x^2}{a^2}}} = \frac{\dfrac{dx}{a}}{\sqrt{1 - \dfrac{x^2}{a^2}}} = \frac{dx}{\sqrt{a^2 - x^2}}.$$

2. Differentiate $y' = a \sin^{-1} \dfrac{x}{a}$.

We have, Art. 33, $\quad dy' = \dfrac{a\,\dfrac{dx}{a}}{\sqrt{1 - \dfrac{x^2}{a^2}}} = \dfrac{a\,dx}{\sqrt{a^2 - x^2}}.$

Geometric illustration of Examples 1 and 2 :

Let $\quad OA = 1, \quad OA' = a, \quad y = \text{arc } AB, \quad y' = \text{arc } A'B',$
$x = M'B'.$

Now $\quad BM = \dfrac{B'M'}{OB'} = \dfrac{x}{a};$

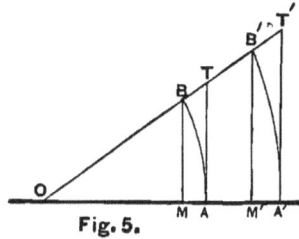
Fig. 5.

$\therefore \quad \text{arc } AB = \sin^{-1} BM$

$\qquad = \sin^{-1} \dfrac{B'M'}{OB'}$

$\qquad = \sin^{-1} \dfrac{x}{a}$

$\qquad = y \ (\text{see Ex. 1}).$

$A'B' = A'O \cdot \text{arc } AB = A'O \cdot \sin^{-1} \dfrac{B'M'}{OB'}$

$\qquad = a \sin^{-1} \dfrac{x}{a} = y' \ (\text{see Ex. 2}).$

Also, $\quad A'B' = \sin^{-1} B'M' = \sin^{-1} x \ (\text{to radius } a)$

$\therefore \quad a \sin^{-1} \dfrac{x}{a} \ (\text{to radius 1}) = \sin^{-1} x \ (\text{to radius } a).$

Hence, in Example 1, y is the arc AB (to radius 1), and is given in terms of the sine $\dfrac{x}{a}$ (to radius 1); while in Example 2, y' is the arc $A'B'$ (to radius a), and is given in terms of the sin $\dfrac{x}{a}$ (to radius 1).

If we give $B'M'$ (which is x in both examples) an increment ($= dx$), the corresponding increment in $A'B'$ will be a times as great as that on AB; that is, dy' in Ex. 2 is a times dy in Ex. 1.

3

3. $y = \cos^{-1} \dfrac{x}{a}.$ $dy = -\dfrac{dx}{\sqrt{a^2 - x^2}}.$

4. $y = \tan^{-1} \dfrac{x}{a}.$ $dy = \dfrac{adx}{a^2 + x^2}.$

5. $y = \cot^{-1} \dfrac{x}{a}.$ $dy = -\dfrac{adx}{a^2 + x^2}.$

6. $y = \sec^{-1} \dfrac{x}{a}.$ $dy = \dfrac{adx}{x\sqrt{x^2 - a^2}}.$

7. $y = \operatorname{cosec}^{-1} \dfrac{x}{a}.$ $dy = -\dfrac{adx}{x\sqrt{x^2 - a^2}}.$

8. $y = \operatorname{vers}^{-1} \dfrac{x}{a}.$ $dy = \dfrac{dx}{\sqrt{2ax - x^2}}.$

9. $y = \operatorname{covers}^{-1} \dfrac{x}{a}.$ $dy = -\dfrac{dx}{\sqrt{2ax - x^2}}.$

10. $y = a \cos^{-1} \dfrac{x}{a}.$

$$dy = -\dfrac{\dfrac{adx}{a}}{\sqrt{1 - \dfrac{x^2}{a^2}}} = -\dfrac{adx}{\sqrt{a^2 - x^2}}.$$

11. $y = a \tan^{-1} \dfrac{x}{a}.$ $dy = \dfrac{a\dfrac{dx}{a}}{1 + \dfrac{x^2}{a^2}} = \dfrac{a^2 dx}{a^2 + x^2}.$

12. $y = a \cot^{-1} \dfrac{x}{a}.$ $dy = -\dfrac{a\dfrac{dx}{a}}{1 + \dfrac{x^2}{a^2}} = -\dfrac{a^2 dx}{a^2 + x^2}.$

13. $y = a \sec^{-1} \dfrac{x}{a}.$ $dy = \dfrac{a\dfrac{dx}{a}}{\dfrac{x}{a}\sqrt{\dfrac{x^2}{a^2} - 1}} = \dfrac{a^2 dx}{x\sqrt{x^2 - a^2}}.$

14. $y = a \cosec^{-1} \dfrac{x}{a}.$

$$dy = -\frac{a\dfrac{dx}{a}}{\dfrac{x}{a}\sqrt{\dfrac{x^2}{a^2}-1}} = -\frac{a^2 dx}{x\sqrt{x^2-a^2}}.$$

15. $y = a \vers^{-1} \dfrac{x}{a}.$ $dy = \dfrac{a\dfrac{dx}{a}}{\sqrt{2\dfrac{x}{a}-\dfrac{x^2}{a^2}}} = \dfrac{adx}{\sqrt{2ax-x^2}}.$

16. $y = a \covers^{-1} \dfrac{x}{a}.$

$$dy = -\frac{a\dfrac{dx}{a}}{\sqrt{2\dfrac{x}{a}-\dfrac{x^2}{a^2}}} = -\frac{adx}{\sqrt{2ax-x^2}}.$$

MISCELLANEOUS EXAMPLES.

1. $y = \dfrac{a+x}{\sqrt{a-x}}.$ $\qquad dy = \dfrac{3a-x}{2(a-x)^{\frac{3}{2}}}dx.$

2. $y = \sqrt{x-\sqrt{a^2-x^2}}.$

$$dy = \frac{\left(x+\sqrt{a^2-x^2}\right)dx}{2\sqrt{a^2-x^2}\left(x-\sqrt{a^2-x^2}\right)^{\frac{1}{2}}}.$$

3. $y = \dfrac{x}{x+\sqrt{1-x^2}}.$ $dy = \dfrac{dx}{2x(1-x^2)+\sqrt{1-x^2}}.$

4. $y = \dfrac{x^3}{\sqrt{(1-x^2)^3}}.$ $\qquad dy = \dfrac{3x^2\,dx}{(1-x^2)^{\frac{5}{2}}}.$

5. $y = \dfrac{a^4}{2\sqrt{a^2x^2-x^4}}.$ $dy = \dfrac{-a^4(a^2-2x^2)\,dx}{2x^2(a^2-x^2)^{\frac{3}{2}}}.$

6. $y = \dfrac{\sqrt{x^2 + 1} - x}{\sqrt{x^2 + 1} + x}.$

In fractions of this form, the student will find it an advantage to rationalize the denominator, by multiplying both terms of the fraction by the *complementary surd form;* that is, in the present case, by $\sqrt{x^2 + 1} - x$. Thus,

$$y = \frac{\sqrt{x^2 + 1} - x}{\sqrt{x^2 + 1} + x} \times \frac{\sqrt{x^2 + 1} - x}{\sqrt{x^2 + 1} - x}$$

$$= \frac{(\sqrt{x^2 + 1} - x)^2}{1}.$$

$$\therefore \; dy = 2\left(\sqrt{x^2 + 1} - x\right)\left(\frac{x}{\sqrt{x^2 + 1}} - 1\right) dx$$

$$= 2\left(2x - \frac{2x^2 + 1}{\sqrt{x^2 + 1}}\right) dx.$$

7. $y = \sqrt{\dfrac{1 - \sqrt{x}}{1 + \sqrt{x}}} = \dfrac{\sqrt{1 - \sqrt{x}}}{\sqrt{1 + \sqrt{x}}} = \dfrac{\sqrt{1 - x}}{1 + \sqrt{x}}.$

$$dy = -\frac{dx}{2\left(1 + \sqrt{x}\right)\sqrt{x - x^2}}.$$

8. $y = \dfrac{\sqrt{1 + x} + \sqrt{1 - x}}{\sqrt{1 + x} - \sqrt{1 - x}}.$

$$dy = -\frac{1 + \sqrt{1 - x^2}}{x^2\sqrt{1 - x^2}}\, dx.$$

9. $y = \sqrt[3]{x} \cdot \sqrt{\sqrt{x} + 1}.$

$$dy = \frac{7\sqrt{x} + 4}{12\sqrt[3]{x^2} \cdot \sqrt{\sqrt{x} + 1}}\, dx.$$

10. $y = \sqrt[4]{\left[a - \dfrac{b}{\sqrt{x}} + \sqrt[3]{(c^2 - x^2)^2} \right]^3}.$

$$dy = \frac{\dfrac{3b}{2x\sqrt{x}} - \dfrac{4x}{\sqrt[3]{(c^2 - x^2)}}}{4 \sqrt[4]{a - \dfrac{b}{\sqrt{x}} + \sqrt[3]{(c^2 - x^2)^2}}} \, dx.$$

11. $y = \sqrt{2x-1-\sqrt{2x-1-\sqrt{2x-1-} \text{ etc.,}}}$ *ad inf.*

Squaring, we have,

$$y^2 = 2x-1-\sqrt{2x-1-\sqrt{2x-1-\sqrt{2x-1-} \text{ etc.,}}} \text{ ad inf.}$$

Hence, $\qquad y^2 = 2x - 1 - y\,;$

and $\qquad\qquad y = -\tfrac{1}{2} \pm \tfrac{1}{2}\sqrt{8x - 3}.$

$$\therefore \cdot dy = \pm \frac{2dx}{\sqrt{8x - 3}}.$$

12. $y = \dfrac{\sqrt{1 + x^2} + \sqrt{1 - x^2}}{\sqrt{1 + x^2} - \sqrt{1 - x^2}}.$

$$dy = -\frac{2}{x^3}\left(1 + \frac{1}{\sqrt{1 - x^4}}\right) dx.$$

13. $y = \log \dfrac{x}{\sqrt{1 + x^2}}.$ $\qquad dy = \dfrac{dx}{x\,(1 + x^2)}.$

14. $y = \log \dfrac{\sqrt{1 + x} + \sqrt{1 - x}}{\sqrt{1 + x} - \sqrt{1 - x}}.$ $dy = -\dfrac{dx}{x\sqrt{1 - x^2}}.$

15. $y = \log \dfrac{x}{\sqrt{x^2 + a^2} - x}.$ $\qquad dy = \dfrac{dx}{x} + \dfrac{dx}{\sqrt{x^2 + a^2}}.$

16.　$y = \log \left[\sqrt{1 + x^2} + \sqrt{1 - x^2} \right].$

$$dy = \frac{dx}{x} - \frac{dx}{x\sqrt{1 - x^4}}.$$

17.　$y = \log (x - a) - \dfrac{a\,(2x - a)}{(x - a)^2}.$　　$dy = \dfrac{x^2 + a^2}{(x - a)^3} dx.$

18.　$y = a^{x^2}.$　　　　　　　　$dy = 2a^{x^2} \log a\; x\,dx.$

19.　$y = e^x (1 - x^3).$　　　　$dy = e^x (1 - 3x^2 - x^3)\, dx.$

20.　$y = \dfrac{e^x - e^{-x}}{e^x + e^{-x}}.$　　　　$dy = \dfrac{4\,dx}{(e^x + e^{-x})^2}.$

21.　$y = \log (e^x + e^{-x}).$　　$dy = \dfrac{e^x - e^{-x}}{e^x + e^{-x}}\, dx.$

22.　$y = x^{\frac{1}{x}}.$　　　　　　　$dy = \dfrac{x^{\frac{1}{x}} (1 - \log x)}{x^2}\, dx.$

23.　$y = 2e^{\sqrt{x}} \left(x^{\frac{3}{2}} - 3x + 6x^{\frac{1}{2}} - 6 \right).$　　$dy = x e^{\sqrt{x}}\, dx.$

24.　$y = \dfrac{(x - 1)^{\frac{5}{2}}}{(x - 2)^{\frac{3}{4}} (x - 3)^{\frac{7}{3}}}.$　　(See Art. 23.)

$$dy = - \frac{(x - 1)^{\frac{3}{2}} (7x^2 + 30x - 97)}{12\,(x - 2)^{\frac{7}{4}} (x - 3)^{\frac{10}{3}}}\, dx.$$

25.　$y = \dfrac{(x + 1)^{\frac{1}{2}} (x + 3)^{\frac{9}{2}}}{(x + 2)^4}.$　　(Art. 23.)

$$dy = \frac{x^2 (x + 3)^{\frac{7}{2}}\, dx}{(x + 2)^5 (x + 1)^{\frac{1}{2}}}.$$

26.　$y = \dfrac{e^{x\sqrt{x^2 - 1}}}{x + \sqrt{x^2 - 1}}.$　　$dy = \dfrac{2\left(\sqrt{x^2 - 1}\right) e^{x\sqrt{x^2 - 1}}\, dx}{x + \sqrt{x^2 - 1}}.$

27.　$y = \sin x - \frac{1}{3} \sin^3 x.$　　　　　$dy = \cos^3 x\, dx.$

28.　$y = \frac{1}{3} \tan^3 x - \tan x + x.$　　　$dy = \tan^4 x\, dx.$

29.　$y = \frac{1}{3} \tan^3 x + \tan x.$　　　　$dy = \sec^4 x\, dx.$

30. $y = \sin e^x.$ $\qquad\qquad dy = e^x \cos e^x \, dx.$

31. $y = \tan^2 x + \log(\cos^2 x).$ $\qquad dy = 2 \tan^3 x \, dx.$

32. $y = \log(\tan x + \sec x).$ $\qquad dy = \sec x \, dx.$

33. $y = \dfrac{\sin x}{1 + \tan x}.$ $\qquad dy = \dfrac{(\cos^3 x - \sin^3 x)}{(\sin x + \cos x)^2} \, dx.$

34. $y = \log \sqrt{\dfrac{a \cos x - b \sin x}{a \cos x + b \sin x}}.$

$$dy = \frac{-ab \, dx}{a^2 \cos^2 x - b^2 \sin^2 x}.$$

35. $y = \tan e^{\frac{1}{x}}.$ $\qquad dy = -\dfrac{e^{\frac{1}{x}} \sec^2 e^{\frac{1}{x}} \, dx}{x^2}.$

36. $y = \tan \sqrt{1 - x}.$ $\qquad dy = -\dfrac{\left(\sec^2 \sqrt{1 - x}\right) dx}{2\sqrt{1 - x}}.$

37. $y = x^{\sin x}.$ $\qquad dy = x^{\sin x}\left(\cos x \cdot \log x + \dfrac{\sin x}{x}\right) dx.$

38. $y = \dfrac{2}{\sin^2 x \cos x} - \dfrac{3 \cos x}{\sin^2 x} + 3 \log \tan \dfrac{x}{2}.$

$$dy = \frac{2dx}{\sin^3 x \cos^2 x}.$$

39. $y = \sin^{-1} \dfrac{x}{\sqrt{1 + x^2}}.$

We have, $\quad dy = d\left(\dfrac{x}{\sqrt{1 + x^2}}\right) \div \sqrt{1 - \dfrac{x^2}{1 + x^2}}$

$$= \frac{dx}{(1 + x^2)^{\frac{3}{2}}} \div \frac{1}{(1 + x^2)^{\frac{1}{2}}} = \frac{dx}{1 + x^2}.$$

40. $y = \cos^{-1} x \sqrt{1 - x^2} = \cos^{-1} \sqrt{x^2 - x^4}.$

We have, $\quad dy = -d\sqrt{x^2 - x^4} \div \sqrt{1 - (x^2 - x^4)}$

$$= -\frac{(1 - 2x^2) \, dx}{\sqrt{1 - x^2}} \div \sqrt{1 - x^2 + x^4}.$$

$$\therefore \quad dy = -\frac{(1 - 2x^2)\, dx}{\sqrt{(1 - x^2)(1 - x^2 + x^4)}}.$$

41. $y = \sin^{-1}\left(2x\sqrt{1 - x^2}\right).$ $dy = \dfrac{2dx}{\sqrt{1 - x^2}}.$

42. $y = \sin^{-1}(3x - 4x^3).$ $dy = \dfrac{3dx}{\sqrt{1 - x^2}}.$

43. $y = \tan^{-1}\dfrac{2x}{1 - x^2}.$ $dy = \dfrac{2dx}{1 + x^2}.$

44. $y = \sin^{-1}\dfrac{1 - x^2}{1 + x^2}.$ $dy = -\dfrac{2dx}{1 + x^2}.$

45. $x = \operatorname{vers}^{-1} y - \sqrt{2ry - y^2}$ (where $\operatorname{vers}^{-1} y$ is taken to radius r).

We may write this (see Art. 40, Exs. 1 and 2),

$$x = r\,\operatorname{vers}^{-1}\frac{y}{r} - \sqrt{2ry - y^2}.$$

$$dx = \frac{rdy}{\sqrt{2ry - y^2}} - \frac{rdy - ydy}{\sqrt{2ry - y^2}}$$

$$= \frac{ydy}{\sqrt{2ry - y^2}}.$$

46. $y = x\sqrt{a^2 - x^2} + a^2\sin^{-1}\dfrac{x}{a}.$ $dy = 2\sqrt{a^2 - x^2}\, dx.$

47. $y = \log\sqrt[4]{\dfrac{1 + x}{1 - x}} + \tfrac{1}{2}\tan^{-1} x.$ $dy = \dfrac{dx}{1 - x^4}.$

48. $y = \operatorname{vers}^{-1}\dfrac{2x}{9}.$ $dy = \dfrac{dx}{\sqrt{9x - x^2}}.$

49. $y = e^{\tan^{-1} x}.$ $dy = e^{\tan^{-1} x}\dfrac{dx}{1 + x^2}.$

50. $y = x^{\sin^{-1} x}$.

$$dy = x^{\sin^{-1} x} \left[\frac{x \log x + (1 - x^2)^{\frac{1}{2}} \sin^{-1} x}{x (1 - x^2)^{\frac{1}{2}}} \right] dx.$$

51. $y = \sec^{-1} nx.$ $\qquad dy = \dfrac{dx}{x\sqrt{n^2 x^2 - 1}}.$

52. $y = \sin^{-1} \dfrac{x}{\sqrt{a^2 + x^2}}.$ $\quad dy = \dfrac{a\,dx}{a^2 + x^2}.$

53. $y = \sin^{-1} \sqrt{\sin x}.$ $\qquad dy = \frac{1}{2}\left(\sqrt{1 + \operatorname{cosec} x}\right) dx.$

54. $y = \tan^{-1} \sqrt{\dfrac{1 - \cos x}{1 + \cos x}}.$ $\qquad dy = \frac{1}{2} dx.$

55. $y = \dfrac{x \sin^{-1} x}{\sqrt{1 - x^2}} + \log \sqrt{1 - x^2}.$ $\; dy = \dfrac{\sin^{-1} x}{(1 - x^2)^{\frac{3}{2}}} dx.$

56. $y = (x + a) \tan^{-1} \sqrt{\dfrac{x}{a}} - \sqrt{ax}.$

$$dy = \tan^{-1} \sqrt{\frac{x}{a}}\, dx.$$

57. $y = \sec^{-1} \dfrac{x\sqrt{5}}{2\sqrt{x^2 + x - 1}}.$ $\quad dy = \dfrac{dx}{x\sqrt{x^2 + x - 1}}.$

58. $y = \tan^{-1} \dfrac{3a^2 x - x^3}{a^3 - 3ax^2}.$ $\qquad dy = \dfrac{3a\,dx}{a^2 + x^2}.$

59. $y = \sin^{-1} \dfrac{x\sqrt{a - b}}{\sqrt{a(1 + x^2)}}.$

$$dy = \frac{\sqrt{a - b}}{(1 + x^2)\sqrt{a + bx^2}}\, dx.$$

60. If two bodies start together from the extremity of the diameter of a circle, the one moving uniformly along the diameter at the rate of 10 feet per second, and the other in the circumference with a variable velocity so as to keep it always directly above the former, what is its velocity in the

circumference when passing the sixtieth degree from the starting-point?

$$Ans. \ \frac{20}{\sqrt{3}} \ \text{feet per second.}$$

61. If two bodies start together from the extremity of the diameter of a circle, the one moving uniformly along the tangent at the rate of 10 feet per second, and the other in the circumference with a variable velocity, so as to be always in the right line joining the first body with the centre of the circle, what is its velocity when passing the forty-fifth degree from the starting-point.   *Ans.* 5 ft. per second.

CHAPTER III

41. Limiting Values.—The rules for differentiation have been deduced, in Chapter II, in accordance with *the method of infinitesimals* explained in Chapter I. We shall now deduce these rules by *the method of limits*.

The limit, or limiting value of a function, is that value toward which the function continually approaches, till it differs from it by less than any assignable quantity, while the independent variable approaches some assigned value. If the assigned value of the independent variable be zero, the limit is called the *inferior limit*; and if the value be infinity, it is called the *superior limit*.

42. Algebraic Illustration.—Take the example,

$$y = \frac{1}{1 + x},$$

and consider the series of values which y assumes when x has assigned to it different positive values. When $x = 0$, $y = 1$, and when x has any positive value, y is a positive proper fraction; as x increases, y decreases, and *can be made smaller than any assignable fraction, however small, by giving to x a value sufficiently great.* Thus, if we wish y to be less than $\frac{1}{1000000}$, we make $x = 1000000$, and get

$$y = \frac{1}{1 + 1000000}$$

which is less than $\frac{1}{1000000}$. If we wish y to be less than one-trillionth, we make $x = 1000000000000$, and the required result is obtained. We see that, however great x may be taken, y can never become *zero*, though it may be

made to differ from it by *as small a quantity as we please.*
Hence, the limit of the function $\dfrac{1}{1+x}$ is zero when x is infinite.

We are accustomed to speak of such expressions thus: "When x is infinite, y equals zero." But both parts of this sentence are abbreviations: "When x is infinite" means, "*When x is continually increased indefinitely,*" and *not,* "When x is *absolute infinity;*" and "y equals zero" means strictly, "y can be made to differ from zero by as small a quantity as we please." Under these circumstances, we say, "the *limit* of y, when x increases indefinitely, is zero."

43. Trigonometric Illustration—An excellent example of a *limit* is found in Trigonometry. To find the values of $\dfrac{\sin\theta}{\tan\theta}$ and $\dfrac{\sin\theta}{\theta}$, when θ diminishes indefinitely. Here we have

$$\frac{\sin\theta}{\tan\theta} = \cos\theta; \quad \text{and when} \quad \theta = 0, \quad \cos\theta = 1.$$

Hence, if θ be diminished indefinitely, the fraction $\dfrac{\sin\theta}{\tan\theta}$ will approach as near as we please to unity. In other words, the *limit* of $\dfrac{\sin\theta}{\tan\theta}$, as θ continually diminishes, is unity. We usually express this by saying, "The limit of $\dfrac{\sin\theta}{\tan\theta}$, when $\theta = 0$, is unity;" or, "$\dfrac{\sin\theta}{\tan\theta} = 1$, when $\theta = 0$;" that is, we use the words "when $\theta = 0$" as an abbreviation for "when θ is *continually diminished toward zero.*"

Since　　　　$\dfrac{\sin\theta}{\tan\theta} = 1$,　when　$\theta = 0$,

we have also　　$\dfrac{\tan\theta}{\sin\theta} = 1$,　when　$\theta = 0$.

It is evident, from geometric considerations, that if θ be the circular measure of an angle, we have

$$\tan \theta > \theta > \sin \theta;$$

or,
$$\frac{\tan \theta}{\sin \theta} > \frac{\theta}{\sin \theta} > 1;$$

but in the limit, *i. e.*, when $\theta = 0$, we have

$$\frac{\tan \theta}{\sin \theta} = 1;$$

and therefore we have, at the same time,

$$\frac{\theta}{\sin \theta} = 1, \qquad \text{and} \qquad \therefore \quad \frac{\sin \theta}{\theta} = 1,$$

which shows that, in a circle, the limit of the ratio of an arc to its chord is unity.

In the expression, "$\dfrac{\sin \theta}{\theta} = 1$, when $\theta = 0$," it is evident that $\dfrac{\sin \theta}{\theta}$ is never equal to 1 so long as θ has a value different from zero; and if we actually make $\theta = 0$, we render the expression $\dfrac{\sin \theta}{\theta}$ meaningless.* That is, while $\dfrac{\sin \theta}{\theta}$ approaches as nearly as we please to the limit unity, *it never actually attains that limit.*

If a variable quantity be supposed to diminish gradually, till it be less than anything finite which can be assigned, it is said in that state to be *indefinitely small*, or an *infinitesimal ;* the cipher 0 is often used as an abbreviation to denote such a quantity, and does not mean absolute zero ; neither does ∞ express absolute infinity.

Rem.—The student may here read Art. 12, which is applicable to this method as well as to that of infinitesimals, which it is not necessary for us to insert again.

---

* See Todhunter's Dif. Cal., p. 6.

44. Derivatives.—The ratio of the increment of u to that of x, when the increments are finite, is denoted by $\dfrac{\Delta u}{\Delta x}$; the ratio of the increment of u to that of x in the *limit*, *i. e.*, when both are infinitely small, is denoted by $\dfrac{du}{dx}$, and is called the *derivative** of u with respect to x.

Thus, let $u = f(x)$; and let x take the increment h ($= \Delta x$), becoming $x + h$, while u takes the corresponding increment Δu; then we have,

$$u + \Delta u = f(x + h);$$

therefore, by subtraction, we have

$$\Delta u = f(x + h) - f(x);$$

and dividing by h ($= \Delta x$), we get

$$\frac{\Delta u}{\Delta x} = \frac{f(x + h) - f(x)}{h}. \tag{1}$$

It may seem superfluous to use both h and Δx to denote the same thing, but in finding the limit of the second member, it will sometimes be necessary to perform several transformations, and therefore a single letter is more convenient. In the first member, we use Δx on account of symmetry.

The limiting value of the expression in (1), *when* h *is infinitely small, is called the derivative of* u *or* $f(x)$ *with respect to* x, *and is denoted by* $f'(x)$.

Therefore, passing to the limit, by making h diminish indefinitely, the second member of (1) becomes $f'(x)$, and the first member becomes, at the same time, $\dfrac{du}{dx}$; hence we have

$$\frac{du}{dx} = f'(x). \tag{2}$$

---

* Called also the *derived function* and the *differential coefficient.*

45. Differential and Differential Coefficient.

Let $u = f(x)$; then. as we have (Art. 44),

$$\frac{du}{dx} = f'(x),$$

we have $\quad du = d f(x) = f'(x) dx,$

where dx and du are regarded as being infinitely small, and are called respectively (Art. 12) the *differential of x* and the corresponding *differential of u.*

$f'(x)$, which represents the ratio of the differential of the function to that of the variable, and called the *derivative of* $f(x)$ (Art. 44), is also called the *differential coefficient* of $f(x)$, because it is the coefficient of dx in the differential of $f(x)$.

Some writers * consider the symbol $\frac{du}{dx}$ only as a *whole,* and do not assign a separate meaning to du and dx; others,† who also consider the symbol $\frac{du}{dx}$ only as a whole, regard it simply as a convenient nota. tion to represent $\frac{0}{0}$, and claim that du and dx are each *absolutely zero.*

46. Differentiation of the Algebraic Sum of a Number of Functions.

Let $\qquad y = au + bv + cw + z + \text{etc.},$

in which y, u, v, w, and z are functions of x. Suppose that when x takes the increment h $(= \Delta x)$, y, u, v, w, and z take the increments Δy, Δu, Δv, Δw, Δz. Then we have,

$$y + \Delta y = a(u + \Delta u) + b(v + \Delta v) + c(w + \Delta w) + (z + \Delta z) + \text{etc.}$$

$$\therefore \quad \Delta y = a \Delta u + b \Delta v + c \Delta w + \Delta z + \text{etc.}$$

Dividing by h or Δx, we have

---

* See Todhunter's Dif. Cal., p. 17 ; also De Morgan's Calculus, p. 14, etc.

† See Young's Dif. Cal., p. 4.

$$\frac{\Delta y}{\Delta x} = a \frac{\Delta u}{\Delta x} + b \frac{\Delta v}{\Delta x} + c \frac{\Delta w}{\Delta x} + \frac{\Delta z}{\Delta x} + \text{etc.,}$$

which becomes in the limit, when h is infinitely small (Art. 12),

$$\frac{dy}{dx} = a \frac{du}{dx} + b \frac{dv}{dx} + c \frac{dw}{dx} + \frac{dz}{dx} + \text{etc. (see Art. 14).}$$

47. Differentiation of the Product of two Functions.

Let $y = uv$, where u and v are both functions of x, and suppose Δy, Δu, Δv to be the increments of y, u, v corresponding to the increment Δx in x. Then we have

$$y + \Delta y = (u + \Delta u)(v + \Delta v)$$
$$= uv + u \Delta v + v \Delta u + \Delta u \Delta v.$$

$$\therefore \quad \Delta y = u \Delta v + v \Delta u + \Delta u \Delta v;$$

or,
$$\frac{\Delta y}{\Delta x} = u \frac{\Delta v}{\Delta x} + v \frac{\Delta u}{\Delta x} + \frac{\Delta u}{\Delta x} \Delta v.$$

Now suppose Δx to be infinitely small, **and**

$$\frac{\Delta y}{\Delta x}, \qquad \frac{\Delta v}{\Delta x}, \qquad \frac{\Delta u}{\Delta x},$$

become in the limit,

$$\frac{dy}{dx}, \qquad \frac{dv}{dx}, \qquad \frac{du}{dx}.$$

Also, since Δv vanishes at the same time, the limit of the last term is zero, and hence in the limit we have

$$\frac{dy}{dx} = u \frac{dv}{dx} + v \frac{du}{dx}. \quad \text{(See Art. 16.)}$$

It can easily be seen that, although the last term vanishes, the remaining terms may have any finite value whatever, since they contain only the *ratios* of vanishing quantities (see Art. 9). For example, $\frac{ax}{x} = \frac{0}{0}$ when $x = 0$; but by canceling x we get $\frac{ax}{x} = a$. But the expression $\frac{ax}{x} \times x$, which equals $\frac{0}{0} \times 0$ when $x = 0$, becomes $\frac{a}{1} \times 0 = 0$ when $x = 0$.

Otherwise thus:

Let $f(x)$ $\phi(x)$ denote the two functions of x, and let

$$u = f(x)\,\phi(x).$$

Change x into $x + h$, and let $u + \Delta u$ denote the new product; then

$$u + \Delta u = f(x + h)\,\phi(x + h)$$

$$\therefore \quad \frac{\Delta u}{\Delta x} = \frac{f(x + h)\,\phi(x + h) - f(x)\,\phi(x)}{h}.$$

Subtract and add $f(x)\,\phi(x + h)$, which will not change the value, and we have

$$\frac{\Delta u}{\Delta x} = \frac{f(x + h) - f(x)}{h}\,\phi(x + h) + f(x)\frac{\phi(x + h) - \phi(x)}{h}.$$

Now in the limit, when h is diminished indefinitely,

$$\frac{f(x + h) - f(x)}{h} = f'(x)\,;$$

$$\frac{\phi(x + h) - \phi(x)}{h} = \phi'(x) \quad \text{(Art. 44)}\,;$$

and $\qquad \phi(x + h) = \phi(x)\,;$

therefore, $\qquad \dfrac{du}{dx} = f'(x)\,\phi(x) + f(x)\,\phi'(x),$

which agrees with the preceding result.

48. Differentiation of the Product of any Number of Functions.

Let $\qquad\qquad y = uvw,$

u, v, w being all functions of x.

Assume $\qquad\qquad z = vw\,;$

then $\qquad\qquad y = uz,$

and by Art. 47 we have

$$\frac{dy}{dx} = \frac{udz}{dx} + \frac{zdu}{dx}.$$

Also, by the same Article,

$$\frac{dz}{dx} = \frac{vdw}{dx} + \frac{wdv}{dx};$$

hence, by substitution, we have

$$\frac{dy}{dx} = uv\frac{dw}{dx} + uw\frac{dv}{dx} + vw\frac{du}{dx}. \quad \text{(See Art. 17.)}$$

The same process can be extended to any number of functions.

49. Differentiation of a Fraction.

Let

$$y = \frac{u}{v}.$$

Then we shall have

$$y + \Delta y = \frac{u + \Delta u}{v + \Delta v};$$

$$\Delta y = \frac{u + \Delta u}{v + \Delta v} - \frac{u}{v}$$

$$= \frac{v\,\Delta u - u\,\Delta v}{v^2 + v\Delta v}.$$

Dividing by Δx and passing to the limit,

$$\frac{dy}{dx} = \frac{v\dfrac{du}{dx} - u\dfrac{dv}{dx}}{v^2}$$

(since vdv vanishes).   (See Art. 18.)

Cor.—If u is a constant, we have

$$\frac{dy}{dx} = \frac{-\dfrac{udv}{dx}}{v^2}.$$

50. Differentiation of any Power of a Single Variable.

1st. *When n is a positive integer.*

Let $$y = x^n;$$

then we have $$y + \Delta y = (x + h)^n;$$

therefore, $$\Delta y = nx^{n-1}h + \frac{n(n-1)}{1 \cdot 2} x^{n-2}h^2 + \text{etc.} \ldots h^n.$$

Dividing by h or Δx, we get

$$\frac{\Delta y}{\Delta x} = nx^{n-1} + \frac{n(n-1)}{2} x^{n-2}h \ldots h^{n-1}.$$

Passing to the limit, we have

$$\frac{dy}{dx} = nx^{n-1}. \quad \text{(See Art. 19, 1st.)} \tag{1}$$

2d. *When n is a positive fraction.*

Let $$y = u^{\frac{m}{n}},$$

where u is a function of x; then

$$y^n = u^m,$$

and $$d(y^n) = d(u^m);$$

hence, by (1), $$ny^{n-1}\frac{dy}{dx} = mu^{m-1}\frac{du}{dx}.$$

$$\therefore \frac{dy}{dx} = \frac{m}{n}\frac{u^{m-1}}{y^{n-1}}\frac{du}{dx}$$

$$= \frac{m}{n}u^{\frac{m}{n}-1}\frac{du}{dx} \quad \text{(Art. 19, 2d).} \tag{2}$$

3d. *When n is a negative exponent, integral or fractional.*

Let $$y = u^{-n};$$

then $$y = \frac{1}{u^n},$$

and by Art. 49, Cor., we have

$$\frac{dy}{dx} = - \frac{nu^{n-1}}{u^{2n}}\frac{du}{dx} = - nu^{-n-1}\frac{du}{dx} \quad \text{(Art. 19, 3d). (3)}$$

51. Differentiation of log x.

Let $\qquad y = \log x$;

therefore, $\qquad y + \Delta y = \log (x + h)$,

and $\qquad \Delta y = \log (x + h) - \log x$

$$= \log \left(\frac{x + h}{x}\right) = \log \left(1 + \frac{h}{x}\right)$$

$$= m \left(\frac{h}{x} - \frac{h^2}{2x^2} + \frac{h^3}{3x^3} - \text{etc.}\right);$$

therefore, $\qquad \dfrac{\Delta y}{\Delta x} = m \left(\dfrac{1}{x} - \dfrac{h}{2x^2} + \text{etc.}\right)$;

therefore, passing to the limit, we get

$$\frac{dy}{dx} = \frac{m}{x} \quad \text{or} \quad \frac{1}{x}$$

(according as the logarithms are *not* or are taken in the Naperian system.   See Art. 20).

52. Differentiation of a^x.

Let $\qquad y = a^x$.

Proceeding exactly as in Art. 21, we get

$$\frac{dy}{dx} = \frac{a^x}{m} \log a \quad \text{or} \quad a^x \log a \quad \text{(Art. 21).}$$

53. Differentiation of sin x.

Let $\qquad y = \sin x$;

therefore $\qquad y + \Delta y = \sin (x + h)$;

hence, $\qquad \Delta y = \sin (x + h) - \sin x$.

But from Trigonometry,

$$\sin A - \sin B = 2 \cos \frac{A + B}{2} \sin \frac{A - B}{2}.$$

$$\therefore \quad \Delta y = \sin (x + h) - \sin x$$

$$= 2 \cos \left(x + \frac{h}{2}\right) \sin \frac{h}{2};$$

hence,

$$\frac{\Delta y}{\Delta x} = \cos \left(x + \frac{h}{2}\right) \frac{\sin \frac{h}{2}}{\frac{h}{2}}.$$

By Art. 43, when h is diminished indefinitely, the limit of $\dfrac{\sin \frac{h}{2}}{\frac{h}{2}} = 1$; also, the limit of $\cos \left(x + \dfrac{h}{2}\right) = \cos x$.

Therefore,

$$\frac{dy}{dx} = \cos x. \quad \text{(See Art. 24.)}$$

54. Differentiation of cos x.

Let

$$y = \cos x;$$

therefore,

$$y + \Delta y = \cos (x + h);$$

hence,

$$\Delta y = \cos (x + h) - \cos x$$

$$= - 2 \sin \left(x + \frac{h}{2}\right) \sin \frac{h}{2},$$

because

$$\cos A - \cos B = - 2 \sin \left(\frac{A + B}{2}\right) \sin \frac{A - B}{2}.$$

Therefore,

$$\frac{\Delta y}{\Delta x} = - \sin \left(x + \frac{h}{2}\right) \frac{\sin \frac{h}{2}}{\frac{h}{2}}.$$

Hence, in the limit,

$$\frac{dy}{dx} = - \sin x. \quad \text{(See Art. 25.)}$$

Of course this differentiation may be obtained directly from Art. 53, in the same manner as was done in the 2d method of Art. 25.

Since tan x, cot x, sec x, and cosec x are all fractional forms, we may find the derivative of each of these functions by Arts. 18 or 49, from those of sin x and cos x, as was done in Arts. 26, 27, 28, and 29; also, the derivatives of vers x and covers x, as well as those of the circular functions, may be found as in Arts. 30, 31, 33 to 40.

From the brief discussion that we have given, the student will be able to compare *the method of limits* with *the method of infinitesimals;* he will see that the *results* obtained by the two methods are identically the same. In discussing by the former method, we restricted ourselves to the use of limiting ratios, which are the proper auxiliaries in this method. It will be observed that, in the former method, very small quantities of higher orders are retained till the end of the calculation, and then neglected in passing to the limit; while in the infinitesimal method such quantities are neglected from the start, from the knowledge that they necessarily disappear in the limit, and therefore cannot affect the final result. As a logical basis of the Calculus, *the method of limits* may have some advantages. In other respects, the superiority is immeasurably on the side of *the method of infinitesimals.*

CHAPTER IV.

SUCCESSIVE DIFFERENTIALS AND DERIVATIVES.

55. Successive Differentials.—The differential obtained immediately from the function is *the first differential.* The differential of the first differential is *the second differential,* represented by d^2y, d^2u, etc., and read, "second differential of y," etc. The differential of the second differential is *the third differential,* represented by d^3y, d^3u, etc., and read, "third differential of y," etc. In like manner, we have the fourth, fifth, etc., differentials. Differentials thus obtained are called *successive differentials.*

Thus, let AB be a right line whose equation is $y = ax + b$; $\therefore dy = adx$. Now regard dx as constant, *i. e.*, let x be equicrescent;* and let MM′, M′M″, and M″M‴ represent the successive equal increments of x, or the dx's, and R′P′, R″P″, R‴P‴ the corresponding increments of y, or the dy's. We see from the figure that $R'P' = R''P'' = R'''P'''$; therefore the dy's are all equal, and hence the difference between any two consecutive dy's being 0, the differential of dy, *i. e.*, $d^2y = 0$. Also, from the equation $dy = adx$ we have $d^2y = 0$, since a and dx are both constants.

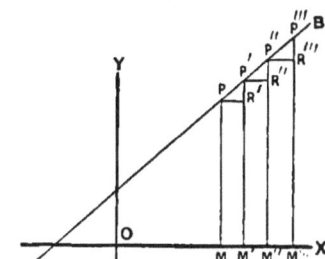

Fig. 6.

Take the case of the parabola $y^2 = 2px$ (Fig. 7), from which we get $dy = \dfrac{pdx}{y}$. Regarding dx as a constant, we

---

* When the variable increases by *equal* increments, *i. e.*, when the differential is *constant*, the variable is called an *equicrescent variable*.

have MM′, M′M″, M″M‴ as the successive equal increments
of x, or the dx's; while we see from
Fig. 7 that R′P′, R″P″, R‴P‴, or
the dy's, are no longer equal, but
diminish as we move towards the
right, and hence the difference between any two consecutive dy's is a
negative quantity (remembering that
the difference is always found by
taking the *first* value from the *second*.
See Art. 12). Also, from the equa-

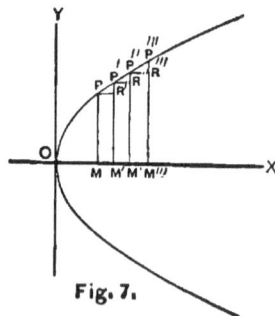

Fig. 7.

tion $dy = \dfrac{p}{y} dx$ we see that dy varies inversely as y.

The student must be careful not to confound d^2y with
dy^2 or $d(y^2)$: the first is "second differential of y;" the
second is "the square of dy;" the third is the differential
of y^2, which equals $2ydy$.

E X A M P L E S.

1. Find the successive differentials of $y = x^5$.

Differentiating, we have $dy = 5x^4 dx$. Differentiating
this, remembering that d of dy is d^2y and that dx is constant, we have $d^2y = 20x^3 dx^2$. In the same way, differentiating again, we have $d^3y = 60x^2 dx^3$. Again, $d^4y = 120x dx^4$.
Once more, $d^5y = 120 dx^5$. If we differentiate again, we
have $d^6y = 0$, since dx is constant.

2. Find the successive differentials of $y = 4x^3 - 3x^2 + 2x$.

$$Ans. \begin{cases} dy = (12x^2 - 6x + 2)\,dx\,; \\ d^2y = (24x - 6)\,dx^2\,; \\ d^3y = 24\,dx^3. \end{cases}$$

3. Find the first six successive differentials of $y = \sin x$.

$$Ans. \begin{cases} dy = \cos x\,dx\,; & d^2y = -\sin x\,dx^2\,; \\ d^3y = -\cos x\,dx^3\,; & d^4y = \sin x\,dx^4\,; \\ d^5y = \cos x\,dx^5\,; & d^6y = -\sin x\,dx^6. \end{cases}$$

4. Find the first six successive differentials of $y = \cos x$.

$$Ans. \begin{cases} dy = -\sin x\, dx\ ; & d^2y = -\cos x\, dx^2\ ; \\ d^3y = \sin x\, dx^3\ ; & d^4y = \cos x\, dx^4\ ; \\ d^5y = -\sin x\, dx^5\ ; & d^6y = -\cos x\, dx^6. \end{cases}$$

5. Find the fourth differential of $y = x^n$.

$$Ans.\ d^4y = n\,(n-1)\,(n-2)\,(n-3)\,x^{n-4}\,dx^4.$$

6. Find the first three successive differentials of $y = a^x$.

$$Ans. \begin{cases} dy = a^x \log a\, dx\ ; \\ d^2y = a^x \log^2 a\, dx^2\ ; \\ d^3y = a^x \log^3 a\, dx^3. \end{cases}$$

7. Find the first four successive differentials of $y = \log x$.

$$Ans. \begin{cases} dy = \dfrac{dx}{x}; & d^2y = -\dfrac{dx^2}{x^2}\ ; \\ d^3y = \dfrac{2dx^3}{x^3}; & d^4y = -\dfrac{6dx^4}{x^4}. \end{cases}$$

8. Find the first four successive differentials of $y = 2a\sqrt{x}$.

$$Ans. \begin{cases} dy = \dfrac{adx}{\sqrt{x}}; & d^2y = -\dfrac{adx^2}{2x^{\frac{3}{2}}}\ ; \\ d^3y = \dfrac{3adx^3}{4x^{\frac{5}{2}}}; & d^4y = -\dfrac{15adx^4}{8x^{\frac{7}{2}}}. \end{cases}$$

9. Find the first four successive differentials of

$y = \log (1 + x)$ in the common system.

$$Ans. \begin{cases} dy = \dfrac{mdx}{1+x}; & d^2y = -\dfrac{mdx^2}{(1+x)^2}\ ; \\ d^3y = \dfrac{2mdx^3}{(1+x)^3}; & d^4y = -\dfrac{6mdx^4}{(1+x)^4}. \end{cases}$$

10. Find the fourth differential of $y = e^x$.

$$Ans.\ d^4y = e^x dx^4.$$

56. Successive Derivatives.—A *first derivative** is the ratio of the differential of a function to the differential of its variable. For example, let

$$y = x^6$$

represent a function of x. Differentiating and dividing by dx, we get

$$\frac{dy}{dx} = 6x^5. \tag{1}$$

The fraction $\frac{dy}{dx}$ is called the *first derivative* of y with respect to x, and represents the ratio of the differential of the function to the differential of the variable, the *value* of which is represented by the second member of the equation.

Clearing (1) of fractions, we have

$$dy = 6x^5 dx \, ;$$

hence, $\frac{dy}{dx}$ or $6x^5$ is also called *the first differential coefficient* of y with respect to x, because it is the coefficient of dx.

A *second derivative* is the ratio of the second differential of a function to the square of the differential of the variable. Thus, differentiating (1) and dividing by dx, we get (since dx is constant, Art. 55),

$$\frac{d^2y}{dx^2} = 30x^4, \tag{2}$$

either member of which is called *the second derivative* of y with respect to x.

A *third derivative* is the ratio of the third differential of a function to the cube of the differential of the variable. Thus, differentiating (2) and dividing by dx, we get

---

* See Arts. 44 and 45.

$$\frac{d^3y}{dx^3} = 120x^3, \tag{3}$$

either member of which is called *the third derivative* of y with respect to x.

In the same way, either member of

$$\frac{d^4y}{dx^4} = 360x^2 \tag{4}$$

is called *the fourth derivative* of y with respect to x, and so on.

Also, $\frac{dy}{dx}$, $\frac{d^2y}{dx^2}$, $\frac{d^3y}{dx^3}$, $\frac{d^4y}{dx^4}$, etc., are called respectively *the first, second, third, fourth*, etc., differential coefficients of y with respect to x, because they are the coefficients of dx, dx^2, dx^3, dx^4, etc., if (1), (2), (3), (4), and so on, be cleared of fractions.

In general, if $y = f(x)$, we have

$$\frac{dy}{dx} = \frac{d f(x)}{dx} = f'(x) \text{ (Art. 45); } \therefore dy = f'(x)\, dx.$$

$$\frac{d^2y}{dx^2} = \frac{d f'(x)}{dx} = f''(x); \qquad \therefore d^2y = f''(x)\, dx^2.$$

$$\frac{d^3y}{dx^3} = \frac{d f''(x)}{dx} = f'''(x); \qquad \therefore d^3y = f'''(x)\, dx^3.$$

$$\frac{d^4y}{dx^4} = \frac{d f'''(x)}{dx} = f^{iv}(x); \qquad \therefore d^4y = f^{iv}(x)\, dx^4.$$

$$\text{etc.} = \text{etc.} = \text{etc.} \qquad \therefore \text{etc.} = \text{etc.}$$

$$\frac{d^ny}{dx^n} = \frac{d f^{(n-1)}(x)}{dx} = f^{(n)}(x); \qquad \therefore d^ny = f^{(n)}(x)\, dx^n.$$

That is, the first, second, third, fourth, etc., derivatives are also represented by $f'(x)$, $f''(x)$, $f'''(x)$, $f^{iv}(x)$, etc.

Strictly speaking, $\frac{dy}{dx}$ or $f'(x)$ are *symbols* representing the ratio of an infinitesimal increment of the function to the corresponding infinitesimal increment of the variable, while the second member expresses its *value*. For example, in the equation $y = ax^4$, we obtain

$$\frac{dy}{dx} = f'(x) = 4ax^3.$$

$\frac{dy}{dx}$ or $f'(x)$ is an *arbitrary symbol*, representing the value of the ratio of the infinitesimal increment of the function (ax^4) to the corresponding infinitesimal increment of the variable (x), while $4ax^3$ *is the value itself.* It is usual, however, to call either the derivative.

56*a*. Geometric Representation of the First Derivative.

—Let AB be any plane curve whose equation is $y = f(x)$. Let P and P′ be consecutive points, and PM and P′M′ consecutive ordinates. The part of the curve PP′, called an *element** of the curve, does not differ from a right line. The line PP′ prolonged is tangent to the curve at the point P (Anal. Geom., Art. 42). Draw PR parallel to XX′, and we have

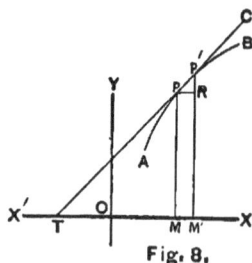

$$MM' = PR = dx, \quad \text{and} \quad RP' = dy.$$

Denote the angle CTX by α, and since $CTX = P'PR$, we have

$$\tan \alpha = \frac{dy}{dx}.$$

And since the tangent has the same direction as the curve

---

* In this work, the word "element" will be used for brevity to denote an "infinitesimal element."

at the tangent point P, α will also denote the inclination of the *curve* to the axis of x.

Hence, *the first derivative of the ordinate of a curve, at any point, is represented by the trigonometric tangent of the angle which the curve at that point, or its tangent, makes with the axis of x.*

In expressing the above differentials and derivatives, we have assumed the independent variable x to be equicrescent (Art. 55), which we are always at liberty to do. This hypothesis greatly simplifies the expressions for the second and higher derivatives and differentials of functions of x, inasmuch as it is equivalent to making all differentials of x above the first vanish. Were we to find the second derivative of y with respect to x, regarding dx as variable, we would have

$$\frac{d^2y}{dx^2} = \frac{d}{dx}\left(\frac{dy}{dx}\right) = \frac{dx\,d^2y - dy\,d^2x}{dx^3},$$

which is much less simple than the expression $\frac{d^2y}{dx^2}$, obtained by supposing dx to be constant.

EXAMPLES.

1. Given $y = x^n$, to find the first four successive derivatives.

$$\frac{dy}{dx} = nx^{n-1};$$

$$\frac{d^2y}{dx^2} = n\,(n-1)\,x^{n-2};$$

$$\frac{d^3y}{dx^3} = n\,(n-1)\,(n-2)\,x^{n-3};$$

$$\frac{d^4y}{dx^4} = n\,(n-1)\,(n-2)\,(n-3)\,x^{n-4}.$$

If n be a positive integer, we have

$$\frac{d^n y}{dx^n} = n(n-1)(n-2)\ldots\ldots 3\cdot 2\cdot 1.$$

and all the higher derivatives vanish.

If n be a negative integer or a fraction, none of the successive derivatives can vanish.

2. Given $y = x^3 \log x$; find $\dfrac{d^4 y}{dx^4}$.

$$\frac{dy}{dx} = 3x^2 \log x + x^2 ;$$

$$\frac{d^2 y}{dx^2} = 6x \log x + 3x + 2x = 6x \log x + 5x ;$$

$$\frac{d^3 y}{dx^3} = 6 \log x + 6 + 5. \qquad \frac{d^4 y}{da^4} = \frac{6}{x}.$$

It can be easily seen that in this case all the terms in the successive derivatives which do not contain $\log x$ will disappear in the final result; thus, the third derivative of x^2 is zero, and therefore that term might have been neglected ; and the same is true of $5x$, its second derivative being zero.

3. $y = \dfrac{1+x}{1-x}$; prove that $\dfrac{d^5 y}{dx^5} = \dfrac{240}{(1-x)^6}.$

4. $y = e^{ax}$;   prove that $\dfrac{d^6 y}{dx^6} = a^6 e^{ax}.$

5. $y = \tan x$; find the first four successive derivatives.

$$\frac{dy}{dx} = \sec^2 x ;$$

$$\frac{d^2 y}{dx^2} = 2 \sec^2 x \tan x ;$$

$$\frac{d^3 y}{dx^3} = 6 \sec^4 x - 4 \sec^2 x ;$$

$$\frac{d^4 y}{dx^4} = 8 \tan x \sec^2 x (3 \sec^2 x - 1).$$

6. $y = \log \sin x$; prove that $\dfrac{d^3y}{dx^3} = 2 \cot x \operatorname{cosec}^2 x$.

7. $y^2 = 2px$; find $\dfrac{d^3y}{dx^3}$.

$$\frac{dy}{dx} = \frac{p}{y};$$

$$\frac{d^2y}{dx^2} = \frac{d}{dx}\left(\frac{p}{y}\right) = -\frac{p\dfrac{dy}{dx}}{y^2} = -\frac{p\dfrac{p}{y}}{y^2} \quad \left(\text{since } \frac{dy}{dx} = \frac{p}{y}\right)$$

$$= -\frac{p^2}{y^3};$$

$$\frac{d^3y}{dx^3} = \frac{d}{dx}\left(-\frac{p^2}{y^3}\right) = \frac{3p^2y^2\dfrac{dy}{dx}}{y^6} = \frac{3p^2\dfrac{p}{y}}{y^4} = \frac{3p^3}{y^5}.$$

8. $y = x^x$; prove that

$$\frac{d^2y}{dx^2} = x^x (1 + \log x)^2 + x^{x-1}.$$

9. $a^2y^2 + b^2x^2 = a^2b^2$; prove that $\dfrac{d^2y}{dx^2} = -\dfrac{b^4}{a^2y^3}$.

10. $y = \dfrac{x^3}{1-x}$;   prove that $\dfrac{d^4y}{dx^4} = \dfrac{24}{(1-x)^5}$.

11. $y^2 = \sec 2x$;   prove that $y + \dfrac{d^2y}{dx^2} = 3y^5$.

12. $y = e^{-x} \cos x$;   prove that $4y + \dfrac{d^4y}{dx^4} = 0$.

13. $y = x^4 \log (x^2)$; prove that $\dfrac{d^5y}{dx^5} = \dfrac{48}{x}$.

14. $y = x^3$; prove that

$$d^3y = 6 (dx)^3 + 18xdxd^2x + 3x^2d^3x,$$

when x is not equicrescent.

15. $y = f(x)$; prove that

$$d^3y = f'''(x)(dx)^3 + 3f''(x) dx d^2x + f'(x) d^3x,$$

when x is not equicrescent.

16. $y = e^x$; prove that

$$d^3y = e^x (dx)^3 + 3e^x dx d^2x + e^x d^3x.$$

CHAPTER V.

DEVELOPMENT OF FUNCTIONS.

57. A function is said to be **developed**, when it is transformed into an equivalent series of terms following some general law.

For example,

$$y = (a + x)^4,$$

when developed by the binomial theorem, becomes

$$y = a^4 + 4a^3x + 6a^2x^2 + 4ax^3 + x^4,$$

which is a *finite* series. Also,

$$y = \frac{1 + x}{1 - x}$$

may be developed by division into the *infinite* series,

$$y = 1 + 2x + 2x^2 + 2x^3 + \text{etc.},$$

in which the terms are arranged according to the ascending powers of x, each coefficient after the first term being 2.

One of the most useful applications of the theory of successive derivatives is the means it gives us of developing functions into series by methods which we now proceed to explain.

MACLAURIN'S THEOREM.

58. **Maclaurin's Theorem**, is a theorem for developing a function of a single variable into a series arranged according to the ascending powers of that variable, with constant coefficients.

Let $$y = f(x)$$

be the function to be developed ; and assume the develop‑ ment of the form

$$y = f(x) = A + Bx + Cx^2 + Dx^3 + Ex^4 + \text{etc.,} \quad (1)$$

in which A, B, C, D, E, etc., are independent of x, and depend upon the *constants* which enter into the given func‑ tion, and upon the *form* of the function. It is now re‑ quired to find such values for the constants A, B, C, etc., as will cause the assumed development to be true for all values of x.

Differentiating (1) and finding the successive derivatives, we have,

$$\frac{dy}{dx} = B + 2Cx + 3Dx^2 + 4Ex^3 + \text{etc.,} \quad (2)$$

$$\frac{d^2y}{dx^2} = 2C + 2\cdot 3Dx + 3\cdot 4Ex^2 + \text{etc.,} \quad (3)$$

$$\frac{d^3y}{dx^3} = 2\cdot 3D + 2\cdot 3\cdot 4Ex + \text{etc.,} \quad (4)$$

$$\frac{d^4y}{dx^4} = 2\cdot 3\cdot 4E + \text{etc.,} \quad (5)$$

Now, as A, B, C, etc., are independent of x, if we can find what they should be for any one value of x, we shall have their values for all values of x. Hence, making $x = 0$ in (1), (2), (3), etc., and representing what y becomes on this hypothesis by (y) ; what $\frac{dy}{dx}$ becomes by $\left(\frac{dy}{dx}\right)$; what $\frac{d^2y}{dx^2}$ becomes by $\left(\frac{d^2y}{dx^2}\right)$; and so on ; we have,

$$(y) = A ; \qquad \therefore \quad A = (y).$$

$$\left(\frac{dy}{dx}\right) = B ; \qquad \therefore \quad B = \left(\frac{dy}{dx}\right).$$

$$\left(\frac{d^2y}{dx^2}\right) = 2C; \qquad \therefore \quad C = \left(\frac{d^2y}{dx^2}\right)\frac{1}{1\cdot 2}.$$

$$\left(\frac{d^3y}{dx^3}\right) = 2\cdot 3D; \qquad \therefore \quad D = \left(\frac{d^3y}{dx^3}\right)\frac{1}{1\cdot 2\cdot 3}.$$

$$\left(\frac{d^4y}{dx^4}\right) = 2\cdot 3\cdot 4E; \quad \therefore \quad E = \left(\frac{d^4y}{dx^4}\right)\frac{1}{1\cdot 2\cdot 3\cdot 4}$$

Substituting these values in (1), we have,

$$y = f(x) = (y) + \left(\frac{dy}{dx}\right)\frac{x}{1} + \left(\frac{d^2y}{dx^2}\right)\frac{x^2}{1\cdot 2} + \left(\frac{d^3y}{dx^3}\right)\frac{x^3}{1\cdot 2\cdot 3}$$
$$+ \left(\frac{d^4y}{dx^4}\right)\frac{x^4}{1\cdot 2\cdot 3\cdot 4} + \text{etc.,} \qquad (6)$$

which is the theorem required.

Hence, by Maclaurin's Theorem, we may develop a function of a single variable, as $y = f(x)$, into a series of terms, the first of which is the value of the function when $x = 0$; the second is the value of the first derivative of the function when $x = 0$ into x; the third is the value of the second derivative when $x = 0$ into $\frac{x^2}{2}$, etc.; the $(n+1)^{th}$ term is

$$\frac{x^n}{1\cdot 2\cdot 3\,..\,n}\left(\frac{d^ny}{dx^n}\right).$$

We may also use the following notation for the function and its successive derivatives : $f(x)$, $f'(x)$, $f''(x)$, $f'''(x)$, $f^{iv}(x)$, etc., as given in Art. 56, and write the above theorem,

$$y = f(x) = f(0) + f'(0)\frac{x}{1} + f''(0)\frac{x^2}{1\cdot 2} + f'''(0)\frac{x^3}{1\cdot 2\cdot 3}$$
$$+ f^{iv}(0)\frac{x^4}{1\cdot 2\cdot 3\cdot 4} + \text{etc.,} \qquad (7)$$

in which $f(0)$, $f'(0)$, $f''(0)$, $f'''(0)$, etc., represent the values which $f(x)$ and its successive derivatives assume when $x = 0$. We shall use this notation instead of $\dfrac{dy}{dx}$, $\dfrac{d^2y}{dx^2}$, etc., for the sake of brevity.

This theorem, which is usually called Maclaurin's Theorem, was previously given by Stirling in 1717; but appearing first in a work on Fluxions by Maclaurin in 1742, it has usually been attributed to him, and has gone by his name. Maclaurin, however, laid no claim to it, for after proving it in his book, he adds, "this theorem was given by Dr. Taylor." See Maclaurin's Fluxions, Vol. 2, Art. 751.

To Develop $y = (a + x)^6$

Here $f(x) = (a + x)^6$;

hence, $f(0) = a^6$.

$$f'(x) = 6(a + x)^5;$$
$$f'(0) = 6a^5.$$
$$f''(x) = 5 \cdot 6(a + x)^4;$$
$$f''(0) = 5 \cdot 6a^4.$$
$$f'''(x) = 4 \cdot 5 \cdot 6(a + x)^3;$$
$$f'''(0) = 4 \cdot 5 \cdot 6a^3.$$
$$f^{iv}(x) = 3 \cdot 4 \cdot 5 \cdot 6(a + x)^2;$$
$$f^{iv}(0) = 3 \cdot 4 \cdot 5 \cdot 6a^2.$$
$$f^{v}(x) = 2 \cdot 3 \cdot 4 \cdot 5 \cdot 6(a + x);$$
$$f^{v}(0) = 2 \cdot 3 \cdot 4 \cdot 5 \cdot 6a;$$
$$f^{vi}(x) = 1 \cdot 2 \cdot 3 \cdot 4 \cdot 5 \cdot 6;$$
$$f^{vi}(0) = 1 \cdot 2 \cdot 3 \cdot 4 \cdot 5 \cdot 6;$$

Substituting in (7), we have,

$$y = (a + x)^6 = a^6 + 6a^5x + 5 \cdot 6a^4\frac{x^2}{1 \cdot 2} + 4 \cdot 5 \cdot 6a^3\frac{x^3}{1 \cdot 2 \cdot 3}$$
$$+ 3 \cdot 4 \cdot 5 \cdot 6\frac{a^2x^4}{1 \cdot 2 \cdot 3 \cdot 4} + \frac{2 \cdot 3 \cdot 4 \cdot 5 \cdot 6ax^5}{1 \cdot 2 \cdot 3 \cdot 4 \cdot 5} + \frac{1 \cdot 2 \cdot 3 \cdot 4 \cdot 5 \cdot 6x^6}{1 \cdot 2 \cdot 3 \cdot 4 \cdot 5 \cdot 6}$$
$$= a^6 + 6a^5x + 15a^4x^2 + 20a^3x^3 + 15a^2x^4 + 6ax^5 + x^6,$$

which is the same result we would obtain by the binomial theorem.

THE BINOMIAL THEOREM.

59. To Develop $y = (a + x)^n$.

Here $\quad f(x) = (a + x)^n$;

hence, $\quad f(0) = a^n$.

$\quad f'(x) = n(a + x)^{n-1}$;

$\quad f'(0) = na^{n-1}$.

$\quad f''(x) = n(n - 1)(a + x)^{n-2}$.

$\quad f''(0) = n(n - 1)a^{n-2}$.

$\quad f'''(x) = n(n - 1)(n - 2)(a + x)^{n-3}$;

$\quad f'''(0) = n(n - 1)(n - 2)a^{n-3}$.

$\quad f^{iv}(x) = n(n - 1)(n - 2)(n - 3)(a + x)^{n-4}$;

$\quad f^{iv}(0) = n(n - 1)(n - 2)(n - 3)a^{n-4}$, etc.

Substituting in (7), Art. 58, we have,

$$y = (a + x)^n = a^n + na^{n-1}x + \frac{n(n - 1)a^{n-2}x^2}{1 \cdot 2}$$
$$+ \frac{n(n - 1)(n - 2)a^{n-3}x^3}{1 \cdot 2 \cdot 3}$$
$$+ \frac{n(n - 1)(n - 2)(n - 3)a^{n-4}x^4}{1 \cdot 2 \cdot 3 \cdot 4} + \text{etc.}$$

Thus the truth of the *binomial theorem* is established, applicable to all values of the exponent, whether positive or negative, integral or fractional, real or imaginary.

60. 1. To Develop $y = \sin x$.

Here $\quad f(x) = \sin x$; $\qquad$ hence, $\quad f(0) = 0.$

" $\quad f'(x) = \cos x$; $\qquad$ " $\quad f'(0) = 1.$

" $\quad f''(x) = -\sin x$; $\qquad$ " $\quad f''(0) = 0.$

Here $f'''(x) = -\cos x$;   hence, $f'''(0) = -1.$
" $f^{iv}(x) = \sin x$;   " $f^{iv}(0) = 0.$
" $f^{v}(x) = \cos x$;   " $f^{v}(0) = 1.$
            Etc., etc.   Etc., etc.

Hence, $y = \sin x = x - \dfrac{x^3}{1 \cdot 2 \cdot 3} + \dfrac{x^5}{1 \cdot 2 \cdot 3 \cdot 4 \cdot 5}$

$$- \frac{x^7}{1 \cdot 2 \cdot 3 \cdot 4 \cdot 5 \cdot 6 \cdot 7} + \text{etc.}$$

2. To Develop $y = \cos x.$

Ans. $y = \cos x = 1 - \dfrac{x^2}{1 \cdot 2} + \dfrac{x^4}{1 \cdot 2 \cdot 3 \cdot 4}$

$$- \frac{x^6}{1 \cdot 2 \cdot 3 \cdot 4 \cdot 5 \cdot 6} + \frac{x^8}{1 \cdot 2 \cdot 3 \cdot 4 \cdot 5 \cdot 6 \cdot 7 \cdot 8} - \text{etc.}$$

The student will observe that by taking the first derivative of the series in (1), we obtain the series in (2), which is clearly as it should be, since the first derivative of sin x is equal to cos x.

Since sin $(-x) = -\sin x$, from Trigonometry we might have inferred at once that the development of sin x in terms of x could contain only *odd* powers of x. Similarly, as cos $(-x) = \cos x$, the development of cos x can contain only *even* powers.

By means of the two formulæ in this Article we may compute the natural sine and cosine of any arc. For example, to compute the natural sine of $20°$, we have $x =$ arc of $20° = \dfrac{\pi}{9} = .3490652$, which substituted in the formulæ, gives sin $20° = .342020$ and cos $20° = .939693$.

THE LOGARITHMIC SERIES.

61. To Develop $y = \log(1 + x)$ **in the system in which the modulus is** m.

Here $f(x) = \log(1 + x)$;   hence, $f(0) = 0.$

" $f'(x) = \dfrac{m}{1 + x}$;   " $f'(0) = m.$

Here $f''(x) = -\dfrac{m}{(1+x)^2}$;   hence, $f''(0) = -m$.

" $f'''(x) = \dfrac{1 \cdot 2m}{(1+x)^3}$;   " $f'''(0) = 1 \cdot 2m$.

" $f^{iv}(x) = -\dfrac{1 \cdot 2 \cdot 3m}{(1+x)^4}$;   " $f^{iv}(0) = -1 \cdot 2 \cdot 3m$.

        Etc.   Etc.

Substituting in (7), Art. 58, we have,

$$y = \log(1+x) = m\left(x - \tfrac{1}{2}x^2 + \tfrac{1}{3}x^3 - \tfrac{1}{4}x^4 + \tfrac{1}{5}x^5 - \text{etc.}\right), \quad (1)$$

which is called the *logarithmic series.*

Since in the Naperian system $m = 1$ (see Art. 20, Cor.), we have,

$$y = \log(1+x) = x - \frac{x^2}{2} + \frac{x^3}{3} - \frac{x^4}{4} + \frac{x^5}{5} - \text{etc.} \quad (2)$$

which is called the *Naperian logarithmic series.*

This formula might be used to compute Naperian logarithms, of *very small fractions;* but in other cases it is useless, as the series in the second number is *divergent* for values of $x > 1$. We therefore proceed to find a formula in which the series is *convergent* for all values of x; *i. e.,* in which the terms will grow smaller as we extend the series.

Substituting $-x$ for x in (2), we have,

$$\log(1-x) = -x - \frac{x^2}{2} - \frac{x^3}{3} - \frac{x^4}{4} - \frac{x^5}{5} - \text{etc.} \quad (3)$$

Subtracting (3) from (2), we have,

$$\log(1+x) - \log(1-x) = 2x + \frac{2x^3}{3} + \frac{2x^5}{5} + \frac{2x^7}{7} + \text{etc.}$$

or $\log\left(\dfrac{1+x}{1-x}\right) = 2\left(x + \dfrac{x^3}{3} + \dfrac{x^5}{5} + \dfrac{x^7}{7} + \text{etc.}\right).$ (4)

Let $x = \dfrac{1}{2z + 1}$; $\therefore$ $\dfrac{1 + x}{1 - x} = \dfrac{z + 1}{z}$.

Substituting in (4), we have,

$$\log \frac{z + 1}{z} = 2\left[\frac{1}{2z + 1} + \frac{1}{3(2z + 1)^3} + \frac{1}{5(2z + 1)^5} + \frac{1}{7(2z + 1)^7} + \text{etc.}\right];$$

$$\text{or}\quad \log(z + 1) = \log z + 2\left[\frac{1}{2z + 1} + \frac{1}{3(2z + 1)^3} + \frac{1}{5(2z + 1)^5} + \frac{1}{7(2z + 1)^7} + \text{etc.}\right]. \quad (5)$$

This series converges for all positive values of z, and more rapidly as z increases. By means of it the Naperian logarithm of any number may be computed when the logarithm of the *preceding number* is known. It is only necessary to compute the logarithms of *prime* numbers from the series, since the logarithm of any other number may be obtained by adding the logarithms of its factors. The logarithm of 1 is 0. Making $z = 1, 2, 4, 6$, etc., successively in (5), we obtain the following

NAPERIAN OR HYPERBOLIC LOGARITHMS.

$$\log 2 = \log 1 + 2\left(\frac{1}{3} + \frac{1}{3 \cdot 3^3} + \frac{1}{5 \cdot 3^5} + \frac{1}{7 \cdot 3^7} + \frac{1}{9 \cdot 3^9} + \frac{1}{11 \cdot 3^{11}}\right.$$
$$\left. + \frac{1}{13 \cdot 3^{13}} + \frac{1}{15 \cdot 3^{15}} + \frac{1}{17 \cdot 3^{17}} + \text{etc.}\right);$$

or, since $\log 1 = 0$,

$$\log 2 = 2 \left\{\begin{array}{l} .33333333 \\ .01234568 \\ .00082305 \\ .00006532 \\ .00000565 \\ .00000051 \\ .00000005 \end{array}\right\} = 2(0.34657359) = 0.69314718$$

$$\log\ 3 = \log 2 + 2\left(\frac{1}{5} + \frac{1}{3\cdot 5^3} + \frac{1}{5\cdot 5^5} + \frac{1}{7\cdot 5^7} + \frac{1}{9\cdot 5^9} + \text{etc.}\right)$$
$$= 1.09861228.$$

$$\log\ 4 = 2\log 2 \qquad\qquad\qquad = 1.38629436.$$

$$\log\ 5 = \log 4 + 2\left(\frac{1}{9} + \frac{1}{3\cdot 9^3} + \frac{1}{5\cdot 9^5} + \frac{1}{7\cdot 9^7} + \frac{1}{9\cdot 9^9} + \text{etc.}\right)$$
$$= 1.60943790.$$

$$\log\ 6 = \log 3 + \log 2 \qquad\qquad = 1.79175946.$$

$$\log\ 7 = \log 6 + 2\left(\frac{1}{13} + \frac{1}{3\cdot 13^3} + \frac{1}{5\cdot 13^5} + \frac{1}{7\cdot 13^7} + \text{etc.}\right)$$
$$= 1.94590996.$$

$$\log\ 8 = 3\log 2 \qquad\qquad\qquad = 2.07944154.$$

$$\log\ 9 = 2\log 3 \qquad\qquad\qquad = 2.19722456.$$

$$\log 10 = \log 5 + \log 2 \qquad\qquad = 2.30258509.$$

In this manner, the Naperian logarithms of all numbers may be computed. Where the numbers are large, their logarithms are computed more easily than in the case of small numbers. Thus, in computing the logarithm of 101, the first term of the series gives the result true to seven places of decimals.

Cor. 1.—From (1) we see that, *the logarithms of the same number in different systems are to each other as the moduli of those systems; and also, that the logarithm of a number in any system is equal to the Naperian logarithm of the same number into the modulus of the given system.*

Cor. 2.—Dividing (1) by (2), we have

$$\frac{\text{Common log } (1 + x)}{\text{Naperian log } (1 + x)} = m. \qquad (6)$$

Hence, *the modulus of the common system is equal to the common logarithm of any number divided by the Naperian logarithm of the same number.*

Substituting in (6) the Naperian logarithm of 10 computed above, and the common logarithm of 10, which is 1, we have

$$m = \frac{1}{2.30258509} = .4342944819032518276511289 \ldots$$

which is *the modulus of the common system*. (See Serret's Calcul Différentiel et Intégral, p. 169.)

Hence, *the common logarithm of any number is equal to the Naperian logarithm of the same number into the modulus of the common system, .43429448.*

COR. 3.—Representing the Naperian base by e (Art. 21, Cor. 2), we have, from Cor. 1 of the present Article,

com. log e : Nap. log e ($=1$) :: .43429448 : 1;

therefore, com. log $e = .43429448$;

and hence, from the table of common logarithms, we have

$$e = 2.718281+.$$

EXPONENTIAL SERIES.

62. To Develop $y = a^x$.

Here $f(x) = a^x$; hence, $f(0) = 1.$
" $f'(x) = a^x \log a$; " $f'(0) = \log a.$
" $f''(x) = a^x (\log a)^2$; " $f''(0) = (\log a)^2.$
" $f'''(x) = a^x (\log a)^3$; " $f'''(0) = (\log a)^3.$

and the development is

$$y = a^x = 1 + \log a \, \frac{x}{1} + \log^2 a \, \frac{x^2}{1\cdot 2} + \log^3 a \, \frac{x^3}{1\cdot 2\cdot 3}$$

$$+ \log^4 a \, \frac{x^4}{1\cdot 2\cdot 3\cdot 4} + \text{etc.} \qquad (1)$$

Cor.—If $a = e$, the Naperian base, the development becomes

$$y = e^x = 1 + \frac{x}{1} + \frac{x^2}{1\cdot 2} + \frac{x^3}{1\cdot 2\cdot 3} + \frac{x^4}{1\cdot 2\cdot 3\cdot 4}$$

$$+ \cdots \cdot \frac{x^n}{1\cdot 2\cdot 3 \ldots n} + \text{etc.} \qquad (2)$$

Putting $x = 1$, we obtain the following series, which enables us to compute the value of the quantity e to any required degree of accuracy:

$$y = e = 2 + \frac{1}{2} + \frac{1}{2\cdot 3} + \frac{1}{2\cdot 3\cdot 4} + \frac{1}{2\cdot 3\cdot 4\cdot 5}$$

$$+ \cdots \cdot \frac{1}{2\cdot 3 \ldots n} + \text{etc.}$$

$$= 2.718281828 + .$$

63. To Develop $y = \tan^{-1} x$.

In the applications of Maclaurin's Theorem, the labor in finding the successive derivatives is often very great. This labor may sometimes be avoided by developing the first derivative by some of the algebraic processes, as follows:

Here $\quad f(x) = \tan^{-1} x$;

hence, $\quad f(0) = 0.$

$$f'(x) = \frac{1}{1 + x^2}$$

$$= \text{(by division)} \ 1 - x^2 + x^4 - x^6 + x^8;$$

$$f'(0) = 1.$$

$$f''(x) = -2x + 4x^3 - 6x^5 + 8x^7 - 10x^9 + \text{etc.};$$

$$f''(0) = 0.$$

$$f'''(x) = -2 + 3\cdot 4x^2 - 5\cdot 6x^4 + 7\cdot 8x^6 - \text{etc.};$$

$$f'''(0) = -2.$$

$$f^{iv}(x) = 2\cdot 3\cdot 4x - 4\cdot 5\cdot 6x^3 + \text{etc.};$$

$$f^{iv}(0) = 0.$$

$$f'(x) = 2 \cdot 3 \cdot 4 - 3 \cdot 4 \cdot 5 \cdot 6x^2 + \text{etc.};$$
$$f'(0) = 2 \cdot 3 \cdot 4.$$
$$f^{vi}(x) = -2 \cdot 3 \cdot 4 \cdot 5 \cdot 6x + \text{etc.};$$
$$f^{vi}(0) = 0.$$
$$f^{vii}(x) = -2 \cdot 3 \cdot 4 \cdot 5 \cdot 6 + \text{etc.};$$
$$f^{vii}(0) = -2 \cdot 3 \cdot 4 \cdot 5 \cdot 6.$$

Substituting in (7) of Art. 58, we get

$$y = \tan^{-1} x = x - \frac{x^3}{3} + \frac{x^5}{5} - \frac{x^7}{7} + \text{etc.}$$

64. It sometimes happens in the application of Maclaurin's Theorem that the function or some of its derivatives become *infinite* when $x = 0$. Such functions cannot be developed by Maclaurin's Theorem, since, in such cases, some of the terms of the series would be *infinite*, while the function itself would be *finite*.

For example, take the function $y = \log x$. Here we have

$$f(x) = \log x; \qquad \text{hence,} \quad f(0) = -\infty.$$

$$f'(x) = \frac{1}{x}; \qquad \text{``} \quad f'(0) = \infty.$$

$$f''(x) = -\frac{1}{x^2}; \qquad \text{``} \quad f''(0) = -\infty.$$

$$\text{etc.} \qquad\qquad\qquad \text{etc.}$$

Substituting in Maclaurin's Theorem, we have

$$y = \log x = -\infty + \infty \frac{x}{1} - \infty \frac{x^2}{2} + \text{etc.}$$

Here we have the absurd result that $\log x = \infty$ for all values of x. Hence, $y = \log x$ cannot be developed by Maclaurin's Theorem.

Similarly, $y = \cot x$ gives, when substituted in Maclaurin's Theorem,

$$y = \cot x = \infty - \infty \frac{x}{1} + \text{etc.} \; ;$$

that is, $\cot x = \infty$ for all values of x, which is an absurd result. Hence, $\cot x$ cannot be developed by Maclaurin's Theorem.

Also, $y = x^{\frac{1}{2}}$ becomes, by Maclaurin's Theorem,

$$y = x^{\frac{1}{2}} = 0 + \infty x + \text{etc.} \; ;$$

that is, $x^{\frac{1}{2}} = \infty$ for all values of x, which is an absurd result.

Whether the failure of Maclaurin's Theorem to develop correctly is due to the fact that the particular function is incapable of *any* development, or whether it is simply because it will not develop in the particular form assumed in this formula, the limits of this book will not allow us to enquire.

TAYLOR'S THEOREM.

65. Taylor's Theorem is a theorem for developing a function of the *sum* of two variables into a series arranged according to the ascending powers of one of the variables, with coefficients that are functions of the other variable and of the constants.

LEMMA.—We have first to prove the following *lemma:* If we have a function of the sum of two variables x and y, the derivative will be the same, whether we suppose x to vary and y to remain constant, or y to vary and x to remain constant. For example, let

$$u = (x + y)^n. \tag{1}$$

Differentiating (1), supposing x to vary and y to remain constant, we have

$$\frac{du}{dx} = n (x + y)^{n-1} \tag{2}$$

Differentiating (1), supposing y to vary and x to remain constant, we have

$$\frac{du}{dy} = n(x+y)^{n-1};$$ (3)

from which we see that the derivative is the same in both (2) and (3).

In general, suppose we have *any* function of $x + y$, as

$$u = f(x+y).$$ (4)

Let $$z = x + y;$$ (5)

$$\therefore \quad u = f(z).$$ (6)

Differentiating (5), supposing x variable and y constant, and also supposing y variable and x constant, we get

$$\frac{dz}{dx} = 1, \quad \text{and} \quad \frac{dz}{dy} = 1.$$

Differentiating (6), we have

$$\frac{du}{dz} = \frac{d f(z)}{dz} = f'(z). \quad \text{(See Art. 45.)}$$

$$\therefore \quad du = f'(z)\, dz.$$

$$\therefore \quad \frac{du}{dx} = f'(z)\frac{dz}{dx} = f'(z)\left(\text{since } \frac{dz}{dx} = 1\right).$$

And similarly,

$$\frac{du}{dy} = f'(z)\frac{dz}{dy} = f'(z)\left(\text{since } \frac{dz}{dy} = 1\right).$$

$$\therefore \quad \frac{du}{dx} = \frac{du}{dy}.$$

That is, *the derivative of u with respect to x, y being constant, is equal to the derivative of u with respect to y, x being constant.*

66. To prove Taylor's Theorem.

Let $u' = f(x + y)$ be the function to be developed, and assume the development of the form

$$u' = f(x + y)$$
$$= A + By + Cy^2 + Dy^3 + Ey^4 + \text{etc.,} \qquad (1)$$

in which A, B, C, etc., are independent of y, but are functions of x and of the constants. It is now required to find such values for A, B, C, etc., as will make the assumed development true for all values of x and y.

Finding the derivative of u', regarding x as constant and y variable, we have

$$\frac{du'}{dy} = B + 2Cy + 3Dy^2 + 4Ey^3 + \text{etc.} \qquad (2)$$

Again, finding the derivative of u', regarding x as variable and y constant, we have

$$\frac{du'}{dx} = \frac{dA}{dx} + \frac{dB}{dx}y + \frac{dC}{dx}y^2 + \frac{dD}{dx}y^3 + \text{etc.} \qquad (3)$$

By Art. 65, we have $\dfrac{du'}{dy} = \dfrac{du'}{dx}$; therefore,

$$B + 2Cy + 3Dy^2 + 4Ey^3 + \text{etc.} = \frac{dA}{dx} + \frac{dB}{dx}y + \frac{dC}{dx}y^2 + \frac{dD}{dx}y^3$$
$$+ \text{etc.} \qquad (4)$$

Since (1) is true for every value of y, it is true when $y = 0$. Making $y = 0$ in (1), and representing what u' becomes on this hypothesis by u, we have

$$u = f(x) = A. \qquad (5)$$

Since (4) is true for every value of y, it follows from the principle of *indeterminate coefficients* (Algebra) that the coefficients of the like powers of y in the two members must be equal. Therefore,

$$B = \frac{dA}{dx}, \qquad\qquad \therefore\ \ B = \frac{du}{dx}, \text{ from (5) ;}$$

$$2C = \frac{dB}{dx}, \qquad\qquad \therefore\ \ C = \frac{1}{1\cdot 2}\cdot\frac{d^2u}{dx^2};$$

$$3D = \frac{dC}{dx}, \qquad\qquad \therefore\ \ D = \frac{1}{1\cdot 2\cdot 3}\cdot\frac{d^3u}{dx^3};$$

$$4E = \frac{dD}{dx}, \qquad\qquad \therefore\ \ E = \frac{1}{1\cdot 2\cdot 3\cdot 4}\cdot\frac{d^4u}{dx^4}.$$

Substituting these values of A, B, C, D, etc., in (1), we have

$$u' = f(x+y) = u + \frac{du}{dx}\frac{y}{1} + \frac{d^2u}{dx^2}\frac{y^2}{1\cdot 2} + \frac{d^3u}{dx^3}\frac{y^3}{1\cdot 2\cdot 3}$$

$$+ \frac{d^4u}{dx^4}\frac{y^4}{1\cdot 2\cdot 3\cdot 4} + \text{etc.} \qquad (6)$$

Or, using the other notation (Art. 56), we have

$$u' = f(x+y) = f(x) + f'(x)\frac{y}{1} + f''(x)\frac{y^2}{1\cdot 2} + f'''(x)\frac{y^3}{1\cdot 2\cdot 3}$$

$$+ f^{\text{iv}}(x)\frac{y^4}{1\cdot 2\cdot 3\cdot 4} + \text{etc.}, \qquad (7)$$

which is *Taylor's Theorem.* It is so called from its discoverer, Dr. Brook Taylor, and was first published by him in 1715, in his *Method of Increments.*

Hence, by Taylor's Theorem, we may develop a function of the sum of two variables, as $u' = f(x + y)$, into a series of terms, the first of which is the value of the function when $y = 0$; the second is the value of the first derivative of the function when $y = 0$, into y ; the third is the value of the second derivative when $y = 0$, into $\frac{y^2}{1\cdot 2}$, etc.

The development of $f(x - y)$ is obtained from (6) or (7), by changing $+ y$ into $- y$; thus,

$$f(x-y) = u - \frac{du}{dx}\frac{y}{1} + \frac{d^2u}{dx^2}\frac{y^2}{1\cdot 2} - \frac{d^3u}{dx^3}\frac{y^3}{1\cdot 2\cdot 3}$$
$$+ \frac{d^4u}{dx^4}\frac{y^4}{1\cdot 2\cdot 3\cdot 4} - \text{etc.}$$

or, $f(x-y) = f(x) - f'(x)\frac{y}{1} + f''(x)\frac{y^2}{1\cdot 2} - f'''(x)\frac{y^3}{1\cdot 2\cdot 3}$
$$+ f^{iv}(x)\frac{y^4}{1\cdot 2\cdot 3\cdot 4} - \text{etc.}$$

Cor.—If we make $x = 0$ in (7), we have

$$u' = f(y) = f(0) + f'(0)\frac{y}{1} + f''(0)\frac{y^2}{1\cdot 2} + f'''(0)\frac{y^3}{1\cdot 2\cdot 3}$$
$$+ f^{iv}(0)\frac{y^4}{1\cdot 2\cdot 3\cdot 4} + \text{etc.},$$

which is Maclaurin's Theorem.   See (7) of Art. 58.

THE BINOMIAL THEOREM.

67. To Develop $u' = (x+y)^n$.

Making $y = 0$, and taking the successive derivatives, we have

$$f(x) = x^n,$$
$$f'(x) = nx^{n-1},$$
$$f''(x) = n(n-1)x^{n-2},$$
$$f'''(x) = n(n-1)(n-2)x^{n-3},$$
$$f^{iv}(x) = n(n-1)(n-2)(n-3)x^{n-4},$$
$$\text{etc.} \qquad \text{etc.}$$

Substituting these values in (7), Art. 66, we have

$$u' = (x+y)^n = x^n + \frac{nx^{n-1}y}{1} + \frac{n(n-1)x^{n-2}y^2}{1\cdot 2}$$
$$+ \frac{n(n-1)(n-2)x^{n-3}y^3}{1\cdot 2\cdot 3} + \text{etc.,}$$

which is the Binomial Theorem (see Art. 59).

5

68. To Develop $u' = \sin(x + y)$.

Here $f(x) = \sin x,$ $f'(x) = \cos x,$

$f''(x) = -\sin x,$ $f'''(x) = -\cos x,$ etc.

Hence,

$u' = \sin(x + y)$

$= \sin x \left(1 - \dfrac{y^2}{1 \cdot 2} + \dfrac{y^4}{1 \cdot 2 \cdot 3 \cdot 4} - \dfrac{y^6}{1 \cdot 2 \cdot 3 \cdot 4 \cdot 5 \cdot 6} + \text{etc.} \right)$

$+ \cos x \left(\dfrac{y}{1} - \dfrac{y^3}{1 \cdot 2 \cdot 3} + \dfrac{y^5}{1 \cdot 2 \cdot 3 \cdot 4 \cdot 5} - \dfrac{y^7}{1 \cdot 2 \cdot 3 \cdot 4 \cdot 5 \cdot 6 \cdot 7} + \text{etc.} \right)$

$= \sin x \cos y + \cos x \sin y.$ (See Art. 60.)

THE LOGARITHMIC SERIES.

69. To Develop $u' = \log(x + y)$.

Here $f(x) = \log x,$ $f'''(x) = \dfrac{2}{x^3},$

$f'(x) = \dfrac{1}{x},$ $f^{\text{iv}}(x) = -\dfrac{6}{x^4},$

$f''(x) = -\dfrac{1}{x^2},$ etc.

Hence, $u' = \log(x + y)$

$= \log x + \dfrac{y}{x} - \dfrac{1}{2}\dfrac{y^2}{x^2} + \dfrac{1}{3}\dfrac{y^3}{x^3} - \dfrac{1}{4}\dfrac{y^4}{x^4} + \text{etc.}$

Cor.—If $x = 1$, this series becomes

$\log(1 + y) = \dfrac{y}{1} - \dfrac{y^2}{2} + \dfrac{y^3}{3} - \dfrac{y^4}{4} + \text{etc.,}$

which is the same as Art. 61.

EXPONENTIAL SERIES.

70. To Develop $u' = a^{x+y}$.

Here $f(x) = a^x,$ $f''(x) = a^x \log^2 a,$

$f'(x) = a^x \log a,$ $f'''(x) = a^x \log^3 a,$ etc.

Hence,

$$u' = a^{x+y}$$

$$= a^x \left(1 + \log a \cdot y + \log^2 a \frac{y^2}{1 \cdot 2} + \log^3 a \frac{y^3}{1 \cdot 2 \cdot 3} + \text{etc.} \right).$$

Cor.—If $x = 0$, this series becomes

$$a^y = 1 + \log a \cdot y + \log^2 a \frac{y^2}{1 \cdot 2} + \log^3 a \frac{y^3}{1 \cdot 2 \cdot 3} + \text{etc.},$$

which is the same as Art. 62.

71. Though Taylor's Theorem in general gives the correct development of every function of the sum of two variables, yet it sometimes happens that, for particular values of one of the variables, the function or some of its derivatives become *infinite;* for these particular values, the theorem fails to give a correct development.

For example, take the function $u' = \sqrt{a + x + y}$.

Here,
$$f(x) = \sqrt{a + x},$$

$$f'(x) = \frac{1}{2\sqrt{a + x}},$$

$$f''(x) = -\frac{1}{4(a + x)^{\frac{3}{2}}},$$

$$f'''(x) = \frac{3}{8(a + x)^{\frac{5}{2}}}, \quad \text{etc.}$$

Substituting in (7) of Art. 66, we have

$$u' = \sqrt{a + x + y}$$

$$= \sqrt{a + x} + \frac{y}{2\sqrt{a + x}} - \frac{y^2}{8(a+x)^{\frac{3}{2}}} + \frac{y^3}{16(a+x)^{\frac{5}{2}}} - \text{etc.}$$

Now when x has the particular value $-a$, this equation becomes

$$u' = \sqrt{y} = 0 + \infty - \infty + \infty \cdot - \text{etc.} ;$$

that is, when $x = -a$, $\sqrt{y} = \infty$. But y is independent of x, and may have any value whatever, irrespective of the value of x, and hence the conclusion that when $x = -a$, $\sqrt{y} = \infty$, cannot be true. For every other value of x, however, all the terms in the series will be finite, and the development true.

Similarly, $u' = a + \sqrt{a - x + y}$ gives, when substituted in Taylor's Theorem,

$$u' = a + \sqrt{a - x + y}$$

$$= a + \sqrt{a - x} - \frac{y}{2\sqrt{a - x}} + \text{etc.},$$

which, when $x = a$, becomes

$$u' = a + \sqrt{y} = a - \infty + \text{etc.};$$

and hence the development fails for the particular value, $x = a$.

It will be seen that when Taylor's Theorem fails to give the true development of a function, the failure is only for *particular* values of the variable, all other values of both variables giving a true development; but when Maclaurin's Theorem fails to develop a function for one value of the variable, it fails for every other value.

Many other formulæ, still more comprehensive than these, have been derived, for the development of functions; but a discussion of them would be out of place in this work.

EXAMPLES.

1. Develop $y = \sqrt{1 + x^2}$.

Put $x^2 = z$, and develop; then replace z by its value.

$$\textit{Ans.} \quad y = \sqrt{1 + x^2}$$

$$= 1 + \frac{x^2}{2} - \frac{x^4}{8} + \frac{x^6}{16} - \frac{5x^8}{128} + \text{etc.}$$

2. $y = \dfrac{1}{1-x}.$

$$y = \frac{1}{1-x} = 1 + x + x^2 + x^3 + x^4 + \text{etc.}$$

3. $y = (a+x)^{-3}.$

$$y = (a+x)^{-3} = a^{-3} - 3a^{-4}x + 6a^{-5}x^2 - 10a^{-6}x^3$$
$$+ \text{etc.}$$

4. $y = e^{\sin x}.$

$$y = e^{\sin x} = 1 + x + \frac{x^2}{2} - \frac{x^4}{2\cdot 4} - \frac{x^5}{3\cdot 5} - \frac{x^6}{2\cdot 4\cdot 5\cdot 6}$$
$$+ \text{etc.}$$

5. $y = xe^x.$

$$y = xe^x = x + x^2 + \frac{x^3}{2} + \frac{x^4}{2\cdot 3} + \text{etc.}$$

6. $y = \sqrt{2x - 1}.$

$$y = \sqrt{2x - 1} = \sqrt{-1}\left(1 - x - \frac{x^2}{2} - \frac{x^3}{2} - \text{etc.}\right).$$

7. $y = (a^2 + x^2)^{\frac{5}{3}}.$

$$y = (a^2 + x^2)^{\frac{5}{3}} = a^{\frac{10}{3}} + \tfrac{5}{3}a^{\frac{4}{3}}x^2 + \tfrac{5}{9}a^{-\frac{2}{3}}x^4 - \tfrac{5}{81}a^{\cdot}x^6$$
$$+ \text{etc.}$$

8. $y = \dfrac{1}{\sqrt[4]{a^4 + x^4}}.$

$$y = (a^4 + x^4)^{-\frac{1}{4}} = \frac{1}{a} - \frac{x^4}{4a^5} + \frac{5x^8}{4\cdot 8a^9} - \frac{5\cdot 9x^{12}}{4\cdot 8\cdot 12a^{13}}$$
$$+ \frac{5\cdot 9\cdot 13x^{16}}{4\cdot 8\cdot 12\cdot 16a^{17}} - \text{etc.}$$

9. $y = (a^5 + a^4x - x^5)^{\frac{1}{5}}.$

Put $a^4x - x^5 = z$, as in Ex. 1.

$$y = a + \frac{x}{5} - \frac{4}{5^2 a}\cdot\frac{x^2}{1\cdot 2} + \frac{4\cdot 9}{5^3 a^2}\cdot\frac{x^3}{1\cdot 2\cdot 3}$$
$$- \frac{4\cdot 9\cdot 14}{5^4 a^3}\cdot\frac{x^4}{1\cdot 2\cdot 3\cdot 4} + \text{etc.}$$

10. $u = (x + y)^{\frac{1}{3}}$.

$u = x^{\frac{1}{3}} + \tfrac{1}{3}x^{-\frac{2}{3}}y - \tfrac{1}{9}x^{-\frac{5}{3}}y^2 + \tfrac{5}{81}x^{-\frac{8}{3}}y^3 - $ etc.

11. $u = \cos(x + y)$.   (See Art. 68.)

$u = \cos(x + y) = \cos x \cos y - \sin x \sin y.$

12. $y = \tan x$.

$$y = \tan x = x + \frac{x^3}{3} + \frac{2x^5}{15} + \text{etc.}$$

13. $y = \sec x$.

$$y = \sec x = 1 + \frac{x^2}{2} + \frac{5x^4}{24} + \frac{61x^6}{720} + \text{etc.}$$

14. $y = \log(1 + \sin x)$.

$$y = \log(1 + \sin x) = x - \frac{x^2}{2} + \frac{x^3}{6} - \frac{x^4}{12} + \text{etc.}$$

CHAPTER VI.

72. Indeterminate Forms.—When an algebraic expression is in the form of a fraction, each of whose terms is variable, it sometimes happens that, for a particular value of the independent variable, the expression becomes indeterminate; thus, if a certain value a when substituted for x makes both terms of the fraction $\dfrac{f(x)}{\phi(x)}$ vanish, then it reduces to the form $\dfrac{0}{0}$, and its value is said to be *indeterminate*.

Similarly, the fraction becomes indeterminate if its terms both become *infinite* for a particular value of x; also the forms $\infty \times 0$ and $\infty - \infty$, as well as certain others whose logarithms assume the form $\infty \times 0$, are *indeterminate forms*. It is the object of this chapter to show how the *true* value of such expressions is to be found. By its true value is meant the *limiting value which the fraction assumes when x differs by an infinitesimal from the particular value which makes the expression indeterminate*. It is evident (Arts. 9, 43) that though the *terms* of the fraction may be infinitesimal, the *ratio* of the terms may have any value whatever.

In many cases, the true values of indeterminate forms can be best found by ordinary algebraic and trigonometric processes.

For example, suppose we have to evaluate $\dfrac{x^3 - 1}{x^2 - 1}$ when $x = 1$. This fraction assumes the form $\dfrac{0}{0}$ when $x = 1$; but if we divide the numerator and denominator by $x - 1$

before making $x = 1$, the fraction becomes $\dfrac{x^2 + x + 1}{x + 1}$;
and now if we make $x = 1$, the fraction becomes

$$\frac{1 + 1 + 1}{1 + 1} = \frac{3}{2},$$

which is its true value when $x = 1$.

73. Hence the first step towards the evaluation of such expressions is to detect, if possible, the factors common to both terms of the fraction, and to divide them out; and then to evaluate the resulting fraction by giving to the variable the assigned value.

EXAMPLES.

1. Evaluate $\dfrac{x^3 - 1}{x^3 - 2x^2 + 2x - 1}$, when $x = 1$.

This fraction may be written

$$\frac{(x - 1)(x^2 + x + 1)}{(x - 1)(x^2 - x + 1)} = \frac{x^2 + x + 1}{x^2 - x + 1} = 3, \text{ when } x = 1.$$

2. The fraction $\dfrac{x}{\sqrt{a + x} - \sqrt{a - x}} = \dfrac{0}{0}$, when $x = 0$.

To find its true value, multiply both terms of the fraction by the *complementary surd*, $\sqrt{a + x} + \sqrt{a - x}$, and it becomes

$$\frac{x(\sqrt{a + x} + \sqrt{a - x})}{2x} \quad \text{or} \quad \frac{\sqrt{a + x} + \sqrt{a - x}}{2};$$

and now making $x = 0$, the fraction becomes $\sqrt{a}$, which is its true value when $x = 0$.

3. $\dfrac{2x - \sqrt{5x^2 - a^2}}{x - \sqrt{2x^2 - a^2}}$, when $x = a$. *Ans.* $\frac{1}{2}$.

4. $\dfrac{a - \sqrt{a^2 - x^2}}{x^2}$, when $x = 0$. *Ans.* $\dfrac{1}{2a}$.

5. $\dfrac{x^5 - 1}{x - 1}$, when $x = 1$. *Ans.* 5.

6. $\sqrt{x^2 + ax} - x$, when $x = \infty$ *Ans.* $\dfrac{a}{2}$.

There are many indeterminate forms in which it is either impossible to detect the factor common to both terms, or else the process is very laborious, and hence the necessity of some general method for evaluating indeterminate forms. Such a method is furnished us by the Differential Calculus, which we now proceed to explain.

METHOD OF THE DIFFERENTIAL CAL-CULUS.

74. To evaluate Functions of the form $\dfrac{0}{0}$.

Let $f(x)$ and $\phi(x)$ be two functions of x such that $f(x) = 0$ and $\phi(x) = 0$, when $x = a$.

Then we shall have $\dfrac{f(a)}{\phi(a)} = \dfrac{0}{0}$.

Let x take an increment h, becoming $x + h$; then the fraction becomes
$$\frac{f(x + h)}{\phi(x + h)}.$$

Now develop $f(x + h)$ and $\phi(x + h)$ by Taylor's Theorem; substituting h for y in (7) of Art. 66, we have

$$\frac{f(x + h)}{\phi(x + h)} = \frac{f(x) + f'(x)\dfrac{h}{1} + f''(x)\dfrac{h^2}{2} + \text{etc.}}{\phi(x) + \phi'(x)\dfrac{h}{1} + \phi''(x)\dfrac{h^2}{2} + \text{etc.}};$$

or when $x = a$,

$$\frac{f(a+h)}{\phi(a+h)} = \frac{f(a) + f'(a) h + f''(a) \dfrac{h^2}{2} + \text{etc.}}{\phi(a) + \phi'(a) h + \phi''(a) \dfrac{h^2}{2} + \text{etc.}}. \quad (1)$$

But by hypothesis $f(a) = 0$, and $\phi(a) = 0$. Hence, dropping the first term in the numerator and denominator, and dividing both by h, we have,

$$\frac{f(a+h)}{\phi(a+h)} = \frac{f'(a) + f''(a) \dfrac{h}{2} + \text{etc.}}{\phi'(a) + \phi''(a) \dfrac{h}{2} + \text{etc.}}.$$

Now when $h = 0$, the numerator and denominator of the second member become $f'(a)$ and $\phi'(a)$ respectively; hence we have,

$$\frac{f(a)}{\phi(a)} = \frac{f'(a)}{\phi'(a)},$$

as the *true* value of the fraction $\dfrac{f(x)}{\phi(x)}$, when $x = a$.

(1.) If $f'(a) = 0$ and $\phi'(a)$ be not 0, the true value of $\dfrac{f(a)}{\phi(a)}$ is zero.

(2.) If $f'(a)$ be not zero and $\phi'(a) = 0$, the true value of $\dfrac{f(a)}{\phi(a)}$ is ∞.

(3.) If $f'(a) = 0$, and $\phi'(a) = 0$, the new fraction $\dfrac{f'(a)}{\phi'(a)}$ is still of the indeterminate form $\dfrac{0}{0}$. Dropping in this case the first *two* terms of the numerator and denominator of (1), dividing both by $\dfrac{h^2}{2}$, and making $h = 0$, we have,

$$\frac{f(a)}{\phi(a)} = \frac{f''(a)}{\phi''(a)},$$

as the true value of the fraction $\frac{f(x)}{\phi(x)}$, when $x = a$.

If this fraction be also of the form $\frac{0}{0}$, we proceed to the next derivative, and thus we proceed till a pair of derivatives is found which do not *both* reduce to zero, when $x = a$. The last result is the true value of the fraction.

EXAMPLES.

1. Evaluate $\frac{\log x}{x-1}$, when $x = 1$.

Here $\quad f(x) = \log x, \qquad \phi(x) = x - 1;$

$\therefore \quad f'(x) = \frac{1}{x}, \qquad$ and $\quad \phi'(x) = 1;$

$\therefore \quad \frac{f(x)}{\phi(x)} = \frac{f'(x)}{\phi'(x)} = \dfrac{\dfrac{1}{x}}{1} = \dfrac{1}{x}\bigg]_{1^*} = 1.$

That is, $\frac{\log x}{x-1} = 1$, when $x = 1$.

2. Evaluate $\frac{1 - \cos x}{x^2}$, when $x = 0$.

$$\frac{f'(x)}{\phi'(x)} = \frac{\sin x}{2x} = \frac{0}{0}, \text{ when } x = 0;$$

$$\frac{f''(x)}{\phi''(x)} = \frac{\cos x}{2}\bigg]_0 = \tfrac{1}{2}.$$

Therefore, $\dfrac{f(x)}{\phi(x)}\bigg]_0 = \tfrac{1}{2}.$

---

* The subscript denotes the value of the independent variable for which the function is evaluated.

3.   Evaluate $\dfrac{x \sin x - \dfrac{\pi}{2}}{\cos x}$, when $x = \dfrac{\pi}{2}$.

Here $\dfrac{f'(x)}{\phi'(x)} = \dfrac{x \cos x + \sin x}{-\sin x}\Big]_{\frac{\pi}{2}} = \dfrac{1}{-1} = -1.$

Hence $\dfrac{f(x)}{\phi(x)} = -1.$

4. $\dfrac{a^x - b^x}{x}$, when $x = 0$.   *Ans.* $\log \dfrac{a}{b}.$

5. $\dfrac{e^{nx} - e^{na}}{(x - a)^s}$, when $x = a$.

Here $\dfrac{f'(x)}{\phi'(x)} = \dfrac{n e^{nx}}{s(x - a)^{s-1}}\Big]_a = \infty \text{ or } 0,$

according as $s >$ or < 1.

6. $\dfrac{x - \sin x}{x^3}$, when $x = 0$.   *Ans.* $\tfrac{1}{6}.$

7. $\dfrac{e^x - e^{-x} - 2x}{x - \sin x}$, when $x = 0$.   *Ans.* 2.

8. $\dfrac{e^x - e^{-x}}{\sin x}$, when $x = 0$.   *Ans.* 2.

9. $\dfrac{e^x - 2 \sin x - e^{-x}}{x - \sin x}$, when $x = 0$.   Take the third
derivative.   *Ans.* 4.

10. $\dfrac{(a^2 - x^2)^{\frac{3}{2}}}{(a - x)^{\frac{3}{2}}}$, when $x = a$. Cancel the factor $(a - x)^{\frac{3}{2}}.$

Ans. $(2a)^{\frac{3}{2}}.$

75. To evaluate Functions of the form $\dfrac{\infty}{\infty}$.

Let $\dfrac{f(x)}{\phi(x)} = \dfrac{\infty}{\infty},$ when $x = a$.

Since the terms of this fraction are infinites when $x = a$, their reciprocals are infinitesimals (Art. 8); that is, $\dfrac{1}{f(x)} = 0$, and $\dfrac{1}{\phi(x)} = 0$, when $x = a$; hence,

$$\frac{f(x)}{\phi(x)} = \frac{\dfrac{1}{\phi(x)}}{\dfrac{1}{f(x)}} = \frac{0}{0}, \text{ when } x = a,$$

and therefore the true values of $\dfrac{\dfrac{1}{\phi(x)}}{\dfrac{1}{f(x)}}$ may be obtained by Art. 74; that is, by taking the derivatives of the terms, thus,

$$\frac{f(x)}{\phi(x)} = \frac{\dfrac{1}{\phi(x)}}{\dfrac{1}{f(x)}} = \frac{\dfrac{\phi'(x)}{[\phi(x)]^2}}{\dfrac{f'(x)}{[f(x)]^2}} = \frac{\phi'(x)\,[f(x)]^2}{f'(x)\,[\phi(x)]^2}, \text{ when } x = a,$$

or

$$\frac{f(a)}{\phi(a)} = \frac{\phi'(a)\,[f(a)]^2}{f'(a)\,[\phi(a)]^2}.$$

Dividing by $\dfrac{f(a)}{\phi(a)}$, we get,

$$1 = \frac{\phi'(a)}{f'(a)} \cdot \frac{f(a)}{\phi(a)};$$

whence

$$\frac{f(a)}{\phi(a)} = \frac{f'(a)}{\phi'(a)}.$$

Hence the true value of the indeterminate form $\dfrac{\infty}{\infty}$ is found in the same manner as that of the form $\dfrac{0}{0}$.

In the above demonstration, in dividing the equation by $\dfrac{f(a)}{\phi(a)}$, when $x = a$, we assumed that $\dfrac{f(a)}{\phi(a)}$ is neither 0 nor ∞, so that the proof would fail in either of these cases.

It may, however, be completed as follows: Suppose the true value of $\dfrac{f(a)}{\varphi(a)}$ to be 0; then the value of $\dfrac{f(a) + h\,\varphi(a)}{\varphi(a)}$ is h, where h may be any constant. But as this latter fraction has a value which is neither 0 nor ∞, its value by the above method is $\dfrac{f'(a) + h\,\varphi'(a)}{\varphi'(a)}$ or $\dfrac{f'(a)}{\varphi'(a)} + h$; and since the value of this fraction is h, the first term $\dfrac{f'(a)}{\varphi'(a)} = 0$; *i. e.*, where $\dfrac{f(a)}{\varphi(a)} = 0$, $\dfrac{f'(a)}{\varphi'(a)}$ is also 0.

Similarly, if the true value of $\dfrac{f(x)}{\varphi(x)}$ be ∞ when $x = a$, then $\dfrac{\varphi(a)}{f(a)} = 0$; and therefore we have $\dfrac{\varphi'(a)}{f'(a)} = 0$, by what has just been shown; $\therefore \dfrac{f'(a)}{\varphi'(a)} = \infty$.

Therefore, in every case the value of $\dfrac{f'(a)}{\varphi'(a)}$ determines the value of $\dfrac{f(a)}{\varphi(a)}$ for either of the indeterminate forms $\dfrac{0}{0}$ or $\dfrac{\infty}{\infty}$. (See Williamson's Dif. Cal., p. 100.)

EXAMPLES.

1. Evaluate $\dfrac{\log x}{x^n}$, when $x = \infty$.

Here $\quad \dfrac{f(x)}{\varphi(x)} = \dfrac{f'(x)}{\varphi'(x)} = \dfrac{\dfrac{1}{x}}{nx^{n-1}} = \dfrac{1}{nx^n}\Big]_\infty = 0.$

2. Evaluate $\dfrac{\log x}{\cot x}$, when $x = 0$.

Here $\quad \dfrac{f'(x)}{\varphi'(x)} = \dfrac{\dfrac{1}{x}}{-\operatorname{cosec}^2 x} = -\dfrac{\sin^2 x}{x}\Big]_0 = \dfrac{0}{0};$

$\dfrac{f''(x)}{\varphi''(x)} = -\dfrac{2\sin x \cos x}{1}\Big]_0 = 0;$

$\therefore \dfrac{\log x}{\cot x} = 0,$ when $x = 0.$

3. $\dfrac{1 - \log x}{e^{\frac{1}{x}}}$, when $x = 0$.   *Ans.* 0.

4. $\dfrac{\dfrac{\pi}{4x}}{\cot \dfrac{\pi x}{2}}$, when $x = 0$.   *Ans.* $\dfrac{\pi^2}{8}$.

5. $\dfrac{\log \tan (2x)}{\log \tan x}$, when $x = 0$.   *Ans.* 1.

76. To evaluate Functions of the form $0 \times \infty$.

Let $\quad f(x) \times \phi(x) = 0 \times \infty$, when $x = a$.

The function in this case is easily reducible to the form $\dfrac{0}{0}$; for if $f(a) = 0$, and $\phi(a) = \infty$, the expression can be

written $\dfrac{f(a)}{\dfrac{1}{\phi(a)}}$, which $= \dfrac{0}{0}$; therefore $f(x) \times \phi(x) = \dfrac{f(x)}{\dfrac{1}{\phi(x)}}$,

may be evaluated by the method of Art. 74.

EXAMPLES.

1. Evaluate $(1 - x) \tan \dfrac{\pi x}{2}$, when $x = 1$.

We may write this

$$\dfrac{1 - x}{\cot \dfrac{\pi x}{2}}.$$

Here $\quad \dfrac{f'(x)}{\phi'(x)} = \dfrac{-1}{-\dfrac{\pi}{2} \operatorname{cosec}^2 \dfrac{\pi x}{2}} = \left.\dfrac{2 \sin^2 \left(\dfrac{\pi x}{2}\right)}{\pi}\right]_1 = \dfrac{2}{\pi}.$

2. Evaluate $x^n \log x$, when $x = 0$.

$$x^n \log x = \left.\dfrac{\log x}{x^{-n}}\right]_0 = \dfrac{\dfrac{1}{x}}{-nx^{-n-1}} = \left.-\dfrac{x^n}{n}\right]_0 = 0.$$

3. $e^{-x} \log x$, when $x = \infty$.   *Ans.* 0.

4. $\sec x \left(x \sin x - \dfrac{\pi}{2} \right)$, when $x = \dfrac{\pi}{2}$.   *Ans.* -1.

77. To evaluate Functions of the form $\infty - \infty$.

Let $f(x)$ and $\phi(x)$ be two functions of x which become infinite when $x = a$. Then $f(x) - \phi(x) = \infty - \infty$, when $x = a$.

The function in this case can be easily reduced to the form $\dfrac{0}{0}$, and may be evaluated as heretofore.

EXAMPLES.

1. Evaluate $\dfrac{2}{x^2 - 1} - \dfrac{1}{x - 1}$, when $x = 1.$

This takes the form $\infty - \infty$, when $x = 1.$

$$\frac{2}{x^2 - 1} - \frac{1}{x - 1} = \frac{2 - x - 1}{x^2 - 1} = \frac{0}{0}, \text{ when } x = 1.$$

$$\frac{f'(x)}{\phi'(x)} = \frac{-1}{2x} = -\tfrac{1}{2}, \quad \text{ when } x = 1.$$

2. Evaluate $\dfrac{1}{x(1 + x)} - \dfrac{\log(1 + x)}{x^2}$, when $x = 0,$

which takes the form $\infty - \infty$, when $x = 0.$

$$\frac{1}{x(1 + x)} - \frac{\log(1 + x)}{x^2} = \frac{x - (1 + x)\log(1 + x)}{x^2(1 + x)}$$

$$= \frac{x - (1 + x)\log(1 + x)}{x^2}$$

(remembering that $1 + x = 1$, when x vanishes).

$$\frac{f'(x)}{\phi'(x)} = \frac{1 - \log(1 + x) - 1}{2x} = \frac{0}{0}, \text{ when } x = 0;$$

$$\frac{f''(x)}{\phi''(x)} = \frac{-\dfrac{1}{1 + x}}{2}\Bigg]_0 = -\tfrac{1}{2}.$$

3. Evaluate $\sec x - \tan x$, when $x = \dfrac{\pi}{2}$.

$$\sec x - \tan x = \frac{1 - \sin x}{\cos x} = \frac{0}{0}, \text{ when } x = \frac{\pi}{2}.$$

$$\frac{f'(x)}{\phi'(x)} = \frac{-\cos x}{-\sin x}\Bigg]_{\frac{\pi}{2}} = 0.$$

Hence, $\sec \dfrac{\pi}{2}$ and $\tan \dfrac{\pi}{2}$ are either absolutely equal, or differ by a quantity which must be neglected in their algebraic sum.*

4. $\dfrac{x}{x - 1} - \dfrac{1}{\log x}$, when $x = 1$.　　　　*Ans.* $\tfrac{1}{2}$.

5. $\dfrac{1}{\log x} - \dfrac{x}{\log x}$, when $x = 1$.　　　　*Ans.* -1.

78. To evaluate Functions of the forms 0^0, ∞^0, and $1^{\pm\infty}$.

Let $f(x)$ and $\phi(x)$ be two functions of x which, when $x = a$, assume such values that $[f(x)]^{\phi(x)}$ is one of the above forms.

Let　　　　　$y = [f(x)]^{\phi(x)}$;

$$\therefore \ \log y = \phi(x) \log f(x).$$

(*1.*) When $f(x) = \infty$ or 0, and $\phi(x) = 0$.

$$\log y = \phi(x) \log f(x) = 0\,(\pm\infty),$$

which is the form of Art. 76.

---

* Price's Infinitesimal Calculus, Vol. I, p. 210.

Hence, $[f(x)]^{\phi(x)}$ becomes indeterminate when it is of the form 0^0 or ∞^0.

(2.) When $f(x) = 1$, and $\phi(x) = \pm \infty$.

$$\log y = \phi(x) \log f(x) = \pm \infty \times 0.$$

Hence, $[f(x)]^{\phi(x)}$ is indeterminate when of the forms

$$1^{\pm \infty}.$$

Hence the indeterminate forms of this class are

$$0^0, {}^* \qquad \infty^0, \qquad 1^{\pm \infty},$$

and may all be evaluated as in Art. 76, by first evaluating their logarithms, which take the form $0 \times \infty$.

EXAMPLES

1. Evaluate x^x, when $x = 0$.

We have $\qquad \log x^x = x \log x = \dfrac{\log x}{x^{-1}};$

$$\frac{f'(x)}{\phi'(x)} = \frac{\dfrac{1}{x}}{-x^{-2}} = -x]_0 = 0;$$

$\therefore \quad \log x^x = 0$, when $x = 0$;

hence, $\qquad x^x = 1$, when $x = 0$.

2. Evaluate $x^{\frac{1}{x}}$, when $x = \infty$.

$$\log x^{\frac{1}{x}} = \frac{1}{x} \log x = \frac{\log x}{x};$$

$$\frac{f'(x)}{\phi'(x)} = \frac{\dfrac{1}{x}}{1} = \frac{1}{x}\Bigg]_\infty = 0;$$

---

* In general, the value of the indeterminate form 0^0 is 1. (See Note on Indeterminate Exponential Forms, by F. Franklin, in Vol. I, No. 4, of American Journal of Mathematics.)

$$\therefore \quad \log x^{\frac{1}{x}} = 0, \quad \text{when} \quad x = \infty;$$

hence, $\qquad x^{\frac{1}{x}} = 1, \quad \text{when} \quad x = \infty.$

3. Evaluate $\left(1 + \dfrac{a}{x}\right)^{x}$, when $x = \infty$.

Let $x = \dfrac{1}{z}$, and denote the function by *u.*

Then $\qquad u = (1 + az)^{\frac{1}{z}}\Big]_{0}$

(since when $x = \infty$, $z = 0$); and

$$\log u = \frac{\log (1 + az)}{z}, \quad \text{when} \quad z = 0.$$

Taking derivatives, we have

$$\log u_{\infty} = \frac{a}{1 + az}\Big]_{0} = a;$$

$$\therefore \quad \log \left(1 + \frac{a}{x}\right)^{x} = a, \quad \text{when} \quad x = \infty;$$

hence, $\quad \left(1 + \dfrac{a}{x}\right)^{x} = e^{a}, \quad \text{when} \quad x = \infty.$

If $a = 1$, we have

$$\left(1 + \frac{1}{x}\right)^{x}_{\infty} = e;$$

that is, as x increases indefinitely, the *limiting value* (Art. 41) of the function $\left(1 + \dfrac{1}{x}\right)^{x}$ is the Naperian base.

4. $\left(\dfrac{1}{x}\right)^{\tan x}$, $\qquad$ when $x = 0$. $\qquad$ *Ans.* **1.**

5. $x^{\sin x}$, $\qquad$ when $x = 0$. $\qquad$ *Ans.* **1.**

6. $\left(2 - \dfrac{x}{a}\right)^{\tan \frac{\pi x}{2a}}$, when $x = a$. $\qquad$ *Ans.* $e^{\frac{2}{\pi}}$.

79. Compound Indeterminate Forms.—If an indeterminate form be the product of two or more expressions, each of which becomes indeterminate for the same value of x, its true value can be found by evaluating each factor separately; also, when the value of any indeterminate form is known, that of any power of it can be determined.

EXAMPLES.

1. Evaluate $\dfrac{x^n}{e^x}$, when $x = \infty$.

This fraction may be written $\left(\dfrac{x}{e^{\frac{x}{n}}}\right)^n$.

We first evaluate $\dfrac{x}{e^{\frac{x}{n}}}$, when $x = \infty$.

Here $\qquad \dfrac{f'(x)}{\phi'(x)} = \dfrac{1}{\dfrac{1}{n}e^{\frac{x}{n}}} \Bigg]_{\infty} = \dfrac{1}{\infty} = \mathbf{0.}$

Hence, $\qquad \dfrac{x^n}{e^x} = 0^n = 0.$

2. Evaluate $x^m \log^n x$, when $x = 0$, and m and n are positive.

Here $\qquad (x^{\frac{m}{n}} \log x)^n = \left(\dfrac{\log x}{x^{-\frac{m}{n}}}\right)^n$.

We first evaluate $\dfrac{\log x}{x^{-\frac{m}{n}}}$, when $x = 0$.

We have $\qquad \dfrac{f'(x)}{\phi'(x)} = \dfrac{\dfrac{1}{x}}{-\dfrac{m}{n}x^{-\frac{m}{n}-1}}$

$$= -\dfrac{n}{m}x^{\frac{m}{n}}\Bigg]_0 = \mathbf{0.}$$

$\therefore \quad x^m \log^n x = 0^n = 0.$

3. $\dfrac{x^m - x^{m+n}}{1 - x^{2p}}$, when $x = 1$.

This function can be written in the form

$$\frac{x^m}{1 + x^p} \cdot \frac{1 - x^n}{1 - x^p}.$$

We have to evaluate only the latter function for $x = 1$, since the former is determinate.

Here $\qquad \dfrac{f'(x)}{\phi'(x)} = \dfrac{-nx^{n-1}}{-px^{p-1}} = \dfrac{n}{p} x^{n-p}$

$$= \frac{n}{p}, \qquad \text{when} \quad x = 1.$$

$$\frac{x^m}{1 + x^p} = \frac{1}{2}, \qquad \text{when} \quad x = 1.$$

$$\therefore \quad \frac{x^n - x^{m+n}}{1 - x^{2p}} = \frac{n}{2p}, \qquad \text{when} \quad x = 1.$$

4. $\dfrac{(x^2 - a^2) \sin \dfrac{\pi x}{2a}}{x^2 \cos \dfrac{\pi x}{2a}}$, when $x = a$.

$$\frac{(x^2 - a^2) \sin \dfrac{\pi x}{2a}}{x^2 \cos \dfrac{\pi x}{2a}} = \frac{x^2 - a^2}{\cos \dfrac{\pi x}{2a}} \cdot \frac{\sin \dfrac{\pi x}{2a}}{x^2}.$$

We have only to evaluate the first factor,

$$\left. \frac{x^2 - a^2}{\cos \dfrac{\pi x}{2a}} \right|_a = \left. \frac{2x}{-\dfrac{\pi}{2a} \sin \dfrac{\pi x}{2a}} \right|_a = -\frac{4a^2}{\pi},$$

and $\qquad \left. \dfrac{\sin \dfrac{\pi x}{2a}}{x^2} \right|_a = \dfrac{1}{a^2}.$

$$\therefore \quad \frac{(x^2 - a^2)\sin\dfrac{\pi x}{2a}}{x^2 \cos\dfrac{\pi x}{2a}} = -\frac{4}{\pi}.$$

EXAMPLES.

1. $\dfrac{e^x - e^{-x}}{\log(1+x)}$,   when $x = 0$.   *Ans.* 2.

2. $\dfrac{x^3 - 5x^2 + 7x - 3}{x^3 - x^2 - 5x - 3}$, when $x = 3$. $\frac{1}{4}$.

3. $\left(\dfrac{\sin nx}{x}\right)^m$,   when $x = 0$. n^m.

4. $\dfrac{e^x - e^{-x} - 2x}{(e^x - 1)^3}$ (dif. three times), when $x = 0$.

                                                           $\frac{1}{8}$.

5. $\dfrac{1 - \sin x + \cos x}{\sin x + \cos x - 1}$, when $x = \dfrac{\pi}{2}$.   1.

6. $\dfrac{\tan x - \sin x}{\sin^3 x}$,   when $x = 0$. $\frac{1}{2}$.

7. $\dfrac{a^{\sin x} - a}{\log \sin x}$,   when $x = \dfrac{\pi}{2}$. $a \log a$.

8. $\dfrac{x^2 + 2\cos x - 2}{x^4}$ (dif. four times), when $x = 0$.

                                                            $\frac{1}{12}$.

9. $x^{\frac{1}{1-x}}$ (pass to logarithms), when $x = 1$.

                                                            $\dfrac{1}{e}$.

10. $\dfrac{x \log(1+x)}{1 - \cos x}$,   when $x = 0$.   2.

11. $x \cdot e^{\frac{1}{x}}$, when $x = 0$.

                                                            ∞.

12. $\left(\dfrac{\log x}{x}\right)^{\frac{1}{x}}$ (pass to logarithms), when $x = \infty$.

$\qquad\qquad\qquad\qquad\qquad\qquad\qquad$ *Ans.* 1.

13. $2x \sin \dfrac{a}{2x}$, when $x = \infty$. $\qquad\qquad$ *a.*

14. $e^{\frac{1}{x}} \sin x$, when $x = 0$. $\qquad\qquad$ ∞.

15. $(\cos ax)^{\operatorname{cosec}^2 cx}$ (pass to logarithms, and dif. twice), when $x = 0$. $\qquad\qquad\qquad\qquad\qquad\qquad e^{-\frac{a^2}{2c^2}}$.

16. $x^m (\sin x)^{\tan x} \left(\dfrac{\pi - 2x}{2 \sin 2x}\right)^3$ (see Art. 79), when $x = \dfrac{\pi}{2}$.

$\qquad\qquad\qquad\qquad\qquad\qquad\qquad\qquad \dfrac{\pi^m}{2^{m+3}}$.

17. $(\sin x)^{\tan x}$, when $x = \dfrac{\pi}{2}$. $\qquad\qquad$ **1.**

CHAPTER VII.

FUNCTIONS OF TWO OR MORE VARIABLES, AND CHANGE OF THE INDEPENDENT VARIABLE.

80. Partial Differentiation.—In the preceding chapters, we have considered only functions of *one independent variable;* such functions are furnished us in *Analytic Geometry of Two Dimensions.* In the present chapter, we are to consider functions of *two or more* variables. *Analytic Geometry of Three Dimensions* introduces us to functions of the latter kind. For example, the equation

$$z = ax + by + c \tag{1}$$

represents a plane; x and y are *two independent* variables, of which z is a function. In this equation, z may be changed by changing either x or y, or by changing them both, as they are entirely independent of each other, and either of them may be considered to change without affecting the other; in this case z, the value of which depends upon the values of x and y, is called a function of the independent variables x and y.

A *partial differential* of a function of several variables is a differential obtained on the hypothesis that only one of the variables changes.

A *total differential* of a function of several variables is a differential obtained on the hypothesis that all the variables change.

A *partial derivative* of a function of several variables is the ratio of a partial differential of the function to the differential of the variable supposed to change.

A *total derivative* of a function of several variables is the ratio of the total differential of the function to the differential of some one of its variables. (See Olney's Calculus, p. 45.)

As all the variables except one are, for the time being, treated as constants, it follows that the *partial differentials and derivatives* of any expression can be obtained by the same rules as the differentials and derivatives in the case of a single variable.

If we differentiate (1), first with respect to x, regarding y as constant, and then with respect to y, regarding x as constant, we get

$$dz = adx, \tag{2}$$

and
$$dz = bdy. \tag{3}$$

Dividing (2) and (3) by dx and dy respectively, we get,

$$\frac{dz}{dx} = a, \tag{4}$$

and
$$\frac{dz}{dy} = b. \tag{5}$$

The expressions in (2) and (3) are called the *partial differentials* of z with respect to x and y, respectively, while $\frac{dz}{dx}$ and $\frac{dz}{dy}$ are called the *partial derivatives* of z with respect to the same variables.

Since a and b in (4) and (5) are the partial derivatives of z with respect to x and y, respectively, we see from (2) that the partial differential of z with respect to x, is equal to the partial derivative of z with respect to x multiplied by dx, and similarly for the partial differential of y.

Hence, generally, if

$$f(x, y, z)$$

denotes a function of three variables, x, y, z, its derivative or differential when x *alone* is supposed to change, is called

6

the partial derivative or differential of the function *with respect to x*, and similarly for the other variables, *y* and *z*. If the function be represented by *u*, its partial derivatives are denoted by

$$\frac{du}{dx}, \qquad \frac{du}{dy}, \qquad \frac{du}{dz},$$

and its partial differentials by

$$\frac{du}{dx}\,dx, \qquad \frac{du}{dy}\,dy, \qquad \frac{du}{dz}\,dz.$$

81. Differentiation of a Function of Two Variables.—Let $u = f(x, y)$, and represent the partial differential of *u* with respect to *x*, by $\frac{du}{dx}\,dx$, and with respect to *y*, by $\frac{du}{dy}\,dy$, while *du* represents the total differential.

Let *x* and *y* receive the infinitesimal increments *dx* and *dy*, and let the corresponding increment of *u* be *du*. Then we have,

$$du = f(x + dx,\ y + dy) - f(x, y).$$

Subtract and add $f(x,\ y + dy)$, and we have

$$du = f(x+dx,\ y+dy) - f(x,\ y+dy) + f(x,\ y+dy)$$
$$- f(x,\ y).$$

Now $f(x+dx,\ y+dy) - f(x,\ y+dy) = \dfrac{du}{dx}\,dx$, because it is the difference between two consecutive states of the function due to a change in *x* alone; that is, whatever difference there is between $f(x+dx,\ y+dy)$ and $f(x,\ y+dy)$ is due solely to the change in *x*, as $y+dy$ is the value of *y* in both of them. For the same reason,

$$f(x,\ y+dy) - f(x, y) = \frac{du}{dy}\,dy\,;$$

and therefore we have $du = \dfrac{du}{dx}\,dx + \dfrac{du}{dy}\,dy,$

in which $\dfrac{du}{dx}\,dx$ and $\dfrac{du}{dy}\,dy$ are the *partial differentials* of u with respect to x and y, respectively, while du is the *total differential* of u when both x and y are supposed to vary. In the same way, we may find the differential of any number of variables.

Hence, the total differential of a function of two or more variables is equal to the sum of its partial differentials.

The student will carefully observe the different meanings given to the infinitely small quantity du in this equation, otherwise the equation will seem to be inconsistent with the principles of algebra. Thus, in $\dfrac{du}{dx}\,dx$, du denotes the infinitely small change in u arising from the increment dx in x, *y being regarded as constant*. Also, in $\dfrac{du}{dy}\,dy$, du denotes the infinitely small change in u arising from the increment dy in y, *x being regarded as constant*, while du in the first member denotes the *total* change in u caused by both x and y changing. If the partial differentials of x and y be represented by $d_x u$ and $d_y u$, respectively, the preceding equation may be written

$$du = d_x u + d_y u.$$

E X A M P L E S.

1. Let $u = ay^2 + bxy + cx^2 + ey + gx + k$, to find the total differential of u.

Differentiating with respect to x, we have

$$d_x u = by dx + 2cx dx + g dx.$$

Differentiating with respect to y, we have

$$d_y u = 2ay dy + bx dy + e dy.$$

Hence, $\quad du = (by + 2cx + g)\, dx + (2ay + bx + e)\, dy.$

2. $u = x^y$.

Here $d_x u = yx^{y-1} dx,$ $d_y u = x^y \log x\, dy.$

Hence, $du = yx^{y-1} dx + x^y \log x\, dy.$

3. $u = \dfrac{x^2}{a^2} + \dfrac{y^2}{b^2}.$

Here $d_x u = \dfrac{2x}{a^2} dx,$ $d_y u = \dfrac{2y}{b^2} dy.$

Hence, $du = \dfrac{2x}{a^2} dx + \dfrac{2y}{b^2} dy.$

4. $u = \tan^{-1} \dfrac{y}{x}.$

Here $d_z u = \dfrac{-\dfrac{ydx}{x^2}}{1 + \dfrac{y^2}{x^2}} = -\dfrac{ydx}{x^2 + y^2},$

$d_y u = \dfrac{\dfrac{dy}{x}}{1 + \dfrac{y^2}{x^2}} = \dfrac{xdy}{x^2 + y^2}.$

Hence, $du = \dfrac{xdy - ydx}{x^2 + y^2}.$

5. $u = \sin^{-1} \dfrac{x}{a} + \sin^{-1} \dfrac{y}{b}.$

Here $d_x u = \dfrac{dx}{\sqrt{a^2 - x^2}},$ $d_y u = \dfrac{dy}{\sqrt{b^2 - y^2}}.$

Hence, $du = \dfrac{dx}{\sqrt{a^2 - x^2}} + \dfrac{dy}{\sqrt{b^2 - y^2}}.$

6. $u = y^{\sin x}.$ $du = y^{\sin x} \log y \cos x\, dx + \dfrac{\sin x\, dy}{y^{\text{covers } x}}.$

7. $u = \text{vers}^{-1}\dfrac{x}{y}$. $\qquad\qquad du = \dfrac{ydx - xdy}{y\sqrt{2xy - x^2}}$.

8. $u = \log x^y$. $\qquad\qquad du = \dfrac{y}{x}dx + \log x\, dy$.

82. To Find the Total Derivative of u with respect to x, when $u = f(y, z)$, and $y = \phi(x)$, $z = \phi_1(x)$.

Since $u = f(y, z)$, we have (Art. 81),

$$du = \frac{du}{dy}dy + \frac{du}{dz}dz; \qquad (1)$$

and since $\qquad y = \phi(x)$, we have $\quad dy = \dfrac{dy}{dx}dx$;

since $\qquad z = \phi_1(x)$, we have $\quad dz = \dfrac{dz}{dx}dx$.

Substituting these values for dy and dz in (1), we get

$$du = \frac{du}{dy}\frac{dy}{dx}dx + \frac{du}{dz}\frac{dz}{dx}dx. \qquad (2)$$

Dividing by dx, and denoting the *total* derivative by ( ), we have

$$\left(\frac{du}{dx}\right) = \frac{du}{dy}\frac{dy}{dx} + \frac{du}{dz}\frac{dz}{dx}. \qquad (3)$$

Cor. 1.—If $z = x$, the proposition becomes $u = f(x, y)$ and $y = \phi(x)$; and since $\dfrac{dz}{dx} = 1$, (3) becomes

$$\left(\frac{du}{dx}\right) = \frac{du}{dx} + \frac{du}{dy}\frac{dy}{dx}.$$

Cor. 2.—If $u = f(x, y, z)$, and $y = \phi(x)$, and $z = \phi_1(x)$, we have

$$du = \frac{du}{dx}dx + \frac{du}{dy}dy + \frac{du}{dz}dz. \qquad (1)$$

$$dy = \frac{dy}{dx}\, dx, \quad \text{and} \quad dz = \frac{dz}{dx}\, dx.$$

Substituting the values of dy and dz in (1), and dividing by dx, we get

$$\left(\frac{du}{dx}\right) = \frac{du}{dx} + \frac{du}{dy}\frac{dy}{dx} + \frac{du}{dz}\frac{dz}{dx}.$$

Cor. 3.—If $u = f(y, z, v)$, and $y = \phi(x)$, and $z = \phi_1(x)$, and $v = \phi_2(x)$, we have,

$$du = \frac{du}{dy}\, dy + \frac{du}{dz}\, dz + \frac{du}{dv}\, dv. \tag{1}$$

$$dy = \frac{dy}{dx}\, dx; \quad dz = \frac{dz}{dx}\, dx \quad dv = \frac{dv}{dx}\, dx.$$

Substituting the values of dy, dz, dv, in (1), and dividing by dx, we get

$$\left(\frac{du}{dx}\right) = \frac{du}{dy}\frac{dy}{dx} + \frac{du}{dz}\frac{dz}{dx} + \frac{du}{dv}\frac{dv}{dx}.$$

Cor. 4.—If $u = f(y)$ and $y = \phi(x)$, to find $\frac{du}{dx}$.

Since $u = f(y)$, we have $du = \frac{du}{dy}\, dy.$

Since $y = \phi(x)$, we have $dy = \frac{dy}{dx}\, dx.$

therefore, $du = \frac{du}{dy}\frac{dy}{dx}\, dx$, and $\quad \therefore \quad \frac{du}{dx} = \frac{du}{dy}\frac{dy}{dx}.$

Sch.—The student must observe carefully the meanings of the terms in this Art. Thus, in the Proposition, u is *indirectly* a function of x through y and z. In Cor. 1, u is *directly* a function of x and *indirectly* a function of x through y. In Cor. 2, u is *directly* a function of x and *indirectly* a function of x through y and z. In Cor. 3, u is *indirectly* a function of x through y, z, and v. In Cor. 4, u is *indirectly* a function of x through y.

The equations in this Article may seem to be inconsistent with the principles of Algebra, and even absurd; but a little reflection will remove the difficulty. The du's must be carefully distinguished from each other. In Cor. 1, for example, the du in $\dfrac{du}{dx}$ is that part of the change in u which results *directly* from a change in x, while y remains constant; and the du in $\dfrac{du}{dy}$ is that part of the change in u which results *indirectly* from a change in x through y; and the du in $\left(\dfrac{du}{dx}\right)$ is the *entire* change in u which results *directly* from a change in x, and *indirectly* from a change in x through y.

EXAMPLES.

1. $u = \tan^{-1}\dfrac{x}{y}$ and $y = (r^2 - x^2)^{\frac{1}{2}}$, to find $\left(\dfrac{du}{dx}\right)$.

Here $\dfrac{du}{dx} = \dfrac{y}{r^2}$, $\dfrac{du}{dy} = -\dfrac{x}{r^2}$, and $\dfrac{dy}{dx} = -\dfrac{x}{y}$.

Substituting in $\left(\dfrac{du}{dx}\right) = \dfrac{du}{dx} + \dfrac{du}{dy}\dfrac{dy}{dx}$ (Art. 82, Cor. 1), we have

$$\left(\dfrac{du}{dx}\right) = \dfrac{y}{r^2} + \left(-\dfrac{x}{r^2}\right)\left(-\dfrac{x}{y}\right)$$

$$= \dfrac{y^2 + x^2}{r^2 y} = \dfrac{1}{\sqrt{r^2 - x^2}};$$

and this value is of course the same that we would obtain if we substituted in $u = \tan^{-1}\dfrac{x}{y}$ for y its value in terms of x, and then differentiated with respect to x.

2. $u = \tan^{-1}(xy)$ and $y = e^x$, to find $\left(\dfrac{du}{dx}\right)$.

Here $\dfrac{du}{dx} = \dfrac{y}{1 + x^2 y^2}$, $\dfrac{du}{dy} = \dfrac{x}{1 + x^2 y^2}$, $\dfrac{dy}{dx} = e^x$.

$\therefore$ (Art. 82, Cor. 1),

$$\left(\frac{du}{dx}\right) = \frac{y + e^x x}{1 + x^2 y^2} = \frac{e^x(1+x)}{1 + x^2 e^{2x}};$$

and this value is of course the same that we would obtain if we differentiated $\tan^{-1}(xe^x)$ with respect to x.

3. $u = z^2 + y^3 + zy$ and $z = \sin x$, $y = e^x$, to find $\left(\dfrac{du}{dx}\right)$.

Here $\dfrac{du}{dy} = 3y^2 + z$, $\dfrac{du}{dz} = 2z + y$,

$\dfrac{dz}{dx} = \cos x$, $\dfrac{dy}{dx} = e^x$.

$\therefore$ (Art. 82),

$$\left(\frac{du}{dx}\right) = (3y^2 + z)\,e^x + (2z + y)\,\cos x$$

$$= (3e^{2x} + \sin x)\,e^x + (2\sin x + e^x)\,\cos x$$

$$= 3e^{3x} + e^x(\sin x + \cos x) + \sin 2x.$$

(See Todhunter's Dif. Cal., p. 150.)

Let the student confirm this result by substituting in u, for y and z, their values in terms of x, thus obtaining

$$u = e^{3x} + e^x \sin x + \sin^2 x,$$

and then differentiate with respect to x.

4. $u = \sin^{-1}(y - z)$, $y = 3x$, $z = 4x^3$.

$$\frac{du}{dy} = \frac{1}{\sqrt{1 - (y - z)^2}};$$

$$\frac{du}{dz} = -\frac{1}{\sqrt{1-(y-z)^2}},$$

$$\frac{dy}{dx} = 3, \qquad \frac{dz}{dx} = 12x^2.$$

$\therefore$ (Art. 82), $\left(\dfrac{du}{dx}\right) = \dfrac{3 - 12x^2}{\sqrt{1-(y-z)^2}}$

$$= \frac{3-12x^2}{\sqrt{1-9x^2+24x^4-16x^6}} = \frac{3}{\sqrt{1-x^2}}.$$

5. $u = \dfrac{e^{ax}(y-z)}{a^2+1}$ and $y = a\sin x,\ z = \cos x.$

$$\frac{du}{dy} = \frac{e^{ax}}{a^2+1}, \qquad \frac{du}{dz} = -\frac{e^{ax}}{a^2+1},$$

$$\frac{dy}{dx} = a\cos x, \qquad \frac{dz}{dx} = -\sin x$$

$$\frac{du}{dx} = \frac{ae^{ax}(y-z)}{a^2+1}.$$

$\therefore$ (Art. 82, Cor. 2),

$$\left(\frac{du}{dx}\right) = \frac{e^{ax}}{a^2+1}(a\cos x + \sin x + a^2\sin x - a\cos x)$$

$$= e^{ax}\sin x. \quad \text{(See Courtenay's Cal., p. 73.)}$$

6. $u = yz$ and $y = e^x,\ z = x^4 - 4x^3 + 12x^2 - 24x + 24.$

$$\left(\frac{du}{dx}\right) = e^x x^4.$$

7.   If $u = f(z)$ and $z = \phi(x, y)$, show that

$$du = \frac{du}{dz}\frac{dz}{dx}dx + \frac{du}{dz}\frac{dz}{dy}dy.$$

8. $u = \dfrac{x^4 y^2}{4} - \dfrac{x^4 y}{8} + \dfrac{x^4}{32}$ and $y = \log x.$

$$\left(\frac{du}{dx}\right) = x^3(\log x)^2.$$

83. Successive Partial Differentiation of Functions of Two Independent Variables.

Let
$$u = f(x, y)$$

be a function of the independent variables x and y; then $\dfrac{du}{dx}$ and $\dfrac{du}{dy}$ are, in general, functions of x and y, and hence may be differentiated with respect to either x or y, thus obtaining a class of *second partial differentials*. Since the partial differentials of u with respect to x and y have been represented by $\dfrac{du}{dx} dx$ and $\dfrac{du}{dy} dy$ (Art. 81), we may represent the successive partial differentials as follows:

The partial differential of $\left(\dfrac{du}{dx} dx\right)$, with respect to x,

$$= \frac{d}{dx}\left(\frac{du}{dx} dx\right) dx,$$

which may be abbreviated into

$$\frac{d^2u}{dx^2} dx^2.$$

The partial differential of $\left(\dfrac{du}{dx} dx\right)$, with respect to y,

$$= \frac{d}{dy}\left(\frac{du}{dx} dx\right) dy,$$

which may be abbreviated into

$$\frac{d^2u}{dy\, dx} dy\, dx.$$

Again, both $\dfrac{d^2u}{dx^2} dx^2$ and $\dfrac{d^2u}{dy\, dx} dy\, dx$ will generally be functions of both x and y, and may be differentiated with respect to x or y, giving us *third partial differentials*, and so on. Hence we use such symbols as

$$\frac{d^2u}{dy\,dx^2}\,dy\,dx^2, \qquad \frac{d^3u}{dx\,dy\,dx}\,dx\,dy\,dx, \quad \text{and} \quad \frac{d^3u}{dy^2\,dx}\,dy^2\,dx,$$

the meaning of which is evident from the preceding remarks. For example, $\dfrac{d^3u}{dx\,dy\,dx}\,dx\,dy\,dx$ denotes that the function u is first differentiated with respect to x, supposing y constant; the resulting function is then differentiated with respect to y, supposing x constant; this last result is then differentiated with respect to x, supposing y constant; and similarly in all other cases.

When $u = f(x, y)$, the partial *derivatives* are denoted by

$$\frac{d^2u}{dx^2}, \quad \frac{d^2u}{dy^2}, \quad \frac{d^2u}{dx\,dy}, \quad \frac{d^3u}{dx^3}, \quad \frac{d^3u}{dx^2\,dy}, \quad \frac{d^3u}{dx\,dy^2}, \quad \text{etc.}$$

84. If u be a Function of x and y, to prove that

$$\frac{d^2u}{dx\,dy}\,dx\,dy = \frac{d^2u}{dy\,dx}\,dy\,dx.$$

Take
$$u = x^2y^3,$$

$$\frac{du}{dx}\,dx = 2xy^3dx,$$

$$\frac{du}{dy}\,dy = 3x^2y^2dy,$$

$$\frac{d^2u}{dy\,dx}\,dy\,dx = 6xy^2\,dy\,dx,$$

$$\frac{d^2u}{dx\,dy}\,dx\,dy = 6xy^2\,dx\,dy.$$

In this particular case,

$$\frac{d^2u}{dx\,dy}\,dx\,dy = \frac{d^2u}{dy\,dx}\,dy\,dx;$$

that is, the values of the partial differentials are independent of the order in which the variables are supposed to change.

To show this generally:

Let $$u = f(x, y);$$

then $$\frac{du}{dx}\, dx = f(x + dx, y) - f(x, y).$$

This expression being regarded as a function of y, let y become $y + dy$, x remaining constant; then

$$\frac{d}{dy}\left(\frac{du}{dx}\, dx\right) dy = f(x + dx, y + dy) - f(x, y + dy)$$
$$- [f(x + dx, y) - f(x, y)]$$
$$= f(x + dx, y + dy) - f(x, y + dy)$$
$$- f(x + dx, y) + f(x, y).$$

In like manner,

$$\frac{du}{dy}\, dy = f(x, y + dy) - f(x, y).$$

$$\frac{d}{dx}\left(\frac{du}{dy}\, dy\right) dx = f(x + dx, y + dy) - f(x + dx, y)$$
$$- [f(x, y + dy) - f(x, y)]$$
$$= f(x + dx, y + dy) - f(x + dx, y)$$
$$- f(x, y + dy) + f(x, y).$$

These two results being identical, we have

$$\frac{d}{dy}\left(\frac{du}{dx}\, dx\right) dy = \frac{d}{dx}\left(\frac{du}{dy}\, dy\right) dx;$$

that is, $$\frac{d^2u}{dy\, dx}\, dy\, dx = \frac{d^2u}{dx\, dy}\, dx\, dy.$$

Dividing by $dy\, dx$, we get

$$\frac{d^2u}{dy\, dx} = \frac{d^2u}{dx\, dy}.$$

In the same manner, it may be shown that

$$\frac{d^3u}{dx^2\, dy}\, dx^2\, dy = \frac{d^3u}{dy\, dx^2}\, dy\, dx^2,$$

or $$\frac{d^3u}{dx^2\, dy} = \frac{d^3u}{dy\, dx^2},$$

and so on to any extent.

E X A M P L E S.

1. Given $u = \sin (x + y)$, to find the successive partial derivatives with respect to x.

$$\frac{du}{dx} = \cos (x + y), \qquad \frac{d^2u}{dx^2} = - \sin (x + y),$$

$$\frac{d^3u}{dx^3} = - \cos (x + y), \qquad \frac{d^4u}{dx^4} = \sin (x + y), \text{ etc.}$$

2. $u = \log (x + y)$, to find the successive partial derivatives with respect to x, and also with respect to y in the common system.

$$\frac{du}{dx} = \frac{m}{x + y}, \qquad \frac{d^2u}{dx^2} = - \frac{m}{(x + y)^2}, \qquad \frac{d^3u}{dx^3} = \frac{2m}{(x + y)^3},$$

$$\frac{du}{dy} = \frac{m}{x + y}, \qquad \frac{d^2u}{dy^2} = - \frac{m}{(x + y)^2}, \qquad \frac{d^3u}{dy^3} = \frac{2m}{(x + y)^3}.$$

(See Art. 65, Lemma.)

3. If $u = x \log y$, verify that $\dfrac{d^2u}{dydx} = \dfrac{d^2u}{dxdy}$.

4. If $u = \tan^{-1} \left(\dfrac{x}{y} \right)$, verify that $\dfrac{d^3u}{dy^2dx} = \dfrac{d^3u}{dxdy^2}$.

5. If $u = \sin (ax^n + by^n)$,

verify that $\dfrac{d^4u}{dx^2dy^2} = \dfrac{d^4u}{dy^2dx^2}$.

85. Successive Differentials of a Function of Two Independent Variables.

Let $u = f(x, y)$.

We have already found the first differential (Art. 81),

$$du = \frac{du}{dx} dx + \frac{du}{dy} dy. \qquad (1)$$

Differentiating this equation, and observing that $\dfrac{du}{dx}, \dfrac{du}{dy},$ are, in general, functions of both x and y (Art. 83), and remembering that x and y are independent, and hence that dx and dy are constant, we have,

$$d^2u = \frac{d\left(\dfrac{du}{dx}\,dx\right)}{dx}\,dx + \frac{d\left(\dfrac{du}{dx}\,dx\right)}{dy}\,dy + \frac{d\left(\dfrac{du}{dy}\,dy\right)}{dx}\,dx$$

$$+ \frac{d\left(\dfrac{du}{dy}\,dy\right)^*}{dy}\,dy.$$

$$d^2u = \frac{d^2u}{dx^2}\,dx^2 + \frac{d^2u}{dydx}\,dydx + \frac{d^2u}{dxdy}\,dxdy + \frac{d^2u}{dy^2}\,dy^2$$

$$= \frac{d^2u}{dx^2}\,dx^2 + 2\,\frac{d^2u}{dxdy}\,dxdy + \frac{d^2u}{dy^2}\,dy^2, \qquad (2)$$

$$\left(\text{since } \frac{d^2u}{dydx}\,dydx = \frac{d^2u}{dxdy}\,dxdy, \text{ Art. 84}\right).$$

Differentiating (2), remembering that each term is a function of x and y, and hence that the total differential of each term is equal to the sum of its partial differentials, we get,

$$d^3u = \frac{d^3u}{dx^3}\,dx^3 + 3\,\frac{d^3u}{dx^2dy}\,dx^2dy + 3\,\frac{d^3u}{dxdy^2}\,dxdy^2 + \frac{d^3u}{dy^3}\,dy^3, (3)$$

and so on. It will be observed that the coefficients and exponents in the different terms of these differentials are the same as those in the corresponding powers of a binomial; and hence any required differential may be written out.

---

* The total differential of each of the terms $\left(\dfrac{du}{dx}\,dx\right)$ and $\left(\dfrac{du}{dy}\,dy\right)$ is equal to the sum of its partial differentials.

EXAMPLES.

1. $u = (x^2 + y^2)^{\frac{1}{2}}$.

$$\frac{du}{dx} = \frac{x}{(x^2 + y^2)^{\frac{1}{2}}} ; \qquad \frac{du}{dy} = \frac{y}{(x^2 + y^2)^{\frac{1}{2}}}.$$

$$\frac{d^2u}{dx^2} = \frac{y^2}{(x^2 + y^2)^{\frac{3}{2}}} ; \qquad \frac{d^2u}{dxdy} = \frac{-xy}{(x^2 + y^2)^{\frac{3}{2}}}.$$

$$\frac{d^2u}{dy^2} = \frac{x^2}{(x^2 + y^2)^{\frac{3}{2}}} ; \qquad \frac{d^3u}{dx^3} = \frac{-3xy^2}{(x^2 + y^2)^{\frac{5}{2}}}.$$

$$\frac{d^3u}{dx^2dy} = y\frac{(2x^2 - y^2)}{(x^2 + y^2)^{\frac{5}{2}}} ; \qquad \frac{d^3u}{dxdy^2} = \frac{x(2y^2 - x^2)}{(x^2 + y^2)^{\frac{5}{2}}}$$

$$\frac{d^3u}{dy^3} = \frac{-3yx^2}{(x^2 + y^2)^{\frac{5}{2}}}.$$

$$\therefore \quad d^3u = [-3xy^2dx^3 + 3y(2x^2 - y^2)\,dx^2dy$$
$$+ 3x(2y^2 - x^2)\,dxdy^2 - 3yx^2dy^3]\,\frac{1}{(x^2 + y^2)^{\frac{5}{2}}}.$$

2. $u = e^{(ax+by)}$.

$$d^2u = [a^2dx^2 + 2abdxdy + b^2dy^2]\,e^{ax+by}$$
$$= [adx + bdy]^2\,e^{ax+by}.$$

86. Implicit Functions (see Art. 6).—Thus far in this Chapter, the methods which we have given, although often convenient, are not absolutely *necessary*, as in every case by making the proper substitutions we may obtain an explicit function of x, and differentiate it by the rules in Chapter II. But the case of *implicit* functions which we are now to consider is one in which a new method is often *indispensable*.

Let $f(x, y) = 0$ be an implicit function of two variables, in which it is required to find $\dfrac{dy}{dx}$. If this equation can be solved with respect to y, giving for example $y = \phi(x)$, then the derivative of y with respect to x can be found by previous rules. But as it is often difficult and sometimes impossible to solve the given equation, it is necessary to investigate a rule for finding $\dfrac{dy}{dx}$ without solving the equation.

87. Differentiation of an Implicit Function.

Let
$$f(x, y) = 0,$$
in which y is an implicit function of x, to find $\dfrac{dy}{dx}$.

Let
$$f(x, y) = u.$$

Then
$$u = f(x, y) = 0.$$

Hence by (Art. 82, Cor. 1), we have,

$$\left(\frac{du}{dx}\right) = \frac{du}{dx} + \frac{du}{dy}\frac{dy}{dx}.$$

But u is always $= 0$, and therefore its total differential $= 0$; hence $\left(\dfrac{du}{dx}\right) = 0$, and therefore,

$$\frac{du}{dx} + \frac{du}{dy}\frac{dy}{dx} = 0,$$

from which we get,

$$\frac{dy}{dx} = -\frac{\dfrac{du}{dx}}{\dfrac{du}{dy}}.$$

SCH.—It will be observed that while $\left(\dfrac{du}{dx}\right) = 0$, neither $\dfrac{du}{dx}$ nor $\dfrac{du}{dy}$ is in general $= 0$. For example,

$$x^2 + y^2 - r^2 = 0$$

is of the form $f(x, y) = 0$. We see that if x changes while y remains constant, the *function* changes, and hence is no longer $= 0$. Also, if y changes while x remains constant, the function does not remain $= 0$. But if when x changes y takes a *corresponding* change by virtue of its dependence on x, the function remains $= 0$.

EXAMPLES.

1. $y^2 - 2xy + a^2 = 0$, to find $\dfrac{dy}{dx}$.

$$\frac{du}{dx} = -2y; \qquad\qquad \frac{du}{dy} = 2y - 2x.$$

therefore, $\dfrac{dy}{dx} = -\dfrac{\dfrac{du}{dx}}{\dfrac{du}{dy}} = -\dfrac{-2y}{2y - 2x} = \dfrac{y}{y - x}.$

2. $a^2y^2 + b^2x^2 - a^2b^2 = 0$, to find $\dfrac{dy}{dx}$.

$$\frac{du}{dx} = 2b^2x; \qquad\qquad \frac{du}{dy} = 2a^2y;$$

therefore, $\dfrac{dy}{dx} = -\dfrac{\dfrac{du}{dx}}{\dfrac{du}{dy}} = -\dfrac{2b^2x}{2a^2y} = -\dfrac{b^2x}{a^2y}.$ \hfill (1)

Since $y = \dfrac{b}{a}\sqrt{a^2 - x^2}$, from the given equation, we may solve this example directly by previous methods, and obtain

$$\frac{dy}{dx} = -\frac{bx}{a\sqrt{a^2 - x^2}}, \tag{2}$$

which agrees with (1) by substituting in it the value of y in terms of x.

In this example we can verify our new rule by comparing the result with that obtained by previous rules. In more complex examples, such as the following one, we can find $\dfrac{dy}{dx}$ only by the new method.

3. $x^5 - ax^3y + bx^2y^2 - y^5 = 0$, to find $\dfrac{dy}{dx}$.

$$\frac{du}{dx} = 5x^4 - 3ax^2y + 2bxy^2;$$

$$\frac{du}{dy} = -ax^3 + 2bx^2y - 5y^4;$$

therefore, $\dfrac{dy}{dx} = \dfrac{5x^4 - 3ax^2y + 2bxy^2}{5y^4 - 2bx^2y + ax^3}.$

4. $ax^3 + x^3y - ay^3 = 0.$ $\dfrac{dy}{dx} = -\dfrac{3ax^2 + 3x^2y}{x^3 - 3ay^2}.$

5. $y^2 - 2axy + x^2 - b^2 = 0.$ $\dfrac{dy}{dx} = \dfrac{ay - x}{y - ax}.$

6. $y^3 - 3y + x = 0.$ $\dfrac{dy}{dx} = \dfrac{1}{3(1 - y^2)}.$

7. $x^3 + 3axy + y^3 = 0.$ $\dfrac{dy}{dx} = -\dfrac{x^2 + ay}{y^2 + ax}.$

88. To Find the Second Derivative of an Im-plicit Function.

Let $u = f(x, y) = 0.$

We have $\dfrac{dy}{dx} = -\dfrac{\dfrac{du}{dx}}{\dfrac{du}{dy}}$ (Art. 87),   (1)

or $\dfrac{du}{dx} + \dfrac{du}{dy}\dfrac{dy}{dx} = 0;$ (2)

it is required to find $\dfrac{d^2y}{dx^2}.$

Differentiating (2), remembering that $\dfrac{du}{dx}, \dfrac{du}{dy},$ are func-tions of x and y, we get

$$\frac{d^2u}{dx^2} + \frac{d^2u}{dy\,dx}\frac{dy}{dx} + \left(\frac{d^2u}{dy^2}\frac{dy}{dx} + \frac{d^2u}{dx\,dy}\right)\frac{dy}{dx} + \frac{du}{dy}\frac{d^2y}{dx^2} = 0.$$

or
$$\frac{d^2u}{dx^2} + 2\frac{d^2u}{dx\,dy}\frac{dy}{dx} + \frac{d^2u}{dy^2}\frac{dy^2}{dx^2} + \frac{du}{dy}\frac{d^2y}{dx^2} = 0. \quad (3)$$

Substituting the value of $\frac{dy}{dx}$ from (1), and clearing of fractions, we get

$$\frac{d^2u}{dx^2}\frac{du^2}{dy^2} - 2\frac{d^2u}{dx\,dy}\frac{du}{dx}\frac{du}{dy} + \frac{d^2u}{dy^2}\frac{du^2}{dx^2} + \frac{du^3}{dy^3}\frac{d^2y}{dx^2} = 0.$$

Solving for $\frac{d^2y}{dx^2}$, we get

$$\frac{d^2y}{dx^2} = -\frac{\frac{d^2u}{dx^2}\left(\frac{du}{dy}\right)^2 - 2\frac{d^2u}{dx\,dy}\frac{du}{dx}\frac{du}{dy} + \frac{d^2u}{dy^2}\left(\frac{du}{dx}\right)^2}{\left(\frac{du}{dy}\right)^3}. \quad (4)$$

Sch.—This equation is so complicated that in practice it is generally more convenient to differentiate the value of the first derivative *immediately* than to substitute in (4). The third and higher derivatives may be obtained in a similar manner, but their forms are very complicated.

Equation (2) is frequently called *the first derived equation* or *the differential equation of the first order ;* and equation (3) is called *the second derived equation* or *the differential equation of the second order.*

EXAMPLES.

1. $y^2 - 2xy + a^2 = 0$, to find $\frac{dy}{dx}$ and $\frac{d^2y}{dx^2}$.

$$\frac{du}{dx} = -2y; \qquad \frac{du}{dy} = 2y - 2x; \qquad \frac{d^2u}{dx^2} = 0;$$

$$\frac{d^2u}{dx\,dy} = -2; \qquad \frac{d^2u}{dy^2} = 2.$$

Therefore, by (1), $\quad \frac{dy}{dx} = \frac{y}{y-x};$

and by (4), $\frac{d^2y}{dx^2} = -\frac{-16y(y-x) + 8y^2}{(2y-2x)^3} = \frac{y(y-2x)}{(y-x)^3}.$

2. $y^2 - 2axy + x^2 - b^2 = 0$, to find $\dfrac{dy}{dx}$, $\dfrac{d^2y}{dx^2}$. (See Sch.)

$$\frac{dy}{dx} = \frac{ay - x}{y - ax}.$$

$$\frac{d^2y}{dx^2} = \frac{(y - ax)\left(a\dfrac{dy}{dx} - 1\right) - (ay - x)\left(\dfrac{dy}{dx} - a\right)}{(y - ax)^2}$$

$$= \frac{(y - ax)(a^2y - y) - (ay - x)(a^2x - x)}{(y - ax)^3}$$

$$\left(\text{by substituting the value of } \frac{dy}{dx}\right)$$

$$= \frac{(a^2 - 1)(y^2 - 2axy + x^2)}{(y - ax)^3} = \frac{b^2(a^2 - 1)}{(y - ax)^3}.$$

3. $x^3 + 3axy + y^3 = 0$, to find $\dfrac{dy}{dx}$ and $\dfrac{d^2y}{dx^2}$.

Differentiating, we have

$$(x^2 + ay)\,dx + (y^2 + ax)\,dy = 0\,;$$

$$\therefore \quad \frac{dy}{dx} = -\frac{x^2 + ay}{y^2 + ax}.$$

$$\therefore \quad \frac{d^2y}{dx^2} = -\frac{\left(2x + a\dfrac{dy}{dx}\right)(y^2 + ax) - \left(2y\dfrac{dy}{dx} + a\right)(x^2 + ay)}{(y^2 + ax)^2}$$

$$= \frac{2a^3xy}{(y^2 + ax)^3}. \quad \text{(See Price's Calculus, Vol. I, p. 142.)}$$

4. $x^2 + y^2 - a^2 = 0.$ $\dfrac{dy}{dx} = -\dfrac{x}{y}$; $\dfrac{d^2y}{dx^2} = -\dfrac{a^2}{y^3}.$

89. Change of the Independent Variable.—Thus far we have employed the derivatives $\dfrac{dy}{dx}$, $\dfrac{d^2y}{dx^2}$, etc., upon the hypothesis that x was the independent variable and y the function.   But in the discussion of expressions containing the successive differentials and derivatives of a function with respect to x, it is frequently desirable to change the expression into its equivalent when y is made the independent variable and x the function ; or to introduce some other variable of which both y and x are functions, and make *it* the independent variable.

90. To Find the Values of $\dfrac{dy}{dx}$, $\dfrac{d^2y}{dx^2}$, $\dfrac{d^3y}{dx^3}$, etc., when neither x nor y is Equicrescent. (Art. 55.)

The value of the first derivative, $\dfrac{dy}{dx}$, will be the same whether x or y, or neither, is considered equicrescent.

The value of the second derivative, $\dfrac{d^2y}{dx^2}$, was obtained in Art. 56 by differentiating $\dfrac{dy}{dx}$ as a fraction with a constant denominator and dividing by dx.

If we now consider that neither x nor y is equicrescent, and hence that both dx and dy are variables, and differentiate $\dfrac{dy}{dx}$, we have

$$\frac{d^2y}{dx^2} = \frac{d}{dx}\left(\frac{dy}{dx}\right) = \frac{d^2y\,dx - d^2x\,dy}{dx^3}, \qquad (1)$$

which is therefore the value of the second derivative when neither variable is equicrescent.

Similarly,

$$\frac{d^3y}{dx^3} = \frac{d}{dx}\left(\frac{d^2y}{dx^2}\right)$$

$$= \frac{(d^3y\,dx - d^3x\,dy)\,dx - 3\,(d^2y\,dx - d^2x\,dy)\,d^2x}{dx^5}. \qquad (2)$$

which is the value of the third derivative when neither variable is equicrescent, and so on for any other derivative.

Cor.—If x is equicrescent, these equations are identical.

If y is equicrescent, $d^2y = d^3y = 0$, and (1) becomes

$$\frac{d^2y}{dx^2} = -\frac{d^2x\,dy}{dx^3}, \qquad (3)$$

and (2) becomes

$$\frac{d^3y}{dx^3} = \frac{3\,(d^2x)^2\,dy - d^3x\,dy\,dx}{dx^5}, \qquad (4)$$

which are the values of the second and third derivatives when y is equicrescent.

Sch. 1.—Hence, if we wish to change an expression when x is equicrescent into its equivalent where neither x nor y is equicrescent, we must replace $\dfrac{d^2y}{dx^2}$, $\dfrac{d^3y}{dx^3}$, etc., by their complete values in (1), (2), etc.; but if we want an equivalent expression in which y is equicrescent, we must replace $\dfrac{d^2y}{dx^2}$, $\dfrac{d^3y}{dx^3}$, etc., by their values in (3), (4), etc.

Sch. 2.—If we wish to change an expression in which x is equicrescent into its equivalent, and have the result in terms of a new independent variable t, of which x is a function, we must replace $\dfrac{d^2y}{dx^2}$, $\dfrac{d^3y}{dx^3}$, etc., by their complete values in (1), (2), etc., and then substitute in the resulting expression, in which neither x nor y is equicrescent, the values of x, dx, d^2x, etc., in terms of the new equicrescent variable.

EXAMPLES.

1. Transform $x\dfrac{d^2y}{dx^2} + \left(\dfrac{dy}{dx}\right)^3 - \dfrac{dy}{dx} = 0$, in which x is equicrescent, into its equivalents, (*1*) when neither x nor y is equicrescent, (*2*) when y is equicrescent.

(*1.*) Replace $\frac{d^2y}{dx^2}$ by its value in (1), and multiply by dx^3, and we have

$$x(d^2y\,dx - d^2x\,dy) + dy^3 - dy\,dx^2 = 0.$$

(*2.*) Put $d^2y = 0$, divide by dy^3 to have the differential of the independent variable in its proper position, the denominator, and change signs, and we have

$$x\frac{d^2x}{dy^2} + \left(\frac{dx}{dy}\right)^2 - 1 = 0.$$

2. Transform $\frac{d^2y}{dx^2} - \frac{x}{1-x^2}\frac{dy}{dx} + \frac{y}{1-x^2} = 0$, in which x is equicrescent, into its equivalent when θ is equicrescent, having given $x = \cos\theta$.

Replacing $\frac{d^2y}{dx^2}$ by its complete value in (1), the given equation becomes

$$\frac{d^2y\,dx - d^2x\,dy}{dx^3} - \frac{x}{1-x^2}\frac{dy}{dx} + \frac{y}{1-x^2} = 0.$$

$$x = \cos\theta;$$

$$\therefore \quad dx = -\sin\theta\,d\theta \quad \text{and} \quad d^2x = -\cos\theta\,d\theta^2.$$

$$1 - x^2 = \sin^2\theta.$$

Substituting, we have

$$\frac{-d^2y\sin\theta\,d\theta + \cos\theta\,d\theta^2\,dy}{-\sin^3\theta\,d\theta^3} + \frac{\cos\theta}{\sin^2\theta}\frac{dy}{\sin\theta\,d\theta} + \frac{y}{\sin^2\theta} = 0.$$

$$\therefore \quad \frac{d^2y}{d\theta^2} + y = 0. \quad \text{(See Price's Calculus, Vol. I, p. 126.)}$$

3. Transform $\frac{d^2y}{dx^2} + \frac{1}{x}\frac{dy}{dx} + y = 0$, in which x is equicrescent, into its equivalent, (*1*) when neither y nor θ is equicrescent; (*2*) when θ is equicrescent; (*3*) when y is equicrescent, having given $x^2 = 4\theta$.

Replacing in this equation the complete value of $\dfrac{d^2y}{dx^2}$, it becomes

$$\frac{d^2y\,dx - d^2x\,dy}{dx^3} + \frac{1}{x}\frac{dy}{dx} + y = 0.$$

$$x = 2\,(\theta)^{\frac{1}{2}}; \qquad \therefore \quad dx = \theta^{-\frac{1}{2}}\,d\theta.$$

$$d^2x = -\frac{d\theta^2}{2\theta^{\frac{3}{2}}} + \frac{d^2\theta}{\theta^{\frac{1}{2}}}.$$

Substituting, we have

(1.) $y\,d\theta^3 + dy\,d\theta^2 + \theta\,d^2y\,d\theta - \theta\,dy\,d^2\theta = 0.$

(2.) When θ is equicrescent, $d^2\theta = 0$; therefore (1) becomes

$$y\,d\theta^3 + dy\,d\theta^2 + 0\,d^2y\,d\theta = 0,$$

or $\theta\,\dfrac{d^2y}{d\theta^2} + \dfrac{dy}{d\theta} + y = 0.$

(3.) When y is equicrescent, $d^2y = 0$; therefore (1) becomes

$$\theta\,\frac{d^2\theta}{dy^2} - \left(\frac{d\theta}{dy}\right)^2 - y\left(\frac{d\theta}{dy}\right)^3 = 0.$$

4. Transform $R = \dfrac{\left[1 + \dfrac{dy^2}{dx^2}\right]^{\frac{3}{2}}}{\dfrac{d^2y}{dx^2}}$ into its equivalent, (1) in

the most general form; (2) when θ is equicrescent; (3) when r is equicrescent, having given $x = r\cos\theta$, and $y = r\sin\theta$.

The complete value of R is

$$R = -\frac{(dx^2 + dy^2)^{\frac{3}{2}}}{d^2x\,dy - d^2y\,dx}.$$

$$dx = dr\cos\theta - r\sin\theta\,d\theta,$$

$$dy = dr\sin\theta + r\cos\theta\,d\theta.$$

$$\therefore \quad d^2x = \cos\theta \, d^2r - 2\sin\theta \, drd\theta - r\cos\theta \, d\theta^2$$
$$- r\sin\theta \, d^2\theta,$$

$$d^2y = \sin\theta \, d^2r + 2\cos\theta \, drd\theta - r\sin\theta \, d\theta^2$$
$$+ r\cos\theta \, d^2\theta \, ;$$

$$\therefore \quad (dx)^2 + (dy)^2 = dr^2 + r^2 d\theta^2,$$

$$d^2x dy - d^2y dx = rd^2rd\theta - 2dr^2d\theta - r^2d\theta^3 - rdrd^2\theta.$$

(1.) $\therefore \quad R = \dfrac{[dr^2 + r^2d\theta^2]^{\frac{3}{2}}}{- rd^2rd\theta + 2dr^2d\theta + r^2d\theta^3 + rdrd^2\theta},$

(2.) $R = \dfrac{\left[\left(\dfrac{dr}{d\theta}\right)^2 + r^2\right]^{\frac{3}{2}}}{r^2 + 2\left(\dfrac{dr}{d\theta}\right)^2 - r\dfrac{d^2r}{d\theta^2}},$

(3.) $R = \dfrac{\left[1 + r^2\left(\dfrac{d\theta}{dr}\right)^2\right]^{\frac{3}{2}}}{r\dfrac{d^2\theta}{dr^2} + r^2\left(\dfrac{d\theta}{dr}\right)^3 + 2\dfrac{d\theta}{dr}}.$

(See Serret's Calcul Différential et Intégral, p. 94.)

5. Transform
$$(dy^2 + dx^2)^{\frac{3}{2}} + adxd^2y = 0,$$

in which x is equicrescent, into its equivalents, (1) when neither x nor y is equicrescent, (2) when y is equicrescent.

(1.) $\quad (dy^2 + dx^2)^{\frac{3}{2}} + a(d^2ydx - d^2xdy) = 0 \, ;$

(2.) $\quad \left(1 + \dfrac{dx^2}{dy^2}\right)^{\frac{3}{2}} - a\dfrac{d^2x}{dy^2} = 0.$

91. General Case of Transformation for Two Inde pendent Variables.—Let u be a function of the independent variables, say $u = f(x, y)$; and suppose x and y functions of two new independent variables r and θ, so that,

7

$$x = \phi\,(r,\,\theta) \quad \text{and} \quad y = \psi\,(r,\,\theta)\,; \tag{1}$$

then u may be regarded as a function of r and θ, through x and y. It is required to find the values of $\dfrac{du}{dx}$ and $\dfrac{du}{dy}$ in terms of derivatives of u, taken with respect to the new variables r and θ.

Since u is a function of r through x and y, we have (Art. 82),

$$\frac{du}{dr} = \frac{du}{dx}\frac{dx}{dr} + \frac{du}{dy}\frac{dy}{dr}. \tag{2}$$

And similarly

$$\frac{du}{d\theta} = \frac{du}{dx}\frac{dx}{d\theta} + \frac{du}{dy}\frac{dy}{d\theta}\,; \tag{3}$$

where the values of $\dfrac{dx}{dr},\,\dfrac{dy}{dr},\,\dfrac{dx}{d\theta},\,\dfrac{dy}{d\theta}$, can be found from (1).

Whenever equations (1) can be solved for r and θ separately, we can find by direct differentiation the values of $\dfrac{dr}{dx},\,\dfrac{dr}{dy},\,\dfrac{d\theta}{dx},\,\dfrac{d\theta}{dy}$, and hence by substituting in

$$\frac{du}{dx} = \frac{du}{dr}\frac{dr}{dx} + \frac{du}{d\theta}\frac{d\theta}{dx}\,,$$

and

$$\frac{du}{dy} = \frac{du}{dr}\frac{dr}{dy} + \frac{du}{d\theta}\frac{d\theta}{dy} \quad \text{(Art. 82)},$$

we can obtain the values of $\dfrac{du}{dx}$ and $\dfrac{du}{dy}$.

When this process is not practicable, we can obtain their values by solving (2) and (3) directly, as follows:

Multiply (2) by $\dfrac{dx}{d\theta}$ and (3) by $\dfrac{dx}{dr}$ and subtract; then multiply (2) by $\dfrac{dy}{d\theta}$ and (3) by $\dfrac{dy}{dr}$ and subtract. We shall then have two equations, from which we obtain,

$$\frac{du}{dx} = \frac{\dfrac{du}{dr}\dfrac{dy}{d\theta} - \dfrac{du}{d\theta}\dfrac{dy}{dr}}{\dfrac{dx}{dr}\dfrac{dy}{d\theta} - \dfrac{dy}{dr}\dfrac{dx}{d\theta}}, \tag{4}$$

and

$$\frac{du}{dy} = \frac{\dfrac{du}{d\theta}\dfrac{dx}{dr} - \dfrac{du}{dr}\dfrac{dx}{d\theta}}{\dfrac{dx}{dr}\dfrac{dy}{d\theta} - \dfrac{dy}{dr}\dfrac{dx}{d\theta}}. \tag{5}$$

The values of $\dfrac{d^2u}{dx^2}$, $\dfrac{d^2u}{dy^2}$, etc., can be obtained from these, but the general formulæ are too complicated to be of much practical use. (See Gregory's Examples, p. 35.)

Cor.—If $x = r \cos \theta$ and $y = r \sin \theta$, (4) and (5) become

$$\frac{du}{dx} = \cos \theta \frac{du}{dr} - \frac{\sin \theta}{r} \frac{du}{d\theta}; \quad \frac{du}{dy} = \frac{\cos \theta}{r} \frac{du}{d\theta} + \sin \theta \frac{du}{dr}.$$

EXAMPLES.

1. $u = \dfrac{x + y}{x - y}$, to find du (Art 81).

$$du = \frac{2\,(x dy - y dx)}{(x - y)^2}.$$

2. $u = \sin ax + \sin by + \tan^{-1}\dfrac{y}{z}$.

$$du = a \cos ax\, dx + b \cos by\, dy + \frac{z dy - y dz}{y^2 + z^4}.$$

3. $u = \sin^{-1}\dfrac{x}{y}$.

$$du = \frac{y dx - x dy}{y \sqrt{y^2 - x^2}}.$$

4. $u = \sin (x + y)$.

$$du = \cos (x + y)\,(dx + dy).$$

5. $u = \dfrac{x^2 y}{z^2 - a^2}.$

$$du = \frac{x(z^2 - a^2)(2y\,dx + x\,dy) - 2x^2 yz\,dz}{(z^2 - a^2)^2}.$$

6. $u = \log \left(\dfrac{x + \sqrt{x^2 - y^2}}{x - \sqrt{x^2 - y^2}} \right).$

$$du = 2\,\frac{y\,dx - x\,dy}{y\sqrt{x^2 - y^2}}.$$

7. $u = \cot x^y$ to find $\left(\dfrac{du}{dx}\right)$ (Art. 82, Cor. 1).

$$\left(\frac{du}{dx}\right) = -x^y \operatorname{cosec}^2 x^y \left(\frac{y}{x} + \log x\,\frac{dy}{dx}\right).$$

8. $u = \sin(y^2 - z)$, and $y = \log x$, $z = x^2$, to find $\left(\dfrac{du}{dx}\right)$
(Art. 82).

$$\left(\frac{du}{dx}\right) = \frac{2(y - x^2)\cos(y^2 - z)}{x}.$$

9. $u = \log \tan \dfrac{x}{y}.$

$$\left(\frac{du}{dx}\right) = \frac{y - x\dfrac{dy}{dx}}{y^2 \sin \dfrac{x}{y} \cos \dfrac{x}{y}}.$$

10. $u = \log(x - a + \sqrt{x^2 - 2ax}).$

$$\left(\frac{du}{dx}\right) = \frac{1}{\sqrt{x^2 - 2ax}}.$$

11.   If $u = x^3 z^4 + e^x y^2 z^3 + x^2 y^2 z^2$, show that

$$\frac{d^4 u}{dx^2\,dy\,dz} = 6e^x yz^2 + 8yz. \quad \text{(Art. 83.)}$$

12.   If $u = \tan^{-1} \dfrac{xy}{\sqrt{1 + x^2 + y^2}}$, show that

$$\frac{d^2 u}{dx\,dy} = \frac{1}{(1 + x^2 + y^2)^{\frac{3}{2}}}, \quad \frac{d^4 u}{dx^2\,dy^2} = \frac{15xy}{(1 + x^2 + y^2)^{\frac{7}{2}}}.$$

13. $u = x^3y^2 + y^3x^2$ to find d^2u (Art. 85).

$$d^2u = 6xy^2dx^2 + 6x^2ydxdy + 2y^3dx^2 + 6xy^2dxdy$$
$$+ 6x^2ydxdy + 2x^3dy^2 + 6xy^2dxdy + 6x^2ydy^2$$
$$= (6xy^2+2y^3)dx^2+12\,(x^2y+y^2x)\,dxdy+(6x^2y+2x^3)dy^2.$$

14. $x^y + y^z - a = 0$, to find $\dfrac{dy}{dx}$. (Art. 87.)

$$\frac{dy}{dx} = -\frac{y.x^{y-1} + y^z \log y}{xy^{z-1} + x^y \log x}.$$

15. $x^y - y^z = 0.$ $\qquad \dfrac{dy}{dx} = \dfrac{y^2 - xy \log y}{x^2 - xy \log x}.$

16. $y^2\,(2a - x) - x^3 = 0.$ $\dfrac{dy}{dx} = \dfrac{3x^2 + y^2}{2y\,(2a - x)}.$

17. $x^3 - 3axy + y^3 = 0.$ $\dfrac{dy}{dx} = \dfrac{ay - x^2}{y^2 - ax}.$

18. $ye^{ny} - ax^m = 0.$ $\qquad \dfrac{dy}{dx} = \dfrac{my}{x\,(1 + ny)}.$

19. $a^{zy} + \sqrt{\sec\,(xy)} = 0.$

$$\frac{dy}{dx} = -\frac{y\,\sqrt{\sec\,(xy)}\,\tan\,(xy) + 2a^{zy}yx^{y-1}\log a}{x\,\sqrt{\sec\,(xy)}\,\tan\,(xy) + 2a^{zy}x^y\,\log a\,\log x}.$$

20. $x^4 + 2ax^2y - ay^3 = 0$, to find $\dfrac{dy}{dx}$ and $\dfrac{d^2y}{dx^2}$.

(Art. 88.) $\qquad \dfrac{dy}{dx} = -\dfrac{4x^3 + 4axy}{2ax^2 - 3ay^2}.$

$$\frac{d^2y}{dx^2} = -\,[(12x^2 + 4ay)\,(2ax^2 - 3ay^2)^2$$
$$-\,8ax\,(4x^3 + 4axy)\,(2ax^2-3ay^2) - (4x^3+4axy)^2\,6ay]$$
$$\div\,(2ax^2 - 3ay^2)^3 = \text{what ?}$$

Show that $\dfrac{dy}{dx} = 0$ or $\pm\sqrt{2}$, when $x = 0$ and $y = 0.$

21. Change the independent variable from x to t in

$$(1 - x^2)\frac{d^2y}{dx^2} - x\frac{dy}{dx} = 0, \text{ when } x = \cos t.$$

$$Ans. \ \frac{d^2y}{dt^2} = 0.$$

22. Change the independent variable from x to θ in

$$\frac{d^2y}{dx^2} + \frac{2x}{1 + x^2}\frac{dy}{dx} + \frac{y}{(1 + x^2)^2} = 0, \text{ when } x = \tan\theta.$$

$$Ans. \ \frac{d^2y}{d\theta^2} + y = 0.$$

23. Change the independent variable from x to r, and eliminate x, y, dx and dy, between

$$t = \frac{xdy - ydx}{ydy + xdx}, \quad x = r\cos\theta, \text{ and } y = r\sin\theta.$$

$$Ans. \ t = \frac{rd\theta}{dr}.$$

24. Change the independent variable from x to z in

$$P = x^2\frac{d^2y}{dx^2} - 2x\frac{dy}{dx} + y\frac{dy}{dx}, \text{ when } x = e^z \text{ and } y = e^z(v + 2)$$

$$Ans. \ P = e^z\left[\frac{d^2v}{dz^2} + (v + 1)\frac{dv}{dz} + v(v + 2)\right].$$

CHAPTER VIII.

92. Definition of a Maximum and a Minimum.—If, while the independent variable increases continuously, a function dependent on it *increases up to a certain value, and then decreases,* the value at the end of the increase is called a *maximum* value of the function.

If while the independent variable increases, the function *decreases to a certain value and then increases,* the value at the end of the decrease is called a *minimum* value of the function. Hence, *a maximum value of a function of a single variable is a value which is greater than the immediately preceding and succeeding values, and a minimum value is less than the immediately preceding and succeeding values.*

For example, sin ϕ *increases* as ϕ increases till the latter reaches 90°, after which sin ϕ *decreases* as ϕ increases; that is, sin ϕ is a maximum when ϕ is 90°, since it is greater than the immediately preceding and succeeding values. Also, cosec ϕ *decreases* as ϕ *increases* till the latter reaches 90°, after which cosec ϕ *increases* as ϕ *increases ;* that is, cosec ϕ is a minimum when ϕ is 90°, since it is less than the immediately preceding and succeeding values.

93. Condition for a Maximum or Minimum.—If y be any function of x, and y be *increasing* as x *increases,* the differential of the function is positive (Art. 12), and hence the first derivative $\dfrac{dy}{dx}$ will be *positive.* If the function be *decreasing* as x increases, the differential of the function is negative, and hence the first derivative $\dfrac{dy}{dx}$ will be *negative*.

Therefore, since at a maximum value the function changes from increasing to decreasing, *the first derivative must change its sign from plus to minus; as the variable increases.* And since, at a minimum value, the function changes from decreasing to increasing, *the first derivative must change its sign from minus to plus.* But as a function which is continuous* can change its sign only by passing through 0 or ∞, it follows that *the only values of the variable corresponding to a maximum or a minimum value of the function, are those which make the first derivative* **0** *or* **∞**.

94. Geometric Illustration. — This result is also evident from geometric considerations ; for, let $y = f(x)$ be the equation of the curve AB. At the points P, P', P'', P^{iv}, the tangents to the curve are parallel to the axis of x,

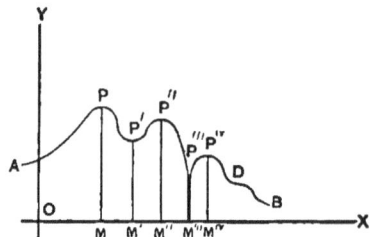

Fig. 9.

and therefore at each of these points the first derivative $f'(x) = 0$, by Art. 56^a.

We see that as x is increasing and y approaching a maximum value, as PM, the tangent to the curve makes an acute angle with the axis of x ; hence, approaching P from the left $\dfrac{dy}{dx}$ is $+$. At P the tangent becomes parallel to the axis of x ; hence, $\dfrac{dy}{dx} = 0$. Immediately after passing P the tangent makes an obtuse angle with the axis of x ; hence, $\dfrac{dy}{dx}$ is $-$.

---

* In this discussion the function is to be regarded as *continuous.*

Also in approaching a minimum value, as P'M', from the left, we see that the tangent makes an obtuse angle with the axis of x, and hence $\frac{dy}{dx}$ is $-$. At the point P', $\frac{dy}{dx} = 0$. After passing P', the angle is acute and $\frac{dy}{dx}$ is $+$.

In passing P''', $\frac{dy}{dx}$ changes sign by passing through ∞, P'''M''' is a minimum ordinate. In approaching it from the left the tangent makes an obtuse angle with the axis of x, and hence $\frac{dy}{dx}$ is $-$. At P''' the tangent is perpendicular to the axis of x, and $\frac{dy}{dx} = \infty$. After passing P'''M''', the angle is acute and $\frac{dy}{dx}$ is $+$.

While the first derivative can change its sign from $+$ to $-$ or from $-$ to $+$ *only* by passing through 0 or ∞, it does not follow that *because* it is 0 or ∞, it therefore *necessarily* changes its sign. The first derivative as the variable increases may be $+$, then 0, and then $+$, or it may be $-$, then 0, and then $-$. This is evident from Fig. 9, where, at the point D, the tangent is parallel to the axis of x, and $\frac{dy}{dx}$ is 0, although just before and just after it is $-$. Hence the values of the variable which make $\frac{dy}{dx} = 0$ or ∞, are simply *critical** values, *i. e.*, values to be examined.

As a maximum value is merely a value greater than that which *immediately* precedes and follows it, *a function may have several maximum values*, and for a like reason *it may have several minimum values*. Also, a maximum value may be *equal to or even less* than a *minimum value* of the same function. For example, in Fig. 9, the minimum P'M' is greater than the maximum P^{iv}M^{iv}.

---

95. Method of Discriminating between Maxima and Minima.—Since the first derivative at a *maximum* state is 0, and at the immediately succeeding state is —, it follows that the *second derivative*, which is the difference between two consecutive first derivatives,* is — at a maximum. Also, since the first derivative at a *minimum* state is 0, and at the immediately succeeding state is +, it follows that the second derivative is + at a minimum. Therefore, for critical values of the variable, *a function is at a maximum or a minimum state according as its second derivative at that state is* — *or* +.

96. Condition for a Maximum or Minimum given by Taylor's Theorem.—Let $u = f(x)$ be any continuous function of one variable; and let a be a value of x corresponding to a maximum or a minimum value of $f(x)$. Then if a takes a small increment and a small decrement each equal to h, in the case of a maximum we must have, for small values of h,

$$f(a) > f(a + h) \quad \text{and} \quad f(a) > f(a - h);$$

and for a minimum,

$$f(a) < f(a + h) \quad \text{and} \quad f(a) < f(a - h).$$

Therefore, in either case,

$$f(a + h) - f(a) \quad \text{and} \quad f(a - h) - f(a)$$

have both the same sign.

By Taylor's Theorem, Art. 66, Eq. 7, and transposing, we have

$$f(a + h) - f(a) = \quad f'(a) h + f''(a) \frac{h^2}{2} + \text{etc.}; \quad (1)$$

$$f(a - h) - f(a) = -f'(a) h + f''(a) \frac{h^2}{2} - \text{etc.} \quad (2)$$

---

* Remembering that the first value is always to be subtracted from the second.

Now if h be taken infinitely small, the first term in the second member of each of the equations (1) and (2) will be greater than the sum of all the rest, and the sign of the second member of each will be the same as that of its first term, and hence $f(a + h) - f(a)$ and $f(a - h) - f(a)$ cannot have the same sign unless the first term of (1) and (2) disappears, which, since h is not 0, requires that $f'(a) = 0$.

Hence, *the values of x which make $f(x)$ a maximum or a minimum are in general roots of the equation, $f'(x) = 0$.*

Also, when $f'(a) = 0$, the second members of (1) and (2), for small values of h, have the same sign as $f''(a)$; that is, the first members of (1) and (2) are both positive when $f''(a)$ is positive, and negative when $f''(a)$ is negative. Therefore, *$f(a)$ is a maximum or a minimum according as $f''(a)$ is negative or positive.*

If, however, $f''(a)$ vanish along with $f'(a)$, the signs of the second members of (1) and (2) will be the same as $f'''(a)$, and since $f'''(a)$ has opposite signs, it follows that in this case *$f(a)$ is neither a maximum nor a minimum unless $f'''(a)$ also vanish.* But if $f'''(a) = 0$, then $f(a)$ is a maximum when $f^{iv}(a)$ is negative, and a minimum when $f^{iv}(a)$ is positive, and so on. If the first derivative which does not vanish is of an odd order, $f(a)$ is neither a maximum nor a minimum; if of an even order, $f(a)$ is a maximum or a minimum, according as the sign of the derivative which does not vanish is negative or positive.

97. Method of Finding Maxima and Minima Values.—Hence, as the result of the preceding investigation we have the following rule for finding the maximum or minimum values of a given function, $f(x)$.

Find its first derivative, $f'(x)$ put it equal to 0, and solve the equation thus formed, $f'(x) = 0$. Sub-

stitute the values of x thus found for x in the second derivative, $f''(x)$. *Each value of x which makes the second derivative negative will, when substituted in the function $f(x)$ make it a maximum; and each value which makes the second derivative positive will make the function a minimum. If either value of x reduces the second derivative to 0, substitute in the third, fourth, etc., until a derivative is found which does not reduce to 0. If this be of an odd order, the value of x will not make the function a maximum or minimum; but if it be of an even order and negative, the function will be a maximum; if positive, a minimum.*

Second Rule.—It is sometimes more convenient to ascertain whether a root a of $f'(x) = 0$ corresponds to a maximum or a minimum value of the function by substituting for x, in $f'(x)$, $a - h$ and $a + h$, where h is infinitesimal. *If the first result is + and the second is —, a corresponds to a maximum; if the first result is — and the second is +, it corresponds to a minimum. If both results have the same sign, it corresponds to neither a maximum nor a minimum.* (See Arts. 93, 94.)

98. Maxima and Minima Values occur alternately. —Suppose that $f(x)$ is a maximum when $x = a$, and also when $x = b$, where $b > a$; then, in passing from a to b, when $x = a + h$ (where h is very small), the function is decreasing, and when $x = b - h$, it is increasing; but in passing from a decreasing to an increasing state, it must pass through a minimum value; hence, between two maxima one minimum at least must exist.

In the same way, it may be shown that between two minima one maximum must exist.

This is also evident from geometric considerations, for in Fig. 9 we see that there is a maximum value at P, a minimum at P′, a maximum at P″, a minimum at P‴, and so on.

99. The Investigation of Maxima and Minima is often facilitated by the following Axiomatic Principles:

1. If u be a maximum or minimum for any value of x, and a be a positive constant, au is also a maximum or minimum for the same value of x. Hence, before applying the rule, *a constant factor or divisor may be omitted.*

2. If any value of x makes u a maximum or minimum, it will make any positive power of u a maximum or minimum, unless u be negative, in which case an even power of a minimum is a maximum, and an even power of a maximum is a minimum. Hence, *the function may be raised to any power; or, if under a radical, the radical may be omitted.*

3. Whenever u is a maximum or a minimum, $\log u$ is a maximum or minimum for the same value of x. Hence, *to examine the logarithm of a function we have only to examine the function itself.* When the function consists of products or quotients of roots and powers, its examination is often facilitated by passing to logarithms, as the differentiation is made easier.

4. When a function is a maximum or a minimum, its reciprocal is at the same time a minimum or a maximum; this principle is of frequent use in maxima and minima.

5. If u is a maximum or minimum, $u \pm c$ is a maximum or minimum. Hence, *a constant connected by $+$ or $-$ may be omitted.*

Other transformations are sometimes useful, but as they depend upon particular forms which but rarely occur, they may be left to the ingenuity of the student who wishes to simplify the solution of the proposed problem.

It is not admissible to assume $x = \infty$ in searching for maxima and minima, for in that case x cannot have a succeeding value.

1. Find the values of x which will make the function $u = 6x + 3x^2 - 4x^3$ a maximum or minimum, and the corresponding values of the function u.

Here $$\frac{du}{dx} = 6 + 6x - 12x^2.$$

Now whatever values of x make u a maximum or minimum, will make $\frac{du}{dx} = 0$ (Art. 97) ; therefore,

$$6 + 6x - 12x^2 = 0, \quad \text{or} \quad x^2 - \tfrac{1}{2}x = \tfrac{1}{2};$$
$$\therefore \quad x = \tfrac{1}{4} \pm \tfrac{3}{4} = +1 \text{ or } -\tfrac{1}{2}.$$

Hence, if u have maximum or minimum values, they must occur when $x = 1$ or $-\tfrac{1}{2}$.

To ascertain whether these values are maxima or minima, we form the second derivative of u; thus,

$$\frac{d^2u}{dx^2} = 6 - 24x.$$

When $x = 1$, $\frac{d^2u}{dx^2} = -18$, which corresponds to a maximum value of u.

When $x = -\tfrac{1}{2}$, $\frac{d^2u}{dx^2} = +18$, which corresponds to a minimum value of u.

Substituting these values of x in the given function, we have

When $x = 1$, $u = 6 + 3 - 4 = 5$, a maximum.

When $x = -\tfrac{1}{2}$, $u = -3 + \tfrac{3}{4} + \tfrac{1}{2} = -\tfrac{7}{4}$, a minimum.

2. Find the maxima and minima values of u in

$$u = x^4 - 8x^3 + 22x^2 - 24x + 12.$$

$$\frac{du}{dx} = 4x^3 - 24x^2 + 44x - 24 = 0,$$

or, $\qquad x^3 - 6x^2 + 11x - 6 = 0.$

By trial, $x = 1$ is found to be a root of this equation; therefore, by dividing the first member of this equation by $x - 1$, we find for the depressed equation,

$$x^2 - 5x + 6 = 0; \qquad \therefore \quad x = 2 \text{ or } 3.$$

Hence the critical values are $x = 1$, $x = 2$, and $x = 3$.

$$\frac{d^2u}{dx^2} = 12x^2 - 48x + 44 = + 8, \quad \text{when} \quad x = 1.$$
$$= - 4, \quad \text{when} \quad x = 2.$$
$$= + 8, \quad \text{when} \quad x = 3.$$

Therefore we have,

when $\quad x = 1, \quad u = 3, \quad$ a minimum;

when $\quad x = 2, \quad u = 4, \quad$ a maximum;

when $\quad x = 3, \quad u = 3, \quad$ a minimum.

3. Find the maxima and minima values of u in

$$u = (x - 1)^4 (x + 2)^3.$$

$$\frac{du}{dx} = 4 (x - 1)^3 (x + 2)^3 + 3 (x + 2)^2 (x - 1)^4$$

$$= (x - 1)^3 (x + 2)^2 [4 (x + 2) + 3 (x - 1)],$$

or $\quad \dfrac{du}{dx} = (x - 1)^3 (x + 2)^2 (7x + 5) = 0 ; \qquad (1)$

$$\therefore \quad (x - 1) = 0, \quad (x + 2) = 0, \quad (7x + 5) = 0.$$

$\therefore \quad x = 1$, $x = - 2$, and $x = - \frac{5}{7}$, as the critical values of x.

In this case, it will be easier to test the critical values by the *second* rule of Art. 97; that is, to see whether $\dfrac{du}{dx}$ changes sign or not in passing through $x = 1$, $- 2$, and $- \frac{5}{7}$ in succession.

If we substitute in the second member of (1), $(1 - h)$ and $(1 + h)$ for x, where h is infinitesimal, we get

$$\frac{du}{dx} = (1 - h - 1)^3 (1 - h + 2)^2 [7 (1 - h) + 5]$$

$$= - h^3 (3 - h)^2 (12 - 7h) = -.$$

and $\quad \dfrac{du}{dx} = (1 + h - 1)^3 (1 + h + 2)^2 [7 (1 + h) + 5]$

$$= h^3 (3 + h)^2 (12 + 7h) = +$$

Therefore, as $\dfrac{du}{dx}$ changes sign from $-$ to $+$ at $x = 1$, the function u at this point is a minimum.

When $x = - 2$, $\dfrac{du}{dx}$ does not change sign; $\therefore u$ has no maximum or minimum at this point.

When $x = - \frac{5}{7}$, $\dfrac{du}{dx}$ changes sign from $+$ to $-$; $\therefore u$, at this point, is a maximum.

Hence,   when $x = 1$, $u = 0$, a minimum.

when $x = - \frac{5}{7}$, $u = \dfrac{124.93}{77}$, a maximum.

It is usually easy to see from inspection whether $\dfrac{du}{dx}$ changes sign in passing through a critical value of x, without actually making the substitution.

4. Examine $u = b + (x - a)^3$ for maxima and minima.

$$\frac{du}{dx} = 3 (x - a)^2 = 0; \quad \therefore \quad x = a, \text{ and } u = b.$$

Since $x = a$ makes $\dfrac{d^2u}{dx^2} = 0$, we must examine it by the second rule of Art. 97, and see whether $\dfrac{du}{dx}$ changes sign at $x = a$.

$\dfrac{du}{dx} = 3\,(a - h - a)^2 = 3h^2$ is the value of $\dfrac{du}{dx}$ immediately preceding $x = a$.

$\dfrac{du}{dx} = 3\,(a + h - a)^2 = 3h^2$ is the value of $\dfrac{du}{dx}$ immediately succeeding $x = a$.

Therefore, as $\dfrac{du}{dx}$ does not change sign at $x = a$, $u = b$ is neither a maximum nor a minimum.

5. Examine $u = b + (x - a)^4$ for maxima and minima.

$$\dfrac{du}{dx} = 4\,(x - a)^3 = 0; \qquad \therefore \quad x = a \quad \text{and} \quad u = b.$$

It is easy to see that $\dfrac{du}{dx}$ changes sign from $-$ to $+$ at $x = a$; $\therefore$ $x = a$ gives $u = b$, a minimum.

6. Examine $u = \dfrac{(x + 2)^3}{(x - 3)^2}$ for maxima and minima.

$$\dfrac{du}{dx} = \dfrac{(x + 2)^2\,(x - 13)}{(x - 3)^3} = 0 \text{ or } \infty;$$

$$\therefore \quad x = -2,\ 13,\ \text{or } 3.$$

We see that when $x = -2$, $\dfrac{du}{dx}$ does not change sign;

$\therefore$ no maximum or minimum;

when $\qquad x = 13$, $\dfrac{du}{dx}$ changes sign from $-$ to $+$;

$\therefore$ a minimum;

when $\qquad x = 3$, $\dfrac{du}{dx}$ changes sign from $+$ to $-$,

$\therefore$ a maximum;

hence when $x = 13$, $u = 1\frac{35}{4}$, a minimum value;

and when $x = 3$, $u = \infty$, a maximum value.

7. Examine $u = b + (x - a)^{\frac{4}{3}}$ for maxima and minima

$$\frac{du}{dx} = \tfrac{4}{3}(x-a)^{\frac{1}{3}} = 0; \qquad \therefore \quad x = a \quad \text{and} \quad u = b.$$

When $x = a$, $\dfrac{du}{dx}$ changes sign from $-$ to $+$.

$$x = a \quad \text{gives} \quad u = b, \quad \text{a minimum.}$$

8. Examine $u = b - (a - x)^{\frac{3}{5}}$ for maxima and minima.

$$\frac{du}{dx} = \tfrac{3}{5}(a-x)^{\frac{3}{5}} = 0; \qquad \therefore \quad x = a \quad \text{and} \quad u = b.$$

When $x = a$, $\dfrac{du}{dx}$ changes sign from $+$ to $-$.

$$\therefore \quad x = a \quad \text{gives} \quad u = b, \quad \text{a maximum.}$$

9. Examine $u = b + \sqrt[3]{a^3 - 2a^2x + ax^2}$ for maxima and minima.

If u is a maxima or minima, $u - b$ will be so; therefore we omit the constant b and the radical by Art. 99, and get

$$u' = a^3 - 2a^2x + ax^2;$$

$$\frac{du'}{dx} = -2a^2 + 2ax = 0; \qquad \therefore \quad x = a \quad \text{and} \quad u = b.$$

When $x = a$, $\dfrac{du'}{dx}$ changes sign from $-$ to $+$.

$$\therefore \quad x = a \quad \text{gives} \quad u = b, \quad \text{a minimum.}$$

10. Examine $u = \dfrac{a^2x}{(a-x)^2}$ for maxima and minima.

Using the reciprocal, since it is more simple, and omitting the constant a^2 (Art. 99), we have

$$u' = \frac{(a-x)^2}{x} = \frac{a^2}{x} - 2a + x;$$

$$\therefore \quad \frac{du'}{dx} = -\frac{a^2}{x^2} + 1 = 0, \quad \text{and} \quad \frac{d^2u'}{dx^2} = \frac{2a^2}{x^3};$$

$$\therefore \quad x = \pm a, \quad \text{and} \quad \therefore \quad \frac{d^2u'}{dx^2} = \pm \frac{2}{a}.$$

Hence, $x = + a$ makes u' a minimum, and $x = - a$ makes it a maximum; therefore, since maxima and minima values of u' correspond respectively to the minima and maxima values of u (Art. 99, 4), we have,

$$\text{when} \quad x = a, \quad u = \infty, \quad \text{a maximum.}$$

$$\text{``} \quad x = - a, \quad u = - \frac{a}{4}, \quad \text{a minimum.}$$

Find the values of x which give maximum and minimum values of the following functions:

1. $u = x^3 - 3x^2 - 24x + 85$.

Ans. $x = - 2$, max.; $x = 4$, min.

2. $u = 2x^3 - 21x^2 + 36x - 20$.

$x = 1$, max.; $x = 6$, min.

3. $u = x^3 - 18x^2 + 96x - 20$.

$x = 4$, max.; $x = 8$, min.

4. $u = \dfrac{(a - x)^2}{a - 2x}.$ $\qquad x = \frac{1}{4}a$, min.

5. $u = \dfrac{1 + 3x}{\sqrt{4 + 5x}}.$ $\quad x = - 1\frac{4}{15}$, max.

6. $u = x^3 - 3x^2 - 9x + 5$.

$x = - 1$, max.; $x = 3$, min.

7. $u = x^3 - 3x^2 + 6x + 7$.

Neither a max. nor a min.

8. $u = (x - 9)^5 (x - 8)^4$.

$x = 8$, max.; $x = 8\frac{5}{9}$, min.

9. $u = \dfrac{x}{1 + x \tan x}.$ $\quad x = \cos x$, max.

10. $u = \sin^3 x \cos x.$ $\quad x = 60°$, max.

11. $u = \dfrac{\sin x}{1 + \tan x}$ $\quad x = 45°$, max.

12. $u = \sin x + \cos x$.

$$x = 45°, \text{ max.}; \quad x = 225°, \text{ min.}$$

13. $u = \dfrac{\log x}{x^n}$. $\qquad\qquad\qquad x = e^{\frac{1}{n}}, \text{ max.}$

GEOMETRIC PROBLEMS.

The only difficulty in the solution of problems in maxima and minima consists in obtaining a convenient algebraic expression for the function whose maximum or minimum value is required. No general rule can well be given by which this expression can be found. Much will depend upon the ingenuity of the student. A careful examination of all the conditions of the problem, and tact in applying his knowledge of principles previously learned in Algebra, Geometry, and Trigonometry, with experience, will serve to guide him in forming the expression for the function. After reducing the expression to its simplest form by the axioms of Art. 99, he must proceed as in Art. 97.

1. Find the maximum cylinder which can be inscribed in a given right cone with a circular base.

Suppose a cylinder inscribed as in the figure. Let $AO = b$, $DO = a$, $CO = x$, $CE = y$.

Then, denoting the volume of the cylinder by v, we have

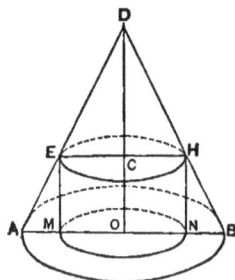

Fig. 10.

$$v = \pi y^2 x. \tag{1}$$

From the similar triangles DOA and DCE, we have

$$DO : AO :: DC : EC,$$

or $\qquad\qquad a : b :: a - x : y;$

$$\therefore \; y = \frac{b}{a}(a - x),$$

which in (1) gives $\qquad v = \pi \dfrac{b^2}{a^2}(a - x)^2 x. \tag{2}$

Dropping constant factors (Art. 99), we have

$$u = (a - x)^2 x = a^2 x - 2ax^2 + x^3;$$

$$\therefore \quad \frac{du}{dx} = a^2 - 4ax + 3x^2 = 0,$$

or $\quad x^2 - \tfrac{4}{3}ax = -\tfrac{1}{3}a^2; \quad \therefore \quad x = a \quad \text{or} \quad \tfrac{1}{3}a.$

$$\frac{d^2u}{dx^2} = -4a + 6x$$

$$= 2a, \quad \text{when } x = a, \quad \therefore \text{ minimum};$$

$$= -2a, \quad \text{when } x = \tfrac{1}{3}a, \quad \therefore \text{ maximum}.$$

Hence the altitude of the maximum cylinder is one-third of the cone.

The second value of x in (2) gives

$$v = \frac{\pi b^2}{a^2}(a - \tfrac{1}{3}a)\frac{a}{3} = \tfrac{4}{27}\pi ab^2.$$

Volume of cone $= \tfrac{1}{3}\pi ab^2.$

$\therefore$ Volume of cylinder $= \tfrac{4}{9}$ volume of cone.

$$y = \frac{b}{a}(a - \tfrac{1}{3}a) = \tfrac{2}{3}b = \text{radius of base of cylinder.}$$

2. What is the altitude of the maximum rectangle that can be inscribed in a given parabola?

Let $AX = a$, $AH = x$, $DH = y$, and $A = $ area of rectangle. Then we have

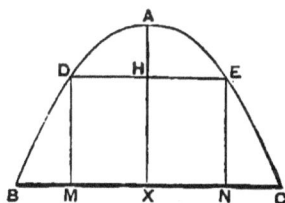

Fig. II.

$$A = 2y(a - x). \tag{1}$$

But from the equation of the parabola, we have

$$y = \sqrt{2px},$$

which in (1) gives $\quad A = 2\sqrt{2px}\,(a - x). \tag{2}$

$$u' = \sqrt{x}\,(a - x) = ax^{\frac{1}{2}} - x^{\frac{3}{2}}.$$

$$\therefore \quad \frac{du'}{dx} = \tfrac{1}{2}ax^{-\frac{1}{2}} - \tfrac{3}{2}x^{\frac{1}{2}} = 0. \quad \therefore \quad x = \tfrac{1}{3}a.$$

Since this value of x makes $\dfrac{du}{dx}$ change sign from $+$ to $-$, it makes the function A a maximum; therefore the altitude of the maximum rectangle is $\frac{2}{3}a$.

3. What is the maximum cone that can be inscribed in a given sphere?

Let ACB be the semicircle, and ACD the triangle which, revolved about AB, generate the sphere and cone respectively. Let $AO = r$, $AD = x$, and $CD = y$, and $v =$ volume of cone.

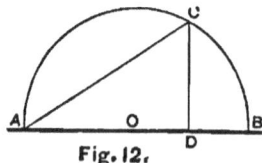

Fig. 12.

Then
$$v = \tfrac{1}{3}\pi y^2 x. \tag{1}$$

But
$$y^2 = AD \times DB = (2r - x)\,x,$$

which in (1) gives
$$v = \tfrac{1}{3}\pi\,(2r - x)\,x^2, \tag{2}$$

or
$$u = 2rx^2 - x^3 ;$$

$$\therefore\ \frac{du}{dx} = 4rx - 3x^2 = 0.$$

$$\therefore\ x = 0 \quad \text{and} \quad \tfrac{4}{3}r.$$

The latter makes $\dfrac{du}{dx}$ change sign from $+$ to $-$; $\therefore$ it makes v a maximum.

Hence the altitude of the maximum cone is $\frac{2}{3}$ of the diameter of the sphere.

The second value of x in (2) gives

$$v = \tfrac{1}{3}\pi\,(2r - \tfrac{4}{3}r)\,(\tfrac{4}{3}r)^2 = \tfrac{32}{81}\pi r^3 = \tfrac{8}{27} \times \tfrac{4}{3}\pi r^3.$$

Volume of sphere $= \tfrac{4}{3}\pi r^3$;

$$\therefore\ \text{the cone} = \tfrac{8}{27} \text{ of the sphere.}$$

4. Find the maximum parabola which can be cut from a given right cone with a circular base, knowing that the area of a parabola is $\frac{2}{3}$ the product of its base and altitude.

Let $AB = a$, $AC = b$, and $BH = x$;
then $AH = a - x$.

$$FE = 2EH = 2\sqrt{AH \times BH}$$
$$= 2\sqrt{(a - x)\, x}.$$

Also, $BA : AC :: BH : HD$,

or $\qquad a : b :: x : HD = \dfrac{b}{a}x.$

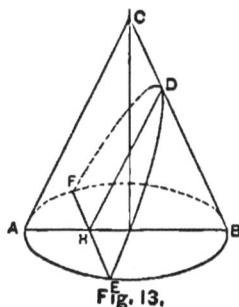

Fig. 13.

Calling the parabola A, we have

$$A = \tfrac{2}{3}FE \times HD = \tfrac{4}{3}\frac{b}{a}x\sqrt{(a - x)\, x},$$

or $\qquad u = ax^3 - x^4.$

$$\frac{du}{dx} = 3ax^2 - 4x^3 = 0;$$

$$\therefore \quad x = 0 \quad \text{and} \quad x = \tfrac{3}{4}a.$$

The second value makes $\dfrac{du}{dx}$ change sign from $+$ to $-$,
and $\therefore$ makes the function A a maximum.

$$\therefore \quad A = \tfrac{4}{3}\cdot\frac{b}{a}\cdot\tfrac{3}{4}a\sqrt{(a - \tfrac{3}{4}a)\tfrac{3}{4}a} = \tfrac{1}{4}ab\sqrt{3},$$

which is the area of the maximum parabola.

REM.—In problems of maxima and minima, it is often more convenient to express the function u in terms of two variables, x and y,
which are connected by some equation, so that either may be regarded
as a function of the other. In this case, either variable of course may
be eliminated, and u expressed in terms of the other, and treated by
the usual process, as in Examples 1, 2, and 3. It is often simpler,
however, to differentiate the function u, and the equation between x
and y, with respect to either of the variables, x, regarding the other,
y, as a function of it, and then eliminate the first derivative, $\dfrac{dy}{dx}$. The
second method of the following example will illustrate the process.

5. To find the maximum rectangle inscribed in a given ellipse.

Let $CM = x$, $PM = y$, and A = area of rectangle. Then we have

$$A = 4xy, \qquad (1)$$

and

$$a^2y^2 + b^2x^2 = a^2b^2. \qquad (2)$$

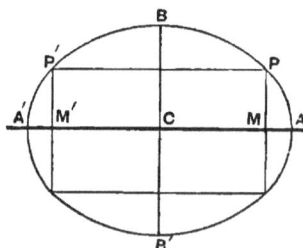

Fig. 14.

1st Method.—From (2) we get

$$y = \frac{b}{a}\sqrt{a^2 - x^2},$$

which in (1) gives $A = 4\dfrac{b}{a}x\sqrt{a^2 - x^2}$,

or

$$u = a^2x^2 - x^4.$$

$$\frac{du}{dx} = 2a^2x - 4x^3 = 0. \quad \therefore x = \pm\frac{a}{\sqrt{2}}.$$

$x = +\dfrac{a}{\sqrt{2}}$ makes $\dfrac{du}{dx}$ change sign from $+$ to $-$; $\therefore$ it makes A a maximum.

Hence, the sides of the maximum rectangle are $a\sqrt{2}$ and $b\sqrt{2}$, and the area is $2ab$.

2d Method.—Differentiate (1) and (2) with respect to x after dropping the factor 4 from (1), and get

$$\frac{dA}{dx} = y + x\frac{dy}{dx} = 0; \qquad \therefore \frac{dy}{dx} = -\frac{y}{x}.$$

$$2a^2y\frac{dy}{dx} + 2b^2x = 0; \qquad \therefore \frac{dy}{dx} = -\frac{b^2x}{a^2y}.$$

$$\therefore \frac{b^2x}{a^2y} = \frac{y}{x}, \quad \text{or} \quad b^2x^2 = a^2y^2;$$

which in (2) gives

$$2a^2y^2 = a^2b^2, \quad \therefore \quad y = \frac{b}{\sqrt{2}} \quad \text{and} \quad x = \frac{a}{\sqrt{2}}.$$

6. Find the cylinder of greatest convex surface that can be inscribed in a right circular cone, whose altitude is h and the radius of whose base is r.

$$\text{Surface} = \frac{\pi h r}{2}.$$

7. Determine the altitude of the maximum cylinder which can be inscribed in a sphere whose radius is r.

Altitude $= \tfrac{2}{3}r \sqrt{3}$.

8. Find the maximum isosceles triangle that can be inscribed in a circle.   An equilateral triangle.

9. Find the area of the greatest rectangle that can be inscribed in a circle whose radius is r.

The sides are each $r \sqrt{2}$.

10. Find the axis of the cone of maximum convex surface that can be inscribed in a sphere of radius r.

The axis $= \tfrac{4}{3}r$.

11. Find the altitude of the maximum cone that can be inscribed in a paraboloid of revolution, whose axis is a, the vertex of the cone being at the middle point of the base of the paraboloid.   Altitude $= \tfrac{1}{2}a$.

12. Find the altitude of the cylinder of greatest convex surface that can be inscribed in a sphere of radius r.

Altitude $= r \sqrt{2}$.

13. From a given surface s, a vessel with circular base and open top is to be made, so as to contain the greatest amount. Find its dimensions. (See Remark under Ex. 4.)

The altitude $=$ radius of base $= \sqrt{\dfrac{s}{3\pi}}.$

14. Find the maximum cone whose convex surface is constant.   The altitude $= \sqrt{2}$ times the radius of base.

15. Find the maximum cylinder that can be inscribed in an oblate spheroid whose semi-axes are a and b.

The radius of base $= a \sqrt{\tfrac{2}{3}}$; the altitude $= b \dfrac{2}{\sqrt{3}}.$

16. Find the maximum difference between the sine and cosine of any angle.   When the angle $= 135°$.

17. Find the number of equal parts into which a must be divided so that their continued product may be a maximum.

Let x be the number of parts, and thus each part equals $\dfrac{a}{x}$, and therefore $u = \left(\dfrac{a}{x}\right)^{x}$, from which we get $x = \dfrac{a}{e}$; therefore each part $= e$, and the product of all $= (e)^{\frac{a}{e}}$.

18. Find a number x such that the xth root shall be a maximum. $x = e = 2.71828+$.

19. Find the fraction that exceeds its mth power by the greatest possible quantity. $\left(\dfrac{1}{m}\right)^{\frac{1}{m-1}}$.

20. A person being in a boat 3 miles from the nearest point of the beach, wishes to reach in the shortest time a place 5 miles from that point along the shore; supposing he can walk 5 miles an hour, but row only at the rate of 4 miles an hour, required the place he must land.

One mile from the place to be reached.

21. A privateer wishes to get to sea unmolested, but has to pass between two lights, A and B, on opposite headlands, the distance between which is c. The intensity of the light A at a unit's distance is a, and the intensity of B at the same distance is b; at what point between the lights must the privateer pass so as to be as little in the light as possible, assuming the principle of optics that the intensity of a light at any distance equals its intensity at the distance one divided by the square of the distance from the light.

$$x = \frac{ca^{\frac{1}{3}}}{a^{\frac{1}{3}} + b^{\frac{1}{3}}}.$$

22. The flame of a candle is directly over the centre of a circle whose radius is r; what ought to be its height above the plane of the circle so as to illuminate the circumference as much as possible, supposing the intensity of the

light to vary directly as the sine of the angle under which it strikes the illuminated surface, and inversely as the square of its distance from the same surface.

$$\text{Height above the plane of the circle} = r\sqrt{\tfrac{1}{2}}.$$

23. Find in the line joining the centres of two spheres, the point from which the greatest portion of spherical surface is visible.

The function to be a maximum is the sum of the two zones whose altitudes are AD and ad; hence

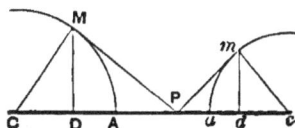

Fig. 15.

we must find an expression for the areas of these zones.

Put CM $= R$ and $cm = r$, Cc $= a$ and CP $= x$.

The area of the zone on the sphere which has R for its radius (from Geometry, or Art. 194) $= 2\pi\text{RAD} = 2\pi R^2 - 2\pi\text{RCD} = 2\pi\left(R^2 - \dfrac{R^3}{x}\right)$, and in the same way for the other zone, from which we readily obtain the solution.

$$x = \frac{aR^{\frac{3}{2}}}{R^{\frac{3}{2}} + r^{\frac{3}{2}}}.$$

24. Find the altitude of the cylinder inscribed in a sphere of radius r, so that its *whole* surface shall be a maximum.

$$\text{Altitude} = r\left[2\left(1 - \frac{1}{\sqrt{5}}\right)\right]^{\frac{1}{2}}.$$

100. Equations of the Tangent and Normal.—Let P, (x', y') be the point of tangency ; the equation of the tangent line at (x', y') will be of the form (Anal. Geom., Art. 25)

$$y - y' = a (x - x'), \qquad (1)$$

in which a is the tangent of the angle which the tangent line makes with the axis of x. It was shown in Article 56a that the value of this tangent is equal to the derivative of the ordinate of the point of tangency, with respect to x,

Fig. 16.

or
$$a = \frac{dy'}{dx'}.$$

Hence
$$y - y' = \frac{dy'}{dx'} (x - x'), \qquad (2)$$

is *the equation of the tangent to the curve at the point* (x', y'), x and y being the *current* co-ordinates of the tangent.

Since the normal is perpendicular to the tangent at the point of tangency, its equation is, from (2),

$$y - y' = - \frac{dx'}{dy'} (x - x'). \qquad (3)$$

(Anal. Geom., Art. 27, Cor. 2.)

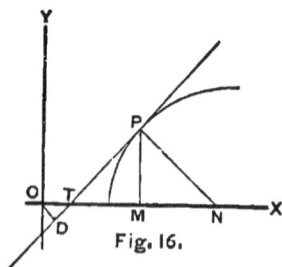

Rem.—To apply (2) or (3) to any particular curve, we substitute for $\dfrac{dy'}{dx'}$ or $\dfrac{dx'}{dy'}$, its value obtained from the equation of the curve and expressed in terms of the co-ordinates of the point of tangency.

EXAMPLES.

1. Find the equations of the tangent and normal to the ellipse

$$a^2y^2 + b^2x^2 = a^2b^2.$$

We find $\quad \dfrac{dy}{dx} = -\dfrac{b^2x}{a^2y}; \quad \therefore \; \dfrac{dy'}{dx'} = -\dfrac{b^2x'}{a^2y'};$

and this value in (2) gives,

$$y - y' = -\frac{b^2x'}{a^2y'}(x - x');$$

which by reduction becomes,

$$a^2yy' + b^2xx' = a^2b^2,$$

which is *the equation of the tangent;* and

$$y - y' = \frac{a^2y'}{b^2x'}(x - x')$$

is *the equation of the normal.*

2. Find the equations of the tangent and normal to the parabola $y^2 = 2px.$

We find $\quad \dfrac{dy}{dx} = \dfrac{p}{y}, \quad \therefore \; \dfrac{dy'}{dx'} = \dfrac{p}{y'},$

and this value in (2) gives

$$y - y' = \frac{p}{y'}(x - x'),$$

or $\qquad\qquad yy' - y'^2 = px - px'.$

But $\qquad\qquad y'^2 = 2px';$

$$\therefore \quad yy' = p\,(x + x'),$$

which is *the equation of the tangent ;* and

$$y - y' = -\frac{y'}{p}\,(x - x')$$

is *the equation of the normal.*

3. Find the equations of the tangent and normal to a hyperbola.

Tangent, $a^2yy' - b^2xx' = -a^2b^2$.

Normal, $\qquad y - y' = -\dfrac{a^2y'}{b^2x'}\,(x - x')$.

4. Find the equation of the tangent to $3y^2 + x^2 - 5 = 0,$ at $x = 1$.

Here $\qquad \dfrac{dy'}{dx'} = -\dfrac{x'}{3y'} = -\dfrac{1}{\pm 3.465} = \mp .29$ about,

which in (2) gives

$$y \mp 1.155 = \mp .29\,(x - 1),$$

or $\qquad\qquad\qquad y = \mp .29x \pm 1.44.$

Hence there are two tangents to this locus at $x = 1$, their equations being

$$y = -.29x + 1.44 \quad \text{and} \quad y = +.29x - 1.44.$$

5. Find the equation of the tangent to the parabola $y^2 = 9x$, at $x = 4$.

At (4, 6)   the equation is $y = \quad \frac{3}{4}x + 3$.
" (4, — 6) "   "   " $y = -\frac{3}{4}x - 3$.

6. Find the equation of the normal to $y^2 = 2x^2 - x^3$, at $x = 1$.

At (1, + 1) the equation is $y = -2x + 3$.
" (1, — 1) "   "   " $y = \quad 2x - 3$.

7. Find the equation of the normal to $y^2 = 6x - 5$, at $y = 5$, and the angle which this normal makes with the axis of x. $y = -\frac{5}{3}x + \frac{40}{3}$; angle $= \tan^{-1}\left(-\frac{5}{3}\right)$.

101. Length of Tangent, Normal. Subtangent, Subnormal, and Perpendicular on the Tangent from the Origin.

Let PT represent the tangent at the point P, PN the normal; draw the ordinate PM ; then

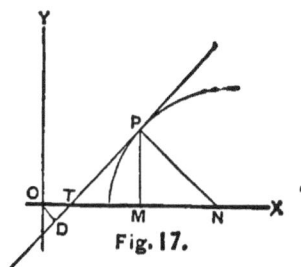

Fig. 17.

    MT is called the *subtangent,*

    MN "   "   "   *subnormal.*

Let α = angle PTM ;

then $\tan \alpha = \dfrac{dy'}{dx'}$ (Art. 56a).

1st. $TM = MP \cot \alpha = y' \dfrac{dx'}{dy'}$;

$$\therefore \; Subtangent = y' \frac{dx'}{dy'}.$$

2d. $MN = MP \tan MPN = y' \tan \alpha;$

$$\therefore \; Subnormal = y' \frac{dy'}{dx'}.$$

3d. $PT = \sqrt{\overline{PM^2 + MT^2}}$

$$= \sqrt{y'^2 + \left(y' \frac{dx'}{dy'}\right)^2} \; ;$$

$$\therefore \; Tangent = y'\sqrt{1 + \left(\frac{dx'}{dy'}\right)^2}.$$

4th. $PN = \sqrt{\overline{PM^2 + MN^2}}$

$$= \sqrt{y'^2 + \left(y' \frac{dy'}{dx'}\right)^2} \; ;$$

$$\therefore \; Normal = y'\sqrt{1 + \left(\frac{dy'}{dx'}\right)^2}.$$

5th. The equation of the tangent at P (x', y') is (Art. 100),

$$y - y' = \frac{dy'}{dx'}(x - x'),$$

or $\qquad x\,dy' - y\,dx' - x'\,dy' + y'\,dx' = 0\,;$

which, written in the normal form, is

$$\frac{x\,dy' - y\,dx' - x'\,dy' + y'\,dx'}{\sqrt{(dx')^2 + (dy')^2}} \quad \text{(Anal. Geom., Art. 24)};$$

hence, $\quad \text{OD} = \dfrac{y'\,dx' - x'\,dy'}{\sqrt{(dx')^2 + (dy')^2}} = 0.$

∴ *Perpendicular on the tangent from the origin*

$$= \frac{y'\,dx' - x'\,dy'}{\sqrt{(dx')^2 + (dy')^2}}.$$

Sch.—In these expressions for the subtangent and subnormal it is to be observed that the subtangent is measured from M towards the *left*, and the subnormal from M towards the *right*. If, in any curve, $y'\dfrac{dy'}{dx'}$ is a *negative* quantity, it denotes that N lies to the *left* of M, and as in that case $y'\dfrac{dx'}{dy'}$ is also negative, T lies to the *right* of M.

EXAMPLES.

1. Find the values of the subtangent, subnormal, and perpendicular from the origin on the tangent, in the ellipse $a^2y^2 + b^2x^2 = a^2b^2$.

Here $\qquad\qquad \dfrac{dy'}{dx'} = -\dfrac{b^2x'}{a^2y'}.$

Hence, the subtangent $= y'\dfrac{dx'}{dy'} = -\dfrac{a^2y'^2}{b^2x'},$

the subnormal $= y'\dfrac{dy'}{dx'} = -\dfrac{b^2}{a^2}x'\,;$

the perpendicular from origin on tangent

$$= \frac{c^2 b^2}{(a^4 y'^2 + b^4 x'^2)^{\frac{1}{2}}}.$$

2. Find the subtangent and subnormal to the Cissoid

$$y^2 = \frac{x^3}{2a - x}. \quad \text{(See Anal. Geom., Art. 149.)}$$

Here
$$\frac{dy'}{dx'} = \pm \frac{x^{\frac{1}{2}} (3a - x)}{(2a - x)^{\frac{3}{2}}}.$$

Hence, the subtangent $= \dfrac{x (2a - x)}{3a - x}.$

the subnormal $= \dfrac{x^2 (3a - x)}{(2a - x)^2}.$

3. Find the value of the subtangent of $y^2 = 3x^2 - 12$, at $x = 4$. Subtangent $= 3$.

4. Find the length of the tangent to $y^2 = 2x$, at $x = 8$. Tangent $= 4\sqrt{17}$.

5. Find the values of the normal and subnormal to the cycloid (Anal. Geom., Art. 156).

$$x = r \operatorname{vers}^{-1} \frac{y}{r} - \sqrt{2ry - y^2};$$

Fig. 18.

$$\therefore \frac{dx}{dy} = \frac{y}{\sqrt{2ry - y^2}} = \frac{\sqrt{2ry - y^2}}{2r - y}.$$

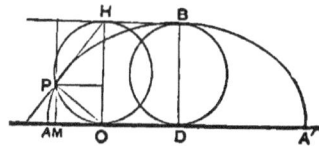

$$\frac{dy}{dx} = \frac{2r - y}{\sqrt{2ry - y^2}}.$$

$\therefore$ Subnormal $= \sqrt{2ry - y^2} = $ MO.

Normal $= \sqrt{2ry} = $ PO.

It can be easily seen that PO is normal to the cycloid at P; for the motion of each point on the generating circle at the instant is one of rotation about the point of contact O, *i.e.*, each point for an instant is describing an infinitely small circular arc whose centre is at O ; and hence PO is normal to the curve, *i.e.*, the normal passes through the foot of the vertical diameter of the generating circle.   Also, since OPH is a right angle, the tangent at P passes through the upper extremity of the vertical diameter.

6. Find the length of the normal in the cycloid, the radius of whose generatrix is 2, at $y = 1$.   Normal $= 2$.

POLAR CURVES

102. Tangents, Normals, Subtangents, Subnormals, and Perpendicular on Tangents.

Let P be any point of the curve APQ, O the pole, OX the initial line.   Denote XOP by θ, and the radius-vector, OP, by r.   Give XOP the infinitesimal increment POQ $= d\theta$, then $OQ = r + dr$.   From the pole O, with the radius OP $= r$, describe the small arc PR, subtending $d\theta$; then, since $d\theta = ab$ is the arc at the unit's distance from the pole O, we have

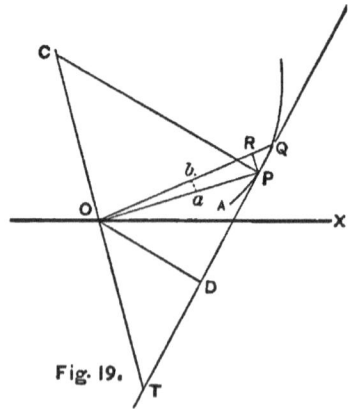

Fig. 19.

$$PR = rd\theta \quad \text{and} \quad RQ = dr. \tag{1}$$

Let PQ, the element* of the arc of the curve, be represented by ds.

$$\therefore \ \ \overline{PQ}^2 = \overline{PR}^2 + \overline{RQ}^2,$$

or $$\overline{ds}^2 = r^2\overline{d\theta}^2 + \overline{dr}^2. \tag{2}$$

Pass through the two points P and Q the right line QPT;

---

* See Art. 56*a*, foot-note.

then, as P and Q are consecutive points, the line QP T is a tangent to the curve at P (Art. 56*a*). Through P draw the normal PC, and through O draw COT perpendicular to OP, and OD perpendicular to PT. The lengths PT and PC are respectively called the *polar tangent* and the *polar normal.* OC is called the *polar subnormal ;* OT the *polar subtangent ;* and OD, the perpendicular from the pole on the tangent, is usually symbolized by *p.* The value of each of these lines is required.

$$\tan RQP = \frac{RP}{RQ} = \frac{rd\theta}{dr}, \text{ from (1).} \qquad (3)$$

Since OPT = OQT + $d\theta$, the two angles OPT and OQT differ from each other by an infinitesimal, and therefore OPT = OQT, and hence,

$$\tan OPT = \frac{rd\theta}{dr}, \text{ from (3),} \qquad (4)$$

$$\sin OPT = \sin OQP = \frac{RP}{QP} = \frac{rd\theta}{ds}, \text{ from (1).} \qquad (5)$$

Hence,

$$OT = \text{polar subtangent} = OP \tan OPT = \frac{r^2 d\theta}{dr},$$
$$\text{from (4).} \qquad (6)$$

$$OC = \text{polar subnormal} = OP \tan OPC = OP \cot OPT$$
$$= \frac{dr}{d\theta}, \text{ from (4).} \qquad (7)$$

$$PT = \text{polar tangent} = \sqrt{\overline{OP}^2 + \overline{OT}^2} = r\sqrt{1 + r^2 \frac{d\theta^2}{dr^2}},$$
$$\text{from (6).} \qquad (8)$$

$$PC = \text{polar normal} = \sqrt{\overline{OP}^2 + \overline{OC}^2} = \sqrt{r^2 + \frac{dr^2}{d\theta^2}},$$
$$\text{from (7).} \qquad (9)$$

$$\text{OD} = p = \text{OP} \sin \text{OPD} = \frac{r^2 d\theta}{ds} \text{ from (5)} = \frac{r^2 d\theta}{\sqrt{r^2 d\theta^2 + dr^2}},$$

$$\text{from (2).} \quad (10)$$

See Price's Calculus, Vol. I, p. 417.

EXAMPLES.

1. The spiral of Archimedes, whose equation is $r = a\theta$. (Anal. Geom., Art. 160.)

Here $\qquad \dfrac{d\theta}{dr} = \dfrac{1}{a}$; $\quad \therefore \quad \text{Subt.} = \dfrac{r^2}{a}$, from (6).

$$\text{Subn.} = a, \text{ from (7).}$$

$$\text{Tangent} = r \sqrt{1 + \frac{r^2}{a^2}}, \text{ from (8),}$$

$$\text{Normal} = \sqrt{r^2 + a^2}, \text{ from (9).}$$

$$p = \frac{r^2}{\sqrt{r^2 + a^2}}, \text{ from (10).}$$

2. The logarithmic spiral $r = a^\theta$. (Anal. Geom., Art. 163.)

Here $\dfrac{dr}{d\theta} = a^\theta \log a = r \log a$;

$$\text{Subt} = \frac{r}{\log a} = mr,$$

(where m is the modulus of the system in which $\log a = 1$).

$$\text{Subn.} = \frac{r}{m}.$$

$$p = \frac{r}{\sqrt{1 + \dfrac{1}{m^2}}} = \frac{mr}{\sqrt{m^2 + 1}}.$$

Fig. 20.

$$\text{Normal} = \left[r^2 + \frac{dr^2}{d\theta^2} \right]^{\frac{1}{2}} = (r^2 + r^2 \log^2 a)^{\frac{1}{2}}.$$

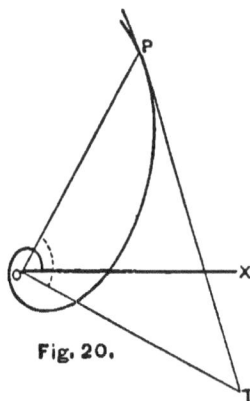

$$\text{Tan. OPT} = r\frac{d\theta}{dr} = \frac{1}{\log a};$$

which is a constant; and therefore the curve cuts every radius-vector at the same angle, and hence it is called the *Equiangular Spiral*.

If $a = e$, the Naperian base, we have,

$$\tan \text{OPT} = \frac{1}{\log e} = 1, \quad \text{and} \quad \therefore \quad \text{OPT} = 45°,$$

and $$\text{OT} = \text{OP} = r.$$

3. Find the subtangent, subnormal, and perpendicular in the Lemniscate of Bernouilli, $r^2 = a^2 \cos 2\theta$. (Anal. Geom., Art. 154.)

$$\text{Subtangent} \quad = \frac{-r^3}{a^2 \sin 2\theta};$$

$$\text{Subnormal} \quad = -\frac{a^2}{r} \sin 2\theta;$$

$$\text{Perpendicular} = \frac{r^3}{\sqrt{r^4 + a^4 \sin^2 2\theta}} = \frac{r^3}{a^2}.$$

4. Find the subtangent and subnormal in the hyperbolic* spiral $r\theta = a$. (Anal. Geom., Art. 161.)

$$\text{Subt.} = -a; \quad \text{Subn.} = -\frac{r^2}{a}.$$

RECTILINEAR ASYMPTOTES.

103. A Rectilinear Asymptote is a line which is continually approaching a curve and becomes tangent to it at an infinite distance from the origin, and yet passes within a finite distance of the origin.

To find whether a proposed curve has an asymptote, we must first ascertain if it has infinite branches, since if it

---

* This curve took its name from the analogy between its equation and that of the hyperbola $xy = a$. (See Strong's Calculus, p. 145; also Young's Dif. Calculus, p. 120.)

has not, there can be no asymptote. If it *has* an infinite branch, we must then ascertain if the intercept on either of the axes is finite. The equation of the tangent (Art. 100) being,

$$y - y' = \frac{dy'}{dx'}(x - x'),$$

if we make successively $y = 0$, $x = 0$, we shall find for the intercepts on the axes of x and y, the following :

$$x_0 = x - y\frac{dx}{dy},$$

(by putting $x = x_0$ and $y = y_0$, and dropping accents),

$$y_0 = y - x\frac{dy}{dx},$$

Now, if for $x = \infty$ both x_0 and y_0 are finite, they will determine two points, one on each axis, through which an asymptote passes. If for $y = \infty$, x_0 is finite and y_0 infinite, the asymptote is parallel to the axis of y. If for $x = \infty$, x_0 is infinite and y_0 finite, the asymptote is parallel to the axis of x. If both x_0 and y_0 are infinite, the curve has no asymptotes corresponding to $x = \infty$. If both x_0 and y_0 are 0, the asymptote passes through the origin, and its direction is obtained by evaluating $\frac{dy}{dx}$ for $x = \infty$.

When there are asymptotes parallel to the axis, they may usually be detected by inspection, as it is only necessary to ascertain what values of x will make $y = \infty$, and what values of y will make $x = \infty$. For example, in the equation $xy = m$, $x = 0$ makes $y = \infty$, and $y = 0$ makes $x = \infty$; hence the two axes are asymptotes. Also in the equation $xy - ay - bx = 0$, which may be put in either of the two forms,

$$y = \frac{bx}{x - a} \quad \text{or} \quad x = \frac{ay}{y - b};$$

$y = \infty$ when $x = a$, and $x = \infty$ when $y = b$;

hence the two lines $x = a$ and $y = b$ are asymptotes to the curve.

In the logarithmic curve $y = a^x$,

$$y = 0 \quad \text{when} \quad x = -\infty,$$

therefore the axis of x is an asymptote to the branch in the second angle.

Also in the Cissoid $y^2 = \dfrac{x^3}{2a - x}$,

$$y = \infty \quad \text{when} \quad x = 2a;$$

hence $x = 2a$ is an asymptote.

EXAMPLES

1. Examine the hyperbola

$$a^2y^2 - b^2x^2 = - a^2b^2, \quad \text{for asymptotes.}$$

Here

$$\frac{dy}{dx} = \frac{b^2x}{a^2y}; \quad \therefore \; x_0 = x - \frac{a^2y^2}{b^2x} = \frac{a^2}{x} = 0 \text{ for } x = \pm \infty.$$

$$y_0 = y - \frac{b^2x^2}{a^2y} = -\frac{b^2}{y} = 0 \text{ for } y = \pm \infty.$$

Hence the hyperbola has two asymptotes passing through the origin.

Also $\quad \dfrac{dy}{dx} = \dfrac{b^2x}{a^2y} = \pm \dfrac{b}{a} \, \dfrac{1}{\sqrt{1 - \dfrac{a^2}{x^2}}} = \pm \dfrac{b}{a} \text{ for } x = \infty.$

Hence the asymptotes make with the axis of x an angle whose tangent is $\pm \dfrac{b}{a}$; that is, they are the produced diagonals of the rectangle of the axes.

2. Examine the parabola $y^2 = 2px$ for asymptotes.

Here

$$\frac{dy}{dx} = \frac{p}{y}; \quad \therefore \quad x_0 = -\frac{y^2}{2p} = -\infty \text{ when } x \text{ or } y = \infty.$$

$$y_0 = \frac{y}{2} = \infty \text{ when } y = \infty \text{ or } x = \infty.$$

Hence the parabola has no asymptotes.

The ellipse and circle have no real asymptotes, since neither has an infinite branch.

3. Examine $y^3 = ax^2 + x^3$ for asymptotes.

When $x = \pm \infty$, $y = \pm \infty$; $\therefore$ the curve has two infinite branches, one in the first and one in the third angle.

$$\frac{dy}{dx} = \frac{2ax + 3x^2}{3y^2};$$

$$\therefore \quad x_0 = x - \frac{3y^3}{2ax + 3x^2} = -\frac{ax^2}{2ax + 3x^2} = -\frac{a}{3},$$

$$\text{when } x = \infty.$$

$$y_0 = y - \frac{2ax^2 + 3x^3}{3y^2} = \frac{3(y^3 - x^3) - 2ax^2}{3y^2}$$

$$= \frac{ax^2}{3(ax^2 + x^3)^{\frac{2}{3}}} = \frac{a}{3}, \text{ when } x = \infty.$$

Hence the asymptote cuts the axis of x at a distance $-\frac{a}{3}$, and that of y at a distance $\frac{a}{3}$ from the origin, and as it is therefore inclined at an angle of 45° to the axis of x, its equation is

$$y = x + \frac{a}{3}.$$

(See Gregory's Examples, p. 153.)

104. Asymptotes Determined by Expansion.—A very convenient method of examining for asymptotes consists in expanding the equation into a series in descending powers of x, by the binomial theorem, or by Maclaurin's theorem, or by division or some other method.

E X A M P L E S .

1. Examine $y^2 = \dfrac{x^3 + ax^2}{x - a}$ for asymptotes.

Then

$$y = \pm x \sqrt{\frac{x + a}{x - a}} = \pm x \left(1 + \frac{a}{x} + \frac{a^2}{2x^2} + \text{etc.}\right) \quad (1)$$

When $x = \infty$ (1) becomes

$$y = \pm (x + a). \tag{2}$$

We see that as x increases, the ordinate of (1) increases, and when x becomes infinitely great, the difference between the ordinate of (1) and that of (2) becomes infinitesimal; that is, the curve (1) is approaching the line (2) and becomes tangent to it when $x = \infty$; therefore, $y = \pm(x+a)$ are the equations of two asymptotes to the curve (1) at right angles to each other.

Another asymptote parallel to the axis of y is given by $x = a$.

2. Examine $x^3 - xy^2 + ay^2 = 0$ for asymptotes.

Here
$$y = \pm \sqrt{\frac{x^3}{x - a}}$$
$$= \pm x \left(1 + \frac{a}{2x} + \frac{3a^2}{8x^2} + \frac{5a^3}{16x^3} + \text{etc.}\right)$$

Hence, $y = \pm \left(x + \dfrac{a}{2}\right)$ are the equations of the two asymptotes.

By inspection, we find that $x = a$ is a third asymptote.

3. Examine $y^2 = x^2 \dfrac{x^2 - 1}{x^2 + 1}$ for asymptotes.

Here
$$y = \pm x \left(1 - \frac{1}{x^2} + \text{etc.} \right)$$

$\therefore$ $y = \pm x$ are the two asymptotes.

105. Asymptotes in Polar Co-ordinates.—When the curve is referred to polar co-ordinates, there will be an asymptote whenever the subtangent is finite for $r = \infty$. Its *position* also will be fixed, since it will be parallel to the radius-vector. Hence, to examine for asymptotes, we find what finite values of θ make $r = \infty$; if the corresponding polar subtangent, $r^2 \dfrac{d\theta}{dr}$, which in this case becomes the perpendicular on the tangent from the pole, is finite or zero, there will be an asymptote parallel to the radius-vector. If for $r = \infty$ the subtangent is ∞, there is no corresponding asymptote.

EXAMPLES.

1. Find the asymptotes of the hyperbola $a^2 y^2 - b^2 x^2 = -a^2 b^2$ by the polar method.

The polar equation is
$$a^2 \sin^2 \theta - b^2 \cos^2 \theta = -\frac{a^2 b^2}{r^2}. \tag{1}$$

When $r = \infty$, (1) becomes, $\tan^2 \theta = \dfrac{b^2}{a^2}$;

$$\therefore \quad \theta = \tan^{-1} \left(\pm \frac{b}{a} \right).$$

Therefore the asymptotes are inclined to the initial line at $\tan^{-1} \left(\pm \dfrac{b}{a} \right)$.

From (1) we get $\dfrac{d\theta}{dr} = \dfrac{a^2 b^2}{r^3 (a^2 + b^2) \sin \theta \cos \theta}$,

and $\qquad r^2 \dfrac{d\theta}{dr} = \pm \dfrac{ab (b^2 \cos^2 \theta - a^2 \sin^2 \theta)^{\frac{1}{2}}}{(a^2 + b^2) \sin \theta \cos \theta}$, (2) which is equal to 0 when $\theta = \tan^{-1}\left(\pm \dfrac{b}{a}\right)$; hence both asymptotes pass through the pole.

2. Find the asymptotes to the hyperbolic spiral $r\theta = a$. (See Anal. Geom., Art. 161.)

Here $\qquad r = \dfrac{a}{\theta}, \qquad \therefore r = \infty$, when $\theta = 0$.

$$\frac{d\theta}{dr} = -\frac{a}{r^2}, \quad \text{and} \quad r^2\frac{d\theta}{dr} = -a.$$

There is an asymptote therefore which passes at a distance a from the pole and is parallel to the initial line.

3. Find the asymptotes to the lituus $r\theta^{\frac{1}{2}} = a$. (Anal. Geom., Art. 162.)

Here $\qquad r = \dfrac{a}{\theta^{\frac{1}{2}}}, \qquad \therefore \quad r = \infty$, when $\theta = 0$.

$$\frac{d\theta}{dr} = -\frac{2a^2}{r^3}, \quad \text{and} \quad r^2\frac{d\theta}{dr} = -2a\theta^{\frac{1}{2}} = 0, \text{ when } \theta = 0.$$

Therefore the initial line is an asymptote to the lituus.

4. Find the asymptotes of the Conchoid of Nicomedes, $r = p \sec\theta + m$. (Anal. Geom., Art. 151.)

Here $r = \infty$ when $\theta = \dfrac{\pi}{2}$; and $r^2\dfrac{d\theta}{dr} = p$ when $\theta = \dfrac{\pi}{2}$.

Therefore the asymptote cuts the initial line at right angles, and at a distance p from the pole.

EXAMPLES.

1. Find the equation of the tangent to $3y^2 - 2x^2 - 10 = 0$, at $x = 4$. $\qquad\qquad$ *Ans.* $y = \pm .7127x \pm .8909$.

2. Find the equation of the tangent to $y^2 = \dfrac{x^3}{4-x}$, at $x = 2$.

$$y = 2x - 2 \text{ and } y = -2x + 2.$$

3. Find the equation of the tangent to the Naperian logarithmic curve.   *Ans.* $y = y'(x - x' + 1)$.

4. At what point on $y = x^3 - 3x^2 - 24x + 85$ is the tangent parallel to the axis of x ?

[Here we must put $\dfrac{dy'}{dx'} = 0$.   See Art. 56*a*.]

                                      At $(4, 5)$ and $(-2, 113)$.

5. At what point on $y^2 = 2x^3$ does the tangent make with the axis of x an angle whose tangent is 3, and where is it perpendicular?   At $(2, 4)$; at infinity.

6. At what angle does the line $y = \frac{1}{2}x + 1$ cut the curve $y^2 = 4x$?   [Find the point of intersection and the tangent to the curve at this point; then find the angle between this tangent and the given line.] $10° \ 14'$ and $33° \ 4'$.

7. At what angle does $y^2 = 10x$ cut $x^2 + y^2 = 144$?
                                            $71° \ 0' \ 58''$.

8. Show that the equation of a perpendicular from the focus of the common parabola upon the tangent is

$$y = -\frac{y'}{p}(x - \tfrac{1}{2}p).$$

9. Show that the length of the perpendicular from the focus of an hyperbola to the asymptote is equal to the semi-conjugate axis.

10. Find the abscissa of the point on the curve

$$y(x - 1)(x - 2) = x - 3$$

at which a tangent is parallel to the axis of x.

$$x = 3 \pm \sqrt{2}.$$

11. Find the abscissa of the point on the curve

$$y^3 = (x - a)^2(x - c)$$

at which a tangent is parallel to the axis of x.

$$x = \frac{2c + a}{3}.$$

12. Find the subtangent of the curve $y = \dfrac{x^{\frac{3}{2}}}{\sqrt{2a - x}}$.

$$Ans. \quad \frac{x(2a - x)}{3a - x}.$$

13. Find the subtangent of the curve $y^3 - 3axy + x^3 = 0$.

$$\frac{2axy - x^3}{ay - x^2}.$$

14. Find the subtangent of the curve $xy^2 = a^2(a - x)$.

$$- \frac{2(ax - x^2)}{a}$$

15. Find the subnormal of the curve $y^2 = 2a^2 \log x$.

$$\frac{a^2}{x}.$$

16. Find the subnormal of the curve $3ay^2 + a^3 = 2x^3$.

$$\frac{x^2}{a}.$$

17. Find the subtangent of the curve $y^2 = \dfrac{x^3}{a - x}$.

$$\frac{2x(a - x)}{3a - 2x}.$$

18. Find the subtangent of the curve

$$x^2y^2 = (a + x)^2(b^2 - x^2).$$

$$- \frac{x(a + x)(b^2 - x^2)}{x^3 + ab^2}.$$

19. Find the subnormal, subtangent, normal, and tangent in the Catenary

$$y = \frac{c}{2}\left(e^{\frac{x}{c}} + e^{-\frac{x}{c}}\right).$$

Subnormal $= \dfrac{c}{4}\left(e^{\frac{2x}{c}} - e^{-\frac{2x}{c}}\right)$; normal $= \dfrac{y^2}{c}$.

Subtangent $= \dfrac{cy}{(y^2 - c^2)^{\frac{1}{2}}}$; tangent $= \dfrac{y^2}{(y^2 - c^2)^{\frac{1}{2}}}$.

20. Find the perpendicular from the pole on the tangent in the lituus $r\theta^{\frac{1}{2}} = a$.

$$p = \frac{2a^2r}{(r^4 + 4a^4)^{\frac{1}{2}}}.$$

21. At what angle does $y^2 = 2ax$ cut $x^3 - 3axy + y^3 = 0$?

$$\cot^{-1} \sqrt[3]{4}.$$

22. Examine $y^2 = 2x + 3x^2$ for asymptotes.

$$y = \sqrt{3}\,x + \frac{1}{\sqrt{3}} \quad \text{is an asymptote.}$$

23. Examine $y^3 = 6x^2 + x^3$ for asymptotes.

$$y = x + 2 \quad \text{is an asymptote.}$$

24. Find the asymptotes of $y^2(x - 2a) = x^3 - a^3$.

$$x = 2a \,; \quad y = \pm (x + a).$$

25. Find the asymptotes of $y = \dfrac{x^3 - 3ax^2 + a^3}{x^2 - 3bx + 2b^2}$.

$$x = b \,; \quad x = 2b \,; \quad y = x - 3(a - b).$$

CHAPTER X.

DIRECTION OF CURVATURE, SINGLE POINTS, TRACING OF CURVES.

106. Concavity and Convexity.—The terms *concavity* and *convexity* are used in mathematics in their ordinary sense. A curve at a point is *concave* towards the axis of x when in passing the point it lies between the tangent and the axis. See Fig. 21. It is *convex* towards the axis of x when its tangent lies between it and the axis. See Fig. 22.

If a curve is concave downwards, as in Fig. 21, it is plain that as x *increases*, α *decreases*, and hence tan α *decreases;* that is, as x increases, $\dfrac{dy}{dx}$ (Art. 56a) *decreases;* and therefore the derivative of $\dfrac{dy}{dx}$ or $\dfrac{d^2y}{dx^2}$ is negative. (Art. 12.)

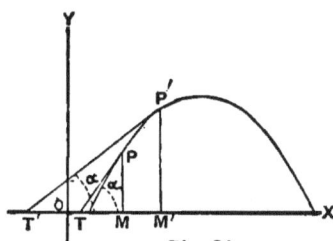

Fig. 21.

In the same way if the curve is *convex* downward, see Fig. 22, it is plain that as x *increases*, α *increases*, and therefore tan α increases; that is, as x increases, $\dfrac{dy}{dx}$ increases, and therefore the derivative of $\dfrac{dy}{dx}$ or $\dfrac{d^2y}{dx^2}$ is positive.

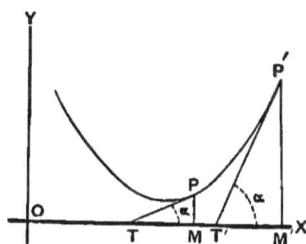

Fig. 22.

Hence the curve is concave or convex downward according as $\dfrac{d^2y}{dx^2}$ is $-$ or $+$.

This is also evident from Fig. 23, where $MM' = M'M''$ $= dx$; PP' is common to the two curves and the common tangent. $PR = P'R' = dx$; and $P'R$ $= P_2R'$. But $P''R' > P_2R' > P_1R'$. Now $P'R$ and $P''R'$ are consecutive values of dy in the upper curve, and $P'R$ and P_1R' are consecutive values of dy in the lower curve, and hence $P''R' - P'R = d(dy) = d^2y$ is $+$, and $P_1R' - P'R = d^2y$ is $-$; that is, d^2y is $-$ or $+$, according as the curve is concave or convex downwards.

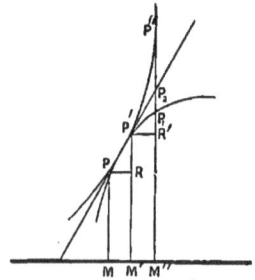
Fig. 23.

The sign of $\dfrac{d^2y}{dx^2}$ is of course the same as that of d^2y, since dx^2 is always positive.

We have supposed in the figures that the curve is *above* the axis of x. If it be below the axis of x, the rule just given still holds, as the student may show by a course of reasoning similar to the above.

If the curve is concave downwards, $\dfrac{d^2y}{dx^2}$ is $-$; if it be above the axis of x, y is $+$; therefore, $y\dfrac{d^2y}{dx^2}$ is $-$; if the curve be concave upwards, $\dfrac{d^2y}{dx^2}$ is $+$; if it be below the axis of x, y is $-$; therefore, $y\dfrac{d^2y}{dx^2}$ is $-$; that is, $y\dfrac{d^2y}{dx^2}$ is $-$ when the curve is concave towards the axis of x. In the same way it may be shown that $y\dfrac{d^2y}{dx^2}$ is $+$, when the curve is convex towards the axis of x.

107. Polar Co-ordinates.—A curve referred to polar co-ordinates is said to be concave or convex to the pole at any point, according as the curve in the neighborhood of that point does or does not lie on the same side of the tangent as the pole.

It is evident from Fig. 24, that when the curve is concave toward the pole O, as r increases p increases also, and
therefore $\dfrac{dr}{dp}$ is positive; and if the curve is convex toward
the pole, as r increases p decreases, and
therefore $\dfrac{dr}{dp}$ is negative. If therefore
the equation of the curve is given in
terms of r and θ, to find whether the
curve is concave or convex towards the
pole, we must transform the equation
into its equivalent between r and p, by

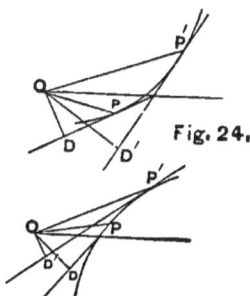

Fig. 24.

means of (10) in Art. 102, and then find $\dfrac{dr}{dp}$.

EXAMPLES.

1. Find the direction of curvature of

$$y = \frac{(x - 1)(x - 3)}{x - 2}.$$

Here $\qquad \dfrac{d^2y}{dx^2} = -\dfrac{2}{(x - 2)^3};$

that is, $\dfrac{d^2y}{dx^2}$ is positive or negative, according as $x <$ or > 2;
and therefore the curve is convex downward for all values
of $x < 2$, and concave downwards for all values of $x > 2$.

2. Find the direction of curvature of

$$y = b + c(x + a)^2 \quad \text{and} \quad y = a^2 \sqrt{x - a}.$$

Ans. The first is concave upward, the second is concave
towards the axis of x.

3. Find the direction of curvature of the lituus $r = \dfrac{a}{\theta^{\frac{1}{2}}}.$

9

Here $\quad \dfrac{dr}{d\theta} = -\dfrac{r}{2\theta} = -\dfrac{r^3}{2a^2}; \quad \therefore \quad \dfrac{dr^2}{d\theta^2} = \dfrac{r^6}{4a^4};$

which in (10) of Art. 102 gives,

$$p = \dfrac{2a^2 r}{(r^4 + 4a^4)^{\frac{1}{2}}}; \quad \therefore \quad \dfrac{dr}{dp} = \dfrac{(4a^4 + r^4)^{\frac{3}{2}}}{2a^2 (4a^4 - r^4)}.$$

Therefore the curve is concave toward the pole for values of $r < a\sqrt{2}$, and convex for $r > a\sqrt{2}$.

4. Find the direction of curvature of the logarithmic spiral $r = a^\theta$.

By Art. 102, Ex. 2,

$$p = \dfrac{mr}{\sqrt{m^2 + 1}}; \quad \therefore \quad \dfrac{dr}{dp} = \dfrac{\sqrt{m^2 + 1}}{m}.$$

which is always positive, and therefore the curve is always concave toward the pole.

SINGULAR POINTS.

108. Singular Points of a curve are those points which have some property peculiar to the curve itself, and not depending on the position of the co-ordinate axes. Such points are : 1st, Points of maxima and minima ordinates ; 2d, Points of inflexion ; 3d, Multiple Points ; 4th, Cusps ; 5th, Conjugate points ; 6th, Stop points ; 7th, Shooting points. We shall not consider any examples of the first kind of points, as they have already been illustrated in Chapter VIII, but will examine very briefly the others.

109. Points of Inflexion.—A *point of inflexion* is a point at which the curve is changing from convexity to concavity, or the reverse ; or it may be defined as the point at which the curve cuts the tangent at that point.

When the curve is convex downwards, $\dfrac{d^2y}{dx^2}$ is $+$ (Art.

106), and when concave downwards, $\frac{d^2y}{dx^2}$ is $-$; therefore, at a point of inflexion $\frac{d^2y}{dx^2}$ is changing from $+$ to $-$, or from $-$ to $+$, and hence it must be 0 or ∞. Hence to find a point of inflexion, we must equate $\frac{d^2y}{dx^2}$ to 0 or ∞, and find the values of x; then substitute for x a value a little greater, and one a little less than the *critical* value ; if $\frac{d^2y}{dx^2}$ changes sign, this is a point of inflexion.

EXAMPLES.

1. Examine $y = b + (x - a)^3$ for points of inflexion.

Here $\qquad\qquad \frac{d^2y}{dx^2} = 6\,(x - a) = 0$;

$$\therefore \quad x = a \quad \text{and hence} \quad y = b.$$

This is a critical point, *i. e.*, one to be examined ; for if there is a point of inflexion it is at $x = a$. For $x > a$, $\frac{d^2y}{dx^2}$ is $+$, and for $x < a$, $\frac{d^2y}{dx^2}$ is $-$. Hence there is a point of inflexion at (a, b).

2. Examine the witch of Agnesi,

$$x^2y = 4a^2\,(2a - y),$$

for points of inflexion.

There are points of inflexion at $\left(\pm \dfrac{2a}{\sqrt{3}},\ \dfrac{3a}{2} \right)$.

3. Examine $y = b + (x - a)^5$ for points of inflexion.
There is a point of inflexion at (a, b).

4. Examine the lituus for points of inflexion.

By Art. 107, Ex. 3, $\frac{dp}{dr}$ is changing sign from $+$ to $-$ when $r = a\sqrt{2}$, indicating that the lituus changes at this point from concavity to convexity, and hence there is a point of inflexion at $r = a\sqrt{2}$.

110. Multiple Points.—A *multiple point* is a point through which two or more branches of a curve pass. If two branches meet at the same point, it is called a *double* point; if three, a *triple* point; and so on. There are two kinds : 1st, a point where two or more branches intersect, their several tangents at that point being inclined to each other ; and 2d, a point where two or more branches are tangent to each other. The latter are sometimes called *points of osculation.*

As each branch of the curve has its tangent, there will be at a multiple point as many tangents, and therefore as many values of $\dfrac{dy}{dx}$ as there are branches which meet in this point. If these branches are all tangent, the values of $\dfrac{dy}{dx}$ will be equal. At a multiple point y will have but one value, while at points near it, it will have two or more values for each value of x. In functions of a simple form, such a point can generally be determined by inspection. After finding a value of x for which y has but one value, and on both sides of which it has two or more values, form $\dfrac{dy}{dx}$. If this has unequal values, the branches of the curve intersect at this point, and the point is of the first kind. If $\dfrac{dy}{dx}$ has but one value, the branches are tangent to each other at this point, and the point is of the second kind.

When the critical points are not readily found by inspection, we proceed as follows :

Let $$f(x, y) = 0 \qquad\qquad (1)$$

be the equation of the locus freed from radicals.   Then

$$\frac{dy}{dx} = -\frac{\dfrac{du}{dx}}{\dfrac{du}{dy}};$$

and as differentiation never introduces radicals when they do not exist in the expression differentiated, the value of $\frac{dy}{dx}$ cannot contain radicals, and therefore cannot have several values, unless by taking the form $\frac{0}{0}$.

Hence we have $\frac{dy}{dx} = \frac{0}{0}$ or $\frac{du}{dx} = 0$, and $\frac{du}{dy} = 0$, from which to determine critical values of x and y. If these values of x and y found from $\frac{du}{dx} = 0$ and $\frac{du}{dy} = 0$ are *real* and satisfy (1), they *may* belong to a multiple point. If y has but one value for the corresponding value of x, and on both sides of it y has two or more real values, this point is a multiple point. We then evaluate $\frac{dy}{dx} = \frac{0}{0}$, and if there are several real and unequal values of $\frac{dy}{dx}$, there will be as many intersecting branches of the curve passing through the point examined. (See Courtenay, p. 190.)

EXAMPLES.

1. Determine whether the curve $y = (x - a)\sqrt{x} + b$ has a multiple point.

Here y has two values for every positive value of $x >$ or $< a$. When $x = 0$ or a, y has but one value, b; hence there are two points to be examined. When $x < 0$, y is imaginary; hence the branches do not pass through the point $(0, b)$, and

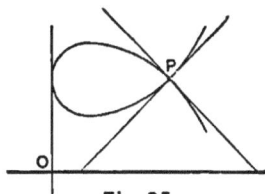

Fig. 25.

therefore it is *not* a multiple point. When $x >$ or $< a$, y has two real values, and therefore (a, b) is a double point.

$$\frac{dy}{dx} = \pm \frac{3x - a}{2\sqrt{x}} = \pm \sqrt{a}, \quad \text{when} \quad x = a.$$

Therefore the point is of the first kind, and the tangents to the curve at the point make with the axis of x angles whose tangents are $+\sqrt{a}$ and $-\sqrt{a}$.

2. Examine $x^4 + 2ax^2y - ay^3 = 0$ for multiple points.

We proceed according to the second method, as all the critical points in this example are not easily found by inspection.

$$\frac{du}{dx} = 4x(x^2 + ay) = 0; \tag{1}$$

$$\frac{du}{dy} = a(2x^2 - 3y^2) = 0; \tag{2}$$

$$\therefore \frac{dy}{dx} = \frac{4x^3 + 4axy}{3ay^2 - 2ax^2}. \tag{3}$$

Solving (1) and (2) for x and y, we find

$$\begin{pmatrix} x = 0 \\ y = 0 \end{pmatrix}; \quad \begin{pmatrix} x = \tfrac{1}{3}a\sqrt{6} \\ y = -\tfrac{2}{3}a \end{pmatrix}; \quad \begin{pmatrix} x = -\tfrac{1}{3}a\sqrt{6} \\ y = -\tfrac{2}{3}a \end{pmatrix}.$$

Only the first pair will satisfy the equation of the curve, and therefore the origin is the only point to be examined.

Evaluating $\frac{dy}{dx}$ in (3) for $x = 0$ and $y = 0$, and representing $\frac{dy}{dx}$ by p, and $\frac{dp}{dx}$ by p', for shortness, we have

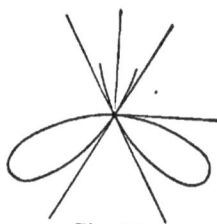

Fig. 26.

$$\frac{dy}{dx} = p = \frac{4x^3 + 4axy}{3ay^2 - 2ax^2} = \frac{0}{0}, \qquad \text{when} \begin{pmatrix} x = 0 \\ y = 0 \end{pmatrix}$$

$$= \frac{12x^2 + 4ay + 4axp}{6ayp - 4ax} = \frac{0}{0}, \quad \text{when} \begin{pmatrix} x = 0 \\ y = 0 \end{pmatrix}$$

$$= \frac{24x + 8ap + 4axp'}{6ap^2 + 6ayp' - 4a}$$

$$= \frac{8ap}{6ap^2 - 4a}, \qquad \text{when} \begin{pmatrix} x = 0 \\ y = 0 \end{pmatrix} .$$

$$\therefore \ p(6ap^2 - 4a) = 8ap;$$

$$\therefore \ p = \frac{dy}{dx} = 0, \ +\sqrt{2}, \text{ or } -\sqrt{2}.$$

Hence the origin is a triple point, the branches being inclined to the axis of x at the angles 0, $\tan^{-1}(\sqrt{2})$, and $\tan^{-1}(-\sqrt{2})$, respectively, as in the figure. (See Courtenay's Calculus, p. 191; or Young's Calculus, p. 151.)

3. Examine $y^2 - x^2(1 - x^2) = 0$ for multiple points.

Ans. There is a double point at the origin, the branches being inclined to the axis of x at angles of 45° and 135° respectively.

4. Show that $ay^3 - x^3y - ax^3 = 0$ has no multiple points.

111. Cusps.—A cusp is a point of a curve at which two branches meet a common tangent, and stop at that point. If the two branches lie on *opposite* sides of the common tangent, the cusp is said to be of the *first* species; if on the *same* side, the cusp is said to be of the *second* species.

Since a cusp is really a multiple point of the second kind, the only difference being that the branches stop at the point, instead of running through it, we examine for cusps as we do for multiple points; and to distin-

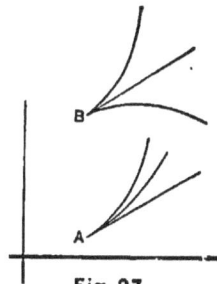

Fig. 27.

guish a cusp from an ordinary multiple point, we trace the curve in the vicinity of the point and see if y is *real* on one side and *imaginary* on the other. To ascertain the *kind* of cusp, we compare the ordinates of the curve, near the point, with the corresponding ordinate of the tangent; or ascertain the direction of curvature by means of the second derivative.

In the particular case in which the common tangent to the two branches is perpendicular to the axis of x, it is best to consider y as the independent variable, and find the values of $\dfrac{dx}{dy}$, etc.

E X A M P L E S.

1. Examine $y = x^2 \pm x^{\frac{5}{2}}$ for cusps.

We see that when $x = 0$, y has but one value, 0; when $x < 0$, y is imaginary; and when $x > 0$, y has two real values; hence, (0, 0) is the point to be examined.

$\dfrac{dy}{dx} = 2x \pm \tfrac{5}{2}x^{\frac{3}{2}} = 0$, when $x = 0$; hence the axis of x is

a common tangent to both branches, and there is a cusp at the origin.

$\dfrac{d^2y}{dx^2} = 2 \pm \tfrac{15}{4}x^{\frac{1}{2}}$ is positive when $x = 0$;

hence the cusp is of the second kind.

The value of $\dfrac{d^2y}{dx^2}$ shows that the upper

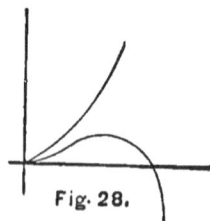

Fig. 28.

branch is always concave upward, while the lower branch has a point of inflexion, when $x = \frac{64}{225}$; from the origin to the point of inflexion this branch is concave upward, after which it is concave downward.

The value of $\dfrac{dy}{dx}$ shows that the branch is horizontal when $x = \frac{16}{25}$. From $y = x^2 - x^{\frac{5}{2}}$, we find that the lower branch cuts the axis of x at $x = 1$. The shape of the curve is given in Fig. 28.

2. Examine $(y - b)^2 = (x - a)^3$ for cusps.

Ans. The point (a, b) is a cusp of the first kind.

3. Examine $cy^2 = x^3$ for cusps.

The origin is a cusp of the first kind.

112. Conjugate Points.—A *conjugate point* is an *isolated* point whose co-ordinates satisfy the equation of the curve, while the point itself is entirely detached from every other point of the curve.

For example, in the equation $y = (a + x)\sqrt{x}$, if x is negative, y is, in general, imaginary but for the particular value $x = -a$, $y = 0$. Hence, P is a point in the curve, and it is entirely detached from all others. When $x = 0$, $y = 0$, which shows that the curve passes through the origin. For positive values of x, there will be two real values of y, numerically equal, with opposite signs. Hence, the curve has two infinite branches on the right, which are symmetrical with respect to the axis of x.

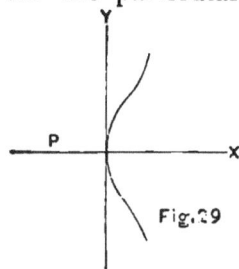

Fig. 29

If the first derivative becomes imaginary for any real values of x and y, the corresponding point will be conjugate, as the curve will then have no direction. It does not follow, however, that at a conjugate point $\dfrac{dy}{dx}$ will be imaginary; for, if the curve $y = f(x)$ have a conjugate point at (x, y), from the definition of a conjugate point, we shall have

$f(x \pm h) =$ an *imaginary* quantity. But

$$f(x \pm h) = y \pm \frac{dy}{dx}\frac{h}{1} + \frac{d^2y}{dx^2}\frac{h^2}{2} \pm \frac{d^3y}{dx^3}\frac{h^3}{6} + \text{etc.},$$

therefore. if either one of the derivatives is imaginary, the first member is imaginary; hence, *at a conjugate point some one or more of the derivatives is imaginary.*

Since at a conjugate point *some* of the derivatives are imaginary, let the nth derivative be the *first* that is imaginary. Suppose the equation of the curve to be freed from radicals, and denoted by $u = f(x, y) = 0$. Take the nth derived equation (Art. 88, Sch.); we have

$$\frac{du}{dy} \frac{d^n y}{dx^n} \cdots + \frac{d^n u}{dx^n} = 0,$$

where the terms omitted contain derivatives of u with respect to x and y, and derivatives of y with respect to x, of lower orders than the nth. If, then, $\frac{du}{dy}$ be not 0, the value of $\frac{d^n y}{dx^n}$ obtained from the derived equation will be *real*, which is contrary to the hypothesis; hence, $\frac{du}{dy} = 0$ is a necessary condition for the existence of a conjugate point. But

$$\frac{du}{dx} + \frac{du}{dy} \frac{dy}{dx} = 0 ;$$

therefore, since $\frac{du}{dy} = 0$, we must have $\frac{du}{dx} = 0$. Hence, at a conjugate point we must have $\frac{du}{dx} = 0$, and $\frac{du}{dy} = 0$.

REM.—Owing to the labor of finding the higher derivatives, it is usually better, if the first derivative does not become imaginary, to substitute successively $a + h$ and $a - h$ for x, in the equation of the curve, where a is the value of x to be tested, and h is very small. If both values of y prove imaginary, the point is a conjugate point.

EXAMPLES.

1. Examine $ay^2 - x^3 + 4ax^2 - 5a^2 x + 2a^3 = 0$ for conjugate points.

$$\frac{du}{dx} = -3x^2 + 8ax - 5a^2 = 0. \tag{1}$$

$$\frac{du}{dy} = 2ay = 0. \tag{2}$$

Solving (1) and (2), we get

$$\begin{pmatrix} x = a \\ y = 0 \end{pmatrix} \quad \text{and} \quad \begin{pmatrix} x = \frac{5}{3}a \\ y = 0 \end{pmatrix}.$$

Only the first pair of values satisfies the equation of the curve, and hence the point $(a, 0)$ is to be examined.

$$\frac{dy}{dx} = p = \frac{3x^2 - 8ax + 5a^2}{2ay} = \frac{6x - 8a}{2ap}$$

$$= -\frac{1}{p}, \quad \text{when} \quad \begin{pmatrix} x = a \\ y = 0 \end{pmatrix};$$

therefore, $\quad p^2 = -1; \quad \therefore \quad p = \pm\sqrt{-1} = \frac{dy}{dx}.$

This result being imaginary, the point $(a, 0)$ is a conjugate point.

2. Show that $x^4 - ax^2y - axy^2 + a^2y^2 = 0$ has a conjugate point at the origin.

3. Examine $(c^2y - x^3)^2 = (x - a)^5 (x - b)^6$ for conjugate points, in which $a > b$.

The point $\left(b, \dfrac{b^3}{c^2}\right)$ is a conjugate point.

The first and second derivatives are real in this example; hence the better method of solving it will be to proceed according to the *Remark* above given

113. Shooting Points are points at which two or more branches of a curve terminate, without having a common tangent.

Stop Points are points in which a single branch of a curve suddenly stops.

These two classes of singular points but rarely occur, and never in curves whose equations are of an *algebraic* form.

EXAMPLES.

1. Examine $y = \dfrac{x}{1 + e^{\frac{1}{x}}}$ for shooting points.

Here $\qquad \dfrac{dy}{dx} = \dfrac{1}{1 + e^{\frac{1}{x}}} + \dfrac{e^{\frac{1}{x}}}{x\left(1 + e^{\frac{1}{x}}\right)^2}.$

If x is $+$ and small, y is $+$; if x is $-$ and small, y is $-$. When x is $+$ and approaches 0, $y = 0$, and $\dfrac{dy}{dx} = 0$; when x is $-$ and approaches 0, $y = 0$, and $\dfrac{dy}{dx} = 1$.

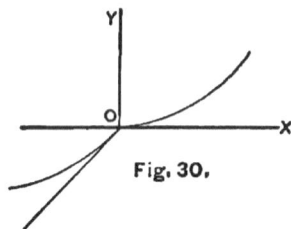

Fig. 30.

Hence, at the origin there is a shooting point, one branch having the axis of x as its tangent, and the other inclined to the axis of x at an angle of $45°$. (See Serret's Calcul Différentiel et Intégral, p. 267.)

2. Examine $y = x \log x$.

When x is $+$, y has one real value; when $x = 0$, $y = 0$; when $x < 0$, y is imaginary; hence there is a stop point at the origin.

3. Examine $y = x \tan^{-1} \dfrac{1}{x}$.

$$\dfrac{dy}{dx} = \tan^{-1} \dfrac{1}{x} - \dfrac{x}{x^2 + 1}.$$

If $\qquad x = + 0$ or $- 0, \quad y = 0;$

$$\dfrac{dy}{dx} = \dfrac{\pi}{2} \quad \text{or} \quad -\dfrac{\pi}{2}.$$

Hence the origin is a shooting point, the tangent being inclined to the axis of x at angles $\tan^{-1}(1.5708)$ and $\tan^{-1}(-1.5708)$.

4. Show that $y = e^{-\frac{1}{x}}$ has a stop point at the origin.

114. Tracing Curves.— We shall conclude this chapter by a brief statement of the mode of tracing curves by means of their equations.

The usual method of tracing curves consists in assigning a series of different values to one of the variables, and calculating the corresponding series of values of the other, thus determining a definite number of points on the curve. By drawing a curve or curves through these points, we are enabled to form a tolerably accurate idea of the shape of the curve. (See Anal. Geometry, Art. 21.)

In the present Article we shall indicate briefly the manner of finding the *general form* of the curve, especially at such points as present any *peculiarity,* so that the mind can conceive the locus, or that it may be sketched without going through the details of substituting a series of values, as was referred to above.

To trace a curve from its equation, the following steps will be found useful:

(1.) If it be possible, solve it with respect to one of its variables, y for example, and observe whether the curve is symmetrical with respect to either axis.

(2.) Find the points in which the curve cuts the axes, also the limits and infinite branches.

(3.) Find the positions of the asymptotes, if any, and at which side of an asymptote the corresponding branches lie.

(4.) Find the value of the first derivative, and thence deduce the maximum and minimum points of the curve, the angles at which the curve cuts the axes, and the multiple points, if any.

(5.) Find the value of the second derivative, and thence the direction of the curvature of the different branches, and the points of inflexion, if any.

(6.) Determine the existence and nature of the singular points by the usual rules.

EXAMPLES.

1. Trace the curve $y = \dfrac{x}{1 + x^2}$.

When $x = 0$, $y = 0$; $\therefore$ the curve passes through the origin.

For all positive values of x, y is positive; and when $x = \infty$, $y = 0$. For negative values of x, y is negative, and when $x = -\infty$, $y = 0$; hence the curve has two infinite branches, one in the first angle and one in the third, and the axis of x is an asymptote to both branches.

$$\frac{dy}{dx} = \frac{1 - x^2}{(1 + x^2)^2}; \qquad \frac{d^2y}{dx^2} = \frac{2x\,(x^2 - 3)}{(1 + x^2)^3}.$$

When $x = \pm 1$, $\dfrac{dy}{dx} = 0$; $\therefore$ there is a maximum ordinate at $x = +1$, and a minimum ordinate at $x = -1$, at which points $y = \frac{1}{2}$ and $-\frac{1}{2}$ respectively.

When $x = 0$, $\dfrac{dy}{dx} = 1$; $\therefore$ the curve cuts the axis of x at an angle of $45°$.

Putting the second derivative equal to 0, we get $x = 0$ or $\pm \sqrt{3}$. Therefore, there are points of inflexion at $(0, 0)$ and at $x = +\sqrt{3}$ and $-\sqrt{3}$, for which we have $y = \frac{1}{4}\sqrt{3}$, $-\frac{1}{4}\sqrt{3}$. From $x = -\sqrt{3}$

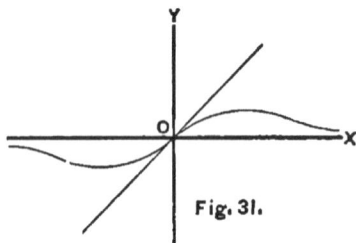

Fig. 31.

to $x = +\sqrt{3}$, the curve is concave towards the axis of x, and beyond them it is convex.

From this investigation the curve is readily constructed, and has the form given in the figure.

2. Trace the curve $y^3 = 2ax^2 - x^3$.

$$y = x^{\frac{2}{3}} (2a - x)^{\frac{1}{3}};$$

$$\frac{dy}{dx} = \frac{4ax - 3x^2}{3y^2}; \qquad \frac{d^2y}{dx^2} = \frac{-8a^2}{9x^{\frac{4}{3}}(2a - x)^{\frac{5}{3}}}.$$

When $x = 0$ or $2a$, $y = 0$; $\therefore$ the curve cuts the axis of x at the origin and at $x = 2a$.

To find the equation of the asymptote, we have

$$y = -x\left(1 - \frac{2a}{x}\right)^{\frac{1}{3}} = -x\left(1 - \frac{2a}{3x} - \ldots\ldots\right);$$

therefore, $y = -x + \frac{2}{3}a$ is the equation of the asymptote, and as the next term of the expression is positive, the curve lies above the asymptote.

Evaluating the first derivative for $x = 0$, $y = 0$, we have

$$\frac{dy}{dx} = \frac{4ax - 3x^2}{3y^2} = \frac{4a - 6x}{6y\,\dfrac{dy}{dx}};$$

$$\therefore \left(\frac{dy}{dx}\right)^2 = \frac{4a}{6y} = \infty, \quad \text{when} \quad x = y = 0;$$

$$\therefore \quad \frac{dy}{dx} = \sqrt{\frac{2a}{3y}} = \pm\,\infty, \quad \text{when} \quad y = 0.$$

Hence, at the origin there are two branches of the curve tangent to the axis of y; and the value of $\frac{dy}{dx}$ shows that if y be negative as it approaches 0, $\frac{dy}{dx}$ will be imaginary; and hence the origin is a cusp of the first species.

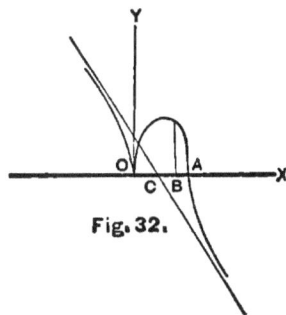

Fig. 32.

When $x = \frac{4}{3}a$, $\frac{dy}{dx} = 0$; $\therefore$ there is a maximum ordinate at $x = \frac{4}{3}a$.

When $x = 2a$, $\frac{dy}{dx} = -\frac{4a^2}{0} = -\infty$; $\therefore$ the curve cuts the axis of x, at the point $x = 2a$, at right angles.

Putting the second derivative equal to ∞, we get $x = 2a$. When $x < 2a$, the second derivative is $-$, and when $> 2a$ it is $+$; hence the left branch is everywhere concave downward, and the right branch is concave downward from $x = 0$ to $x = 2a$. At this last point it cuts the axis of x at right angles, and changes its curvature to concave upward; the two branches touch the asymptote at $x = +\infty$ and $-\infty$, respectively, *i. e.*, they have a common asymptote.

In the figure, $OA = 2a$, $OB = \frac{4}{3}a$, $OC = \frac{2}{3}a$.

3. Trace the curve $y = x\left(\dfrac{x - 2a}{x - a}\right)$.

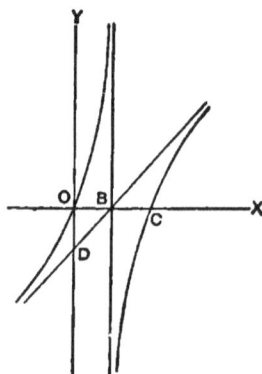

Fig. 32. a

$$
\begin{array}{lll}
\text{Let} & x = 0; & \therefore\ y = 0. \\
& x < a; & \therefore\ y \text{ is positive.} \\
& x = a; & y = \infty. \\
& x > a < 2a; & y \text{ is negative.} \\
& x = 2a; & y = 0. \\
& x > 2a; & y \text{ is positive.} \\
& x = \infty; & y = \infty.
\end{array}
$$

When x is $-$, y is always negative.

To find the asymptote, we have

$$
y = x\left\{\frac{1 - \dfrac{2a}{x}}{1 - \dfrac{a}{x}}\right\} = x\left(1 - \frac{2a}{x}\right)\left(1 + \frac{a}{x} + \text{etc.}\right)
$$

$$
= x\left(1 - \frac{a}{x} - \frac{2a^2}{x^2} - \text{etc.}\right) = x - a - \frac{a^2}{x} - \text{etc.}
$$

$\therefore\ y = x - a$ is the equation of the asymptote.

Hence, take $OB = a = OD$, and the line BD produced is the asymptote; also take $OC = 2a$. Then, since $y = 0$, both when $x = 0$ and $x = 2a$, the curve cuts the axis of x

at O and C. Between O and B, the curve is above the axis; at B the ordinate is infinite; from B to C, the curve is below; from C to infinity, it is above OX. Also, if x is negative, y is negative; therefore the branch on the left of O is entirely below the axis.

Also, $$\frac{dy}{dx} = \frac{x^2 - 2ax + 2a^2}{(x-a)^2}.$$

Let $x = a$; $\therefore \frac{dy}{dx} = \infty$; and the infinite ordinate at the distance a to the right of the origin is an asymptote.

If $x = 0$, $\frac{dy}{dx} = 2$; if $x = 2a$, $\frac{dy}{dx} = 2$; *i.e.*, the curve cuts the axis of x at the origin and the distance $2a$ to the right, at the same angle, $\tan^{-1}(2)$.

If $x^2 - 2ax + 2a^2$ or $(x - a)^2 + a^2 = 0$, x is impossible; hence there is no maximum or minimum ordinate.

Again, $$\frac{d^2y}{dx^2} = \frac{2(x-a)^2 - 2[(x-a)^2 + a^2]}{(x-a)^3}$$
$$= \frac{-2a^2}{(x-a)^3};$$

$\therefore \frac{d^2y}{dx^2}$ is $+$ if $x < a$, and is $-$ if $x > a$.

But $x < a$, y is $+$; and $x > a < 2a$, y is $-$; and $x > 2a$, y is $+$; therefore, from O to B, and B to C, the curve is convex, and from C to infinity, it is concave to the axis of x.

If x be $-$, $\frac{d^2y}{dx^2} = \frac{2a^2}{(x+a)^3}$ is $+$, but y is $-$; therefore the branch from the origin to the left is concave to the axis of x. (See Hall's Calculus, pp. 182, 183.)

4. Trace the curve $y^2 = a^2x^3$.

The curve passes through the origin; is symmetrical with respect to the axis of x; has a cusp of the first kind at the origin; both branches are tangent to the axis of x; are convex towards it; are infinite in the direction of positive abscissas, and the curve has no asymptote or point of inflexion.

115. On Tracing Polar Curves.—Write the equation, if possible, in the form $r = f(\theta)$; give to θ such values as to make r easily found, as for example, 0, $\frac{1}{2}\pi$, π, $\frac{3}{2}\pi$, etc.

Putting $\frac{dr}{d\theta} = 0$, we find the values of θ for which r is a maximum or minimum, *i. e.*, where the radius vector is perpendicular to the curve.

Find the asymptotes and direction of curvature, and points of inflexion. After this there will generally be but little difficulty in finding the form of the curve.

EXAMPLES.

1. Trace the lituus $r = \dfrac{a}{\theta^{\frac{1}{2}}}$.

When $\theta = 0$, $r = \infty$; when $\theta = 1 \; (= 57°.3)$,[*] $r = \pm a$; when $\theta = 2 \; (= 114°.6)$, $r = \pm .7a$; when $\theta = 3$, $r = \pm .58a$, etc.; when $\theta = \infty$, $r = 0$.

$\dfrac{dr}{d\theta} = -\dfrac{r^3}{2a^2}$, and when $\dfrac{dr}{d\theta} = 0$, $r = 0$; hence, r and θ are decreasing functions of each other throughout all their values;[†] and the curve starts from infinity, when $\theta = 0$, and makes an infinite number of revolutions around the pole, cutting every radius-vector at an oblique angle, and reaching the pole only when $\theta = \infty$.

The subtangent $r^2 \dfrac{d\theta}{dr} = -\dfrac{2a^2}{r} = 0$, when $r = \infty$; hence the initial line is an asymptote (Art. 105).

---

* The unit angle is that whose arc is equal to the radius, and is about 57°.29578

† If we consider alone the branch generated by the positive radius-vector.

$$\frac{dp}{dr} = \frac{2a^2(4a^4 - r^4)}{(4a^4 + r^4)^{\frac{3}{2}}}$$ (see Art. 107, Ex. 3) ; hence there is a point of inflexion at $r = a\sqrt{2}$; from $r = 0$ to $r = a\sqrt{2}$ the curve is concave toward the pole, and from $r = a\sqrt{2}$ to $r = \infty$ it is convex.

2. Trace the curve $r = a \sin 3\theta$.

$r = 0$, when $\theta = 0$, 60°, 120°, 180°, 240°, and 300°. When $\theta = 2\pi$, or upwards, the same series of values recur.

If $\theta = 30°$, 90°, 150°, 210°, 270°, and 330°, $r = a, -a, a, -a, a$, and $-a$, successively.

$\frac{dr}{d\theta} = 3a \cos 3\theta$, showing that r begins at 0 when $\theta = 0$, increases till it is a when $\theta = 30°$, diminishes to 0 as θ passes from 30° to 60°, continues to diminish and becomes $-a$ when θ becomes 90°, and so on.

$\frac{dp}{dr} = \frac{18a^2 r - 8r^3}{(9a^2 - 8r^2)^{\frac{3}{2}}}$, which shows that the curve is always concave towards the pole. There is no asymptote, as r is never ∞.

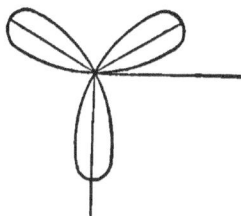

Fig. 33.

Hence the curve consists of three equal loops arranged symmetrically around the pole, each loop being traced twice in each revolution of r. A little consideration will show that the form of the curve is that given in the figure. (See Gregory's Examples, p. 185 ; also Price's Calculus, Vol. I, p. 427.)

3. Trace the Chordel $r = a \operatorname{cosec}\left(\frac{\theta}{2n}\right)$.

If $\theta = 0$, $n\pi$, $2n\pi$, $3n\pi$, $4n\pi$, $5n\pi$, etc., successively, $r = \infty$, a, ∞, $-a$, $-\infty$, a, etc.

$$\frac{dr}{d\theta} = -\frac{a}{2n} \operatorname{cosec} \frac{\theta}{2n} \cot \frac{\theta}{2n} = \frac{a}{2n} \operatorname{cosec}^2 \frac{\theta}{2n}\left(-\cos\frac{\theta}{2n}\right),$$

which is negative from $\theta = 0$ to $\theta = n\pi$, positive from

$\theta = n\pi$ to $\theta = 3n\pi$, negative from $\theta = 3n\pi$ to $\theta = 5n\pi$, etc. Hence we see that r begins at ∞ when $\theta = 0$; diminishes till it becomes a when $\theta = n\pi$; increases as θ passes from $n\pi$ to $2n\pi$; becomes ∞ when $\theta = 2n\pi$; when θ passes $2n\pi$, r changes from $+\infty$ to $-\infty$; when θ increases from $2n\pi$ to $3n\pi$, r increases from $-\infty$ to $-a$; when θ increases from $3n\pi$ to $4n\pi$, r diminishes from $-a$ to $-\infty$; when θ passes $4n\pi$, r changes from $-\infty$ to $+\infty$. When θ increases beyond 4π, the same values of r recur, showing that the curve is complete.

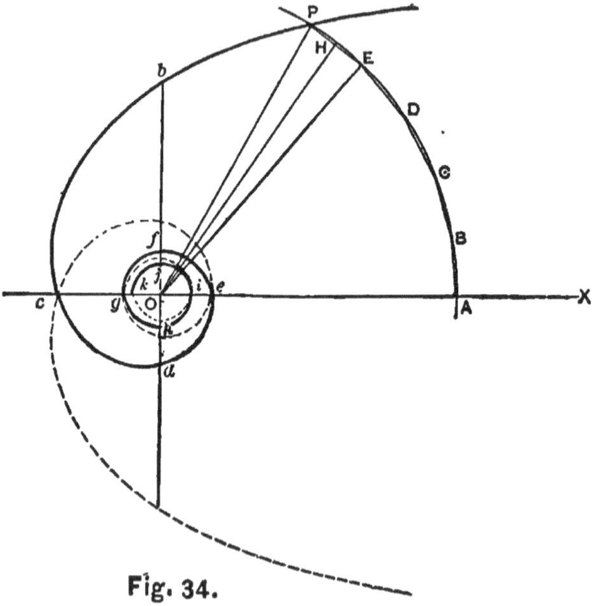

Fig. 34.

$$\frac{dr}{d\theta} = \frac{a}{2n} \operatorname{cosec}^2 \frac{\theta}{2n} \left(-\cos \frac{\theta}{2n} \right) = 0 \quad \text{gives} \quad \theta = n\pi,\ 3n\pi,$$

$5n\pi$, etc.; *i.e.*, the radius-vector is a minimum at $\theta = n\pi$, $3n\pi$, $5n\pi$, etc.

$$\text{The subtangent} = r^2 \frac{d\theta}{dr} = -\frac{2na}{\cos \dfrac{\theta}{2n}}$$

$$= -2na \quad \text{when} \quad \theta = 0;$$

$$\text{and} \quad = +2na \quad \text{when} \quad \theta = 2n\pi;$$

therefore the curve has two asymptotes parallel to the initial line, at the distances $\pm 2na$ from the pole.

$$p = \frac{r^2}{\left(\dfrac{dr^2}{d\theta^2} + r^2\right)^{\frac{1}{2}}} = \frac{2anr}{(4a^2n^2 - a^2 + r^2)^{\frac{1}{2}}};$$

$$\therefore \frac{dp}{dr} = \frac{2a^3n\,(4n^2 - 1)}{[a^2\,(4n^2 - 1) + r^2]^{\frac{3}{2}}};$$

$\therefore$ the curve is always concave towards the pole.

Thus it appears that while θ is increasing from 0 to $2n\pi$, the positive end of the radius-vector traces the branch drawn in Fig. 34; and while θ increases from $2n\pi$ to $4n\pi$, the negative end of the radius-vector traces a second branch (not drawn), the two branches being symmetrical with respect to the vertical line through the pole O.

EXAMPLES.

1. Find the direction of curvature of the Witch of Agnesi

$$x^2y = 4a^2\,(2a - y).$$

The curve is concave downward for all values of y between $2a$ and $\frac{4}{3}a$, and convex for all values of y between $\frac{4}{3}a$ and 0.

2. Find the direction of curvature of $y = b + (x - a)^3$.

Convex towards the axis of x from $x > a$ to $x = \infty$; and from $x = a - b^{\frac{1}{3}}$ to $x = -\infty$; concave towards the axis of x from $x < a$ to $x = a - b^{\frac{1}{3}}$.

3. Examine $y = (a - x)^{\frac{5}{3}} + ax$ for points of inflexion.

There is a point of inflexion at $x = a$.

4. Examine $y = x + 36x^2 - 2x^3 - x^4$ for points of inflexion. Points of inflexion at $x = 2,\ x = -3$.

5. Find the co-ordinates of the point of inflexion of the curve

$$y = \frac{x^2 (a^2 - x^2)}{a^3}.$$

$$x = \pm \frac{a}{\sqrt{6}}; \quad y = \tfrac{5}{36}a.$$

6. Examine $r = \dfrac{a\theta^2}{\theta^2 - 1}$ for points of inflexion.

Here $\dfrac{dr^2}{d\theta^2} = \dfrac{4r (r - a)^3}{a^2};$

$$\therefore \ p = \frac{ar^2}{(4r^4 - 12ar^3 + 13a^2r^2 - 4a^3r)^{\frac{1}{2}}}; \quad \therefore \text{ etc.}$$

There are points of inflexion at $r = \tfrac{3}{2}a$ and $r = \tfrac{2}{3}a$.

7. Examine $y^2 = (x - 1)^2 x$ for multiple points.
        There is a multiple point at $x = 1$.

8. Examine $y^2 = \dfrac{x^2 (a^2 - x^2)}{a^2 + x^2}$ for multiple points.

There is a multiple point at the origin, and the curve is composed of two loops, one on the right and the other on the left of the origin, the tangents bisecting the angles between the axes of co-ordinates.

9. Show that $x^4 + x^2y^2 - 6ax^2y + a^2y^2 = 0$ has a multiple point of the second kind at the origin.

10. Show that $y = a + x + bx^2 \pm cx^{\frac{5}{3}}$ has a cusp of the second kind at the point $(0, a)$, and that the equation of the tangent at the cusp is $y = x + a$.

11. Show that $y^3 = ax^2 + x^3$ has a cusp of the first kind at the origin.

12. Show that $ay^2 - x^3 + bx^2 = 0$ has a conjugate point at the origin, and a point of inflexion at $x = \dfrac{4b}{3}.$

13. Trace the curve $y^3 = a^3 - x^3$.

The curve cuts the axes at $(a, 0)$ and $(0, a)$.

It has an asymptote which passes through the origin.

The points where the curve cuts the axes are points of inflexion.

14. Trace the curve $y = ax^2 \pm \sqrt{bx} \sin x$.

For every positive value of x there are two values of y, and therefore two points, except when $\sin x = 0$, in which case the two points reduce to one. These points form a series of loops like the links of a chain, and have for a

Fig. 35.

diametral curve the parabola $y = ax^2$, from which, when x is positive, the loops recede and approach, meeting the parabola whenever $x = 0$ or π, or any multiple of π. But when x is negative, y is imaginary except when $\sin x = 0$, in which case $y = ax^2$, so that on the negative side there is an infinite number of conjugate points, each one on the parabola opposite a double point of the curve. (See De Morgan's Cal., p. 382; also, Price's Cal., Vol. I, p. 396.)

CHAPTER XI.

116. Curvature.—*The curvature of a curve is its rate of deviation from a tangent*, and is measured by the external angle between the tangents at the extremities of an indefinitely small arc ; that is, by the angle between any infinitesimal element and the prolongation of the preceding element. This angle is called *the angle of contingence* of the arc. Of two curves, that which departs most rapidly from its tangent has the greatest curvature. In the same or in equal circles, the curvature is the same at every point ; but in unequal circles, the greater the radius the less the curvature ; that is, in different circles the curvature varies inversely as their radii.

Whatever be the curvature at any point of a plane curve, it is clear that a circle may be found which has the same curvature as the curve at the given point, and this circle can be placed tangent to the curve at that point, with its radius coinciding in direction with the normal to the curve at the same point. This circle is called the *osculating* circle, or *the circle of curvature* of that point of the curve. *The radius of curvature* is the radius of the osculating circle. *The centre of curvature* is the centre of the osculating circle.

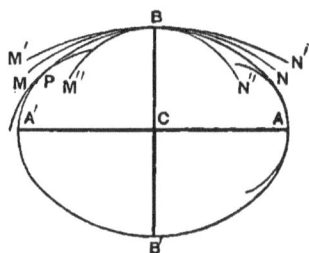

Fig. 36.

For example, let ABA′B′ be an ellipse. If different circles be passed through B with their centres on BB′, it is clear that they will coincide with the ellipse in very different degrees, some falling within and others without. Now, that one which coincides with the ellipse the most nearly of all of them, as in this case MN, is the osculating circle of the ellipse at B, and is entirely exterior to the ellipse. The osculating circle at A or A', is entirely within the ellipse ; while at any other point, as P, it cuts the ellipse, as will be shown hereafter.

117. Order of Contact of Curves. — Let $y = f(x)$ and $y = \phi(x)$ be the equations of the two curves, AB and ab, referred to the axes OX and OY. Giving to x an infinitesimal increment h, and expanding by Taylor's theorem, we have,

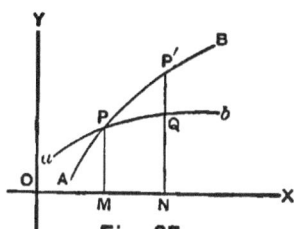

Fig. 37.

$$y_1 = f(x + h) = f(x) + f'(x)\,h + f''(x)\frac{h^2}{2}$$
$$+ f'''(x)\frac{h^3}{2\cdot 3} + \text{etc.} \qquad (1)$$

$$y_2 = \phi(x + h) = \phi(x) + \phi'(x)\,h + \phi''(x)\frac{h^2}{2}$$
$$+ \phi'''(x)\frac{h^3}{2\cdot 3} + \text{etc.} \qquad (2)$$

Now if, when $x = a = $ OM, we have $f(a) = \phi(a)$, the two curves *intersect* at P, *i. e.*, have one point in common. If in addition we have $f'(a) = \phi'(a)$, the curves have a *common tangent* at P, *i. e.*, have two consecutive points in common ; in this case they are said to have a *contact of the first order*. If also we have, not only $f(a) = \phi(a)$ and $f'(a) = \phi'(a)$, but $f''(a) = \phi''(a)$: *i. e.*, in passing along one of the curves to the next consecutive point, $\frac{d^2y}{dx^2}$ (*i. e.*, the curvature), remains the same in both curves, and the new point

10

is also a point of the second curve; *i. e.*, the curves have three consecutive points in common; in this case the curves are said to have a *contact of the second order*. If $f(a) = \phi(a), f'(a) = \phi'(a), f''(a) = \phi''(a), f'''(a) = \phi'''(a)$, the *contact is of the third order*, and so on. It is plain that the higher the order of contact, the more nearly do the curves agree; if *every* term in (1) is equal to the corresponding term in (2), then $y_1 = y_2$, and the two curves become coincident.

118. The Order of Contact depends on the number of Arbitrary Constants.—In order that a curve may have contact of the nth order with a given curve, it follows from Art. 117 that $n + 1$ equations must be satisfied. Hence, if the equation to a species of curve contains $n + 1$ constants, we may by giving suitable values to those constants, find the particular curve of the species that has contact of the nth order with a given curve at a given point. For example, the general equation of the right line has two constants, and hence two conditions can be formed, $f(x) = \phi(x)$ and $f'(x) = \phi'(x)$, from which the values of the constants may be determined so as to find the particular right line which has contact of the *first* order with a given curve at a given point. *In general*, the right line cannot have contact of a higher order than the first.

Contact of the *second* order requires *three* conditions, $f(x) = \phi(x)$, $f'(x) = \phi'(x)$, and $f''(x) = \phi''(x)$, and hence in order that a curve may have contact of the second order with a given curve, its equation must contain *three* constants, and so on. The general equation of the circle has three constants; hence, at any point of a curve a circle may be found which has contact of the *second* order with the curve at that point; this circle is called the *osculating circle* or *circle* of *curvature* of that point; *in general*, the circle cannot have contact of a higher order than the second. The parabola can have contact of the third order, and the ellipse and hyperbola of the fourth.

In this discussion we have assumed that the *given curve* is of such nature as to allow of any order of contact. Of course the order of contact is limited as much by one of the curves as by the other. For example, if the given curve were a right line and the other a circle, the contact could not in general be above the first order, although the circle may have a contact of the second order with curves whose equations have at least three constants. Also, we have used the phrase *in general*, since exceptions occur at particular points, some of which will be noticed hereafter.

119. *To find the radius of curvature of a given curve at a given point, and the co-ordinates of the centre of curvature.*

Let the equation of the given curve be

$$y = f(x), \qquad (1)$$

and that of the required circle be

$$(x' - m)^2 + (y' - n)^2 = r^2; \qquad (2)$$

it is required to determine the values of m, n, and r.

Since (2) has three arbitrary constants, we may impose three conditions, and determine the values of these constants that fulfil them, and the contact will be of the second order (Art. 118).

From (2), by differentiating twice, we have,

$$x' - m + (y' - n)\frac{dy'}{dx'} = 0; \qquad (3)$$

$$1 + \frac{dy'^2}{dx'^2} + (y' - n)\frac{d^2y'}{dx'^2} = 0. \qquad (4)$$

If (2) is the circle of curvature at the point (x, y) of (1), we must have,

$$x' = x, \qquad y' = y;$$

$$\frac{dy'}{dx'} = \frac{dy}{dx}, \qquad \frac{d^2y'}{dx'^2} = \frac{d^2y}{dx^2}.$$

Substituting these values in (2), (3), and (4), we have,

$$(x - m)^2 + (y - n)^2 = r^2 ; \tag{5}$$

$$x - m + (y - n)\frac{dy}{dx} = 0 \tag{6}$$

$$1 + \frac{dy^2}{dx^2} + (y - n)\frac{d^2y}{dx^2} = 0. \tag{7}$$

Therefore,
$$y - n = -\frac{1 + \dfrac{dy^2}{dx^2}}{\dfrac{d^2y}{dx^2}}. \tag{8}$$

$$x - m = \frac{\left(1 + \dfrac{dy^2}{dx^2}\right)\dfrac{dy}{dx}}{\dfrac{d^2y}{dx^2}}. \tag{9}$$

By (5), (8), and (9), we have

$$r = \pm \frac{\left(1 + \dfrac{dy^2}{dx^2}\right)^{\frac{3}{2}}}{\dfrac{d^2y}{dx^2}}. \tag{10}$$

From (9) and (8) we have

$$m = x - \frac{\left(1 + \dfrac{dy^2}{dx^2}\right)\dfrac{dy}{dx}}{\dfrac{d^2y}{dx^2}}. \tag{11}$$

$$n = y + \frac{1 + \dfrac{dy^2}{dx^2}}{\dfrac{d^2y}{dx^2}}. \tag{12}$$

120. Second Method.—Let ds denote an infinitely small element of a curve at a point, and ϕ the angle which the tangent at this point makes with the axis of x. Imagine two normals to be drawn at the extremities of this elementary arc, *i. e.*, at two consecutive points of the curve ; these normals will generally meet at a finite distance. Let r be the distance from the curve to the point of intersection of these consecutive normals. Then the angle included between these consecutive normals is equal to the corresponding angle of contingence (Art. 116), *i. e.*, equal to $d\phi$. Since $d\phi$ is the arc between the two normals at the unit's distance of the point of intersection, we have

$$ds = r\,d\phi, \quad \text{or} \quad r = \frac{ds}{d\phi}. \tag{1}$$

Now this value of r evidently represents the radius of the circle, which has the same curvature as that of the given curve at the given point, and hence is *the radius of curvature* for the given point, while *the centre of curvature* may be defined as *the point of intersection of two consecutive normals.*

To find the value of r, we have (Art. 56*a*),

$$\tan\phi = \frac{dy}{dx}; \qquad \therefore \quad \phi = \tan^{-1}\frac{dy}{dx};$$

and hence $\quad d\phi = \dfrac{\dfrac{d^2y}{dx}}{1 + \dfrac{dy^2}{dx^2}};\quad$ also, $\quad ds = \sqrt{dx^2 + dy^2}.$

Substituting in (1), we have

$$r = \frac{\left(1 + \dfrac{dy^2}{dx^2}\right)^{\frac{3}{2}}}{\dfrac{d^2y}{dx^2}}, \tag{2}$$

which is the same as (10) of Art. 119.

As the expression $\left(1 + \dfrac{dy^2}{dx^2}\right)^{\frac{3}{2}}$ has always two values, the one positive and the other negative, while the curve can generally have only one definite circle of curvature at any point, it will be necessary to agree upon which sign is to be taken. We shall adopt the positive sign, and regard r as positive when the second derivative is positive, *i.e.*, when the curve is convex downwards. (Usage is not uniform on this point. See Price's Calculus, Vol. I, p. 435. Todhunter's Calculus, p. 339, etc.)

121. To Find the Radius of Curvature in Terms of Polar Co-ordinates.

We may obtain this by transforming (2) of Art. 120 to polar co-ordinates, from which we find

$$R = \frac{\left(r^2 + \frac{dr^2}{d\theta^2}\right)^{\frac{3}{2}}}{r^2 + 2\frac{dr^2}{d\theta^2} - r\frac{d^2r}{d\theta^2}} = \frac{N^3}{r^2 + 2\left(\frac{dr}{d\theta}\right)^2 - r\frac{d^2r}{d\theta^2}},$$

where N is the normal. See Art. 102, Eq. 9. [See (2) of Ex. 4, Art. 90.]

122. At a Point where the Radius of Curvature is a Maximum or a Minimum, the Circle of Curvature has Contact of the Third Order with the Curve.

Since r is to be a maximum or a minimum, we must have $\frac{dr}{dx} = 0$.

Differentiating (2) of Art. 120 with respect to x, we have

$$\frac{dr}{dx} = \frac{\frac{3}{2}\left(1 + \frac{dy^2}{dx^2}\right)^{\frac{1}{2}} \times 2\frac{dy}{dx}\left(\frac{d^2y}{dx^2}\right)^2 - \frac{d^3y}{dx^3}\left(1 + \frac{dy^2}{dx^2}\right)^{\frac{3}{2}}}{\left(\frac{d^2y}{dx^2}\right)^2} = 0;$$

$$\therefore \quad \frac{d^3y}{dx^3} = \frac{3\frac{dy}{dx}\left(\frac{d^2y}{dx^2}\right)^2}{1 + \frac{dy^2}{dx^2}}. \tag{1}$$

Differentiating (8) of Art. 119, we have

$$\frac{d^3y}{dx^3} = \frac{3\frac{dy}{dx}\left(\frac{d^2y}{dx^2}\right)^2}{1 + \frac{dy^2}{dx^2}}. \qquad (2)$$

Hence the third derivative at a point of maximum or minimum curvature is the same as it is in the circle of curvature, and therefore the contact at this point is of the third order (Art. 117).

Cor.—*The contact of the osculating circles at the vertices of the conic sections is closer than at other points.*

123. Contact of Different Orders.—Let $y = f(x)$ and $y = \phi(x)$ represent two curves, and let x_1 be the abscissa of a point of their intersection; then we have

$$f(x_1) = \phi(x_1).$$

Substituting $x_1 \pm h$ for x_1 in both equations, and supposing y_1 and y_2 the corresponding ordinates of the two curves, we have

$$y_1 = f(x_1 \pm h) = f(x_1) + f'(x_1)(\pm h) + f''(x_1)\frac{(\pm h)^2}{2}$$

$$+ f'''(x_1)\frac{(\pm h)^3}{2 \cdot 3} + \text{etc.} \qquad (1)$$

$$y_2 = \phi(x_1 \pm h) = \phi(x_1) + \phi'(x_1)(\pm h) + \phi''(x_1)\frac{(\pm h)^2}{2}$$

$$+ \phi'''(x_1)\frac{(\pm h)^3}{2 \cdot 3} + \text{etc.} \qquad (2)$$

Subtracting (2) from (1), we get, for the difference of their ordinates, corresponding to $x_1 \pm h$,

$$y_1 - y_2 = [f'(x_1) - \phi'(x_1)](\pm h) + [f''(x_1) - \phi''(x_1)]\frac{(\pm h)^2}{2}$$

$$+ [f'''(x_1) - \phi'''(x_1)]\frac{(\pm h)^3}{2 \cdot 3} + \text{etc.} \quad (3)$$

Now if these curves have contact of the first order, the first term of (3) reduces to zero (Art. 117). If they have contact of the second order, the first two terms reduce to zero. If they have contact of the third order, the first three terms reduce to zero, and so on. Hence, when the order of contact is *odd*, the first term of (3) that does not reduce to zero must contain an even power of $\pm h$, and $y_1 - y_2$ does not change sign with h, and therefore the curves *do not intersect*, the one lying entirely above the other; but when the order of contact is *even*, the first term of (3) that does not reduce to zero must contain an odd power of $\pm h$, and $y_1 - y_2$ changes sign with h, and therefore the curves *intersect*, the one lying alternately above and below the other.

Cor. 1.—At a point of inflexion of a curve, the second derivative equals 0; also, the second derivative of any point of a right line equals 0. Hence, *at a point of inflexion, a rectilinear tangent to a curve has contact of the second order, and therefore intersects the curve.*

Cor. 2.—Since the circle of curvature has a contact of the second order with a curve, it follows that *the circle of curvature, in general, cuts the curve as well as touches it.*

Cor. 3.—At the points of maximum and minimum curvature, as for example at any of the four vertices of an ellipse, the osculating circle does not cut the curve at its point of contact.

EXAMPLES.

1. Find the radius of curvature of an ellipse,

$$\frac{x^2}{a^2} + \frac{y^2}{b^2} = 1.$$

$$\frac{d^2y}{dx^2} = -\frac{a^2b^2y - a^2b^2x\frac{dy}{dx}}{a^4y^2} = -\frac{b^2(a^2y^2 + b^2x^2)}{a^4y^3} = -\frac{b^4}{a^2y^3}$$

∴ (Art. 120),

$$r = \frac{\left(1 + \frac{dy^2}{dx^2}\right)^{\frac{3}{2}}}{\frac{d^2y}{dx^2}}$$

$$= \frac{(a^4y^2 + b^4x^2)^{\frac{3}{2}}}{a^4b^4} \quad \text{(neglecting the sign).}$$

At the extremity of the major axis,

$$x = a, \qquad y = 0, \qquad \therefore \quad r = \frac{b^2}{a}.$$

At the extremity of the minor axis,

$$x = 0, \qquad y = b, \qquad \therefore \quad r = \frac{a^2}{b}.$$

2. Find the radius of curvature of the common parabola,

$$y^2 = 2px.$$

Here

$$\frac{dy}{dx} = \frac{p}{y}, \qquad \frac{d^2y}{dx^2} = -\frac{p^2}{y^3};$$

$$\therefore \quad r = \frac{(y^2 + p^2)^{\frac{3}{2}}}{p^2} = \frac{(\text{normal})^3}{p^2}.$$

At the vertex, $y = 0$; $\qquad \therefore \quad r = p.$

3. Find the radius of curvature of the cycloid

$$x = r \,\text{vers}^{-1}\frac{y}{r} - \sqrt{2ry - y^2}.$$

Here $\dfrac{dx}{dy} = \dfrac{y}{\sqrt{2ry - y^2}};$ $\qquad \therefore \quad 1 + \dfrac{dy^2}{dx^2} = \dfrac{2r}{y};$

$$\frac{d^2y}{dx^2} = -\frac{r}{y^2}; \qquad \therefore \quad r = 2\sqrt{2ry},$$

which equals twice the normal (Art. 101, Ex. 5).

4. Find the radius of curvature of the parabola whose latus-rectum is 9, at $x = 3$, and the co-ordinates of the centre of curvature. $r = 16.04$; $m = 13\frac{1}{2}$, $n = -6.91$.

5. Find the radius of curvature of the ellipse whose axes are 8 and 4, at $x = 2$, and the co-ordinates of the centre of curvature. $r = 5.86$; $m = .38$, $n = -3.9$.

6. Find the radius of curvature of the logarithmic spiral

$$r = a^\theta.$$

$$\frac{dr}{d\theta} = a^\theta \log a; \qquad \frac{d^2r}{d\theta^2} = a^\theta \log^2 a;$$

$$\therefore \quad R = \frac{(r^2 + r^2 \log^2 a)^{\frac{3}{2}}}{r^2 + 2r^2 \log^2 a - r^2 \log^2 a} = (r^2 + r^2 \log^2 a)^{\frac{1}{2}} = N.$$

(See Ex. 2, Art. 102.)

7. Find the radius of curvature of the spiral of Archimedes, $r = a\theta$. $R = \dfrac{(a^2 + r^2)^{\frac{3}{2}}}{2a^2 + r^2}.$

8. Find the radius of curvature of the hyperbolic spiral, $r\theta = a$. $R = \dfrac{r (a^2 + r^2)^{\frac{3}{2}}}{a^3}.$

124. Evolutes and Involutes.—The curve which is the locus of the centres of all the osculating circles of a given curve, is called the *evolute* of that curve; the latter curve is called the *involute* of the former.

Let P_1, P_2, P_3, etc., represent a series of consecutive points on the curve MN, and C_1, C_2, C_3, etc., the corresponding centres of curvature; then the curve C_1, C_2, C_3, etc., is the *evolute* of MN, and MN is the *involute* of C_1, C_2, C_3, etc. Also, since the lines C_1P_1, C_2P_2, etc., are normals to the involute at the consecutive points, the points C_1, C_2, C_3, etc., may be regarded as

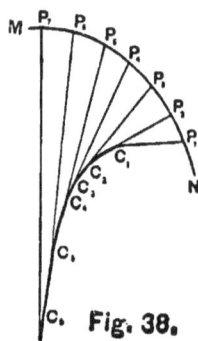

Fig. 38.

consecutive points of the evolute; and since each of the normals P_1C_1, P_2C_2, etc., passes through two consecutive points on the evolute, they are tangents to it.

Let r_1, r_2, r_3, etc., denote the lengths of the radii of curvature at P_1, P_2, P_3, etc., and we have,

$$r_2 - r_1 = P_2C_2 - P_1C_1 = P_2C_2 - P_2C_1 = C_1C_2;$$

also $$r_3 - r_2 = P_3C_3 - P_2C_2 = P_3C_3 - P_3C_2 = C_2C_3;$$

and $r_4 - r_3 = C_3C_4$, and so on to r_n;

hence by addition we have,

$$r_n - r_1 = C_1C_2 + C_2C_3 + \ldots \ldots C_{n-1}C_n.$$

This result holds when the number n is increased indefinitely, and we infer that the *length of any arc of the evolute is equal, in general, to the difference between the radii of curvature at its extremities.*

It is evident that the involute may be generated from its evolute by winding a string round the evolute, holding it tight, and then unwinding it, each point in the string will describe a different involute. It is from this property that the names evolute and involute are given. While a curve can only have *one* evolute, it can have an infinite number of involutes.

The involutes described by two different points in the moving string, are said to be *parallel;* each curve being got from the other by cutting off a constant length on its normal, measured from the involute. (Williamson's Differential Calculus, p. 295.)

125. To find the Equation of the Evolute of a Given Curve.—The co-ordinates of the centre of curvature are the co-ordinates of the evolute (Art. 124). Hence, if we combine (11) and (12) of Art. 119 with the equation of the curve, and eliminate x and y, there will result an equation expressing a relation between m and n, the co-ordinates of the required evolute, which is therefore the required equation; the method can be best illustrated by examples.

The eliminations are often quite difficult; the following are comparatively simple examples.

EXAMPLES.

1. Find the equation of the evolute of the parabola,

$$y^2 = 2px. \tag{1}$$

Here $\quad \dfrac{dy}{dx} = \dfrac{p}{y}; \quad \dfrac{d^2y}{dx^2} = -\dfrac{p^2}{y^3}.$

Substituting in (11) and (12) of Art. 119, we have,

$$m = x + \frac{y^2 + p^2}{y^2} \cdot \frac{p}{y} \cdot \frac{y^3}{p^2} = 3x + p;$$

$$\therefore \quad x = \frac{m - p}{3}.$$

$$n = y - \frac{y^2 + p^2}{y^2} \cdot \frac{y^3}{p^2} = -\frac{y^3}{p^2};$$

$$\therefore \quad y = -p^{\frac{2}{3}} n^{\frac{1}{3}}.$$

And these values of x and y in (1) give,

$$p^{\frac{4}{3}} n^{\frac{2}{3}} = \tfrac{2}{3} p\, (m - p);$$

$$\therefore \quad n^2 = \frac{8}{27p}\, (m - p)^3; \tag{2}$$

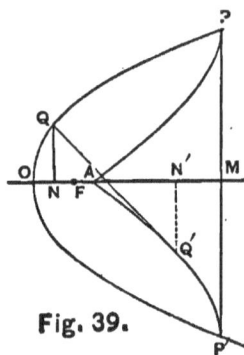

which *is the equation required*, and is called the *semi-cubical parabola*. Tracing the curve, we find its form as given in Fig. 39, where AO = p.

If we transfer the origin from O to A, (2) becomes

$$n^2 = \frac{8}{27p}\, m^3.$$

2. Find the length of the evolute AQ', Fig. 39, in terms of the co-ordinates of its extremities.

Let $ON = x$, $NQ = y$; $ON' = m$, $N'Q' = n$.

Then by Art. 123, Ex. 2, we have,

$$r = \frac{(y^2 + p^2)^{\frac{3}{2}}}{p^2}.$$

Therefore, by Art. 124, we have,

Length of $AQ' = Q'Q - AO = \dfrac{(y^2 + p^2)^{\frac{3}{2}}}{p^2} - p$

$= \left(n^{\frac{2}{3}} + p^{\frac{2}{3}}\right)^{\frac{3}{2}} - p.$ (Since $y^2 = p^{\frac{4}{3}}n^{\frac{2}{3}}$, by Ex. 1.)

3. Find the equation of the evolute of the cycloid,

$$x = r \, \mathrm{vers}^{-1} \frac{y}{r} - \sqrt{2ry - y^2}. \tag{1}$$

Here $\quad \dfrac{dy}{dx} = \dfrac{\sqrt{2ry - y^2}}{y}, \quad \dfrac{d^2y}{dx^2} = -\dfrac{r}{y^2}.$

$\therefore \quad m = x + 2\sqrt{2ry - y^2} \quad$ and $\quad n = -y;$

or $\qquad x = m - 2\sqrt{-2rn - n^2} \quad$ and $\quad y = -n;$

which, in the equation of the cycloid, gives

$m = r \, \mathrm{vers}^{-1}\left(-\dfrac{n}{r}\right) + \sqrt{-2rn - n^2}, (2)$

which is the equation of a cycloid equal to the given cycloid; the origin being at the highest point. This will appear by transforming

$x = {}^{*}r \, \mathrm{versin}^{-1}\dfrac{y}{r} + \sqrt{2ry - y^2}, (3)$

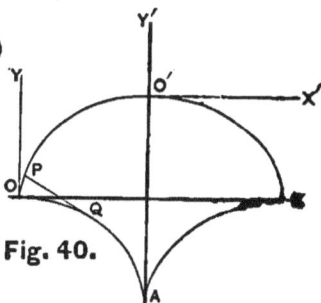

Fig. 40.

---

* $\mathrm{Versin}^{-1}\frac{y}{r}$ is not restricted to first quadrant.

(which gives points between O' and X), to parallel axes through O'.

Denoting by m and n the new co-ordinates, the formulæ for transforming from O to O' are

$$x = \pi r + m, \qquad y = 2r + n ;$$

which in (3) gives

$$\pi r + m = r \text{ vers}^{-1}\left(2 + \frac{n}{r}\right) + \sqrt{2r(2r+n)-(2r+n)^2}$$

$$= r \text{ vers}^{-1} 2 + r \text{ vers}^{-1}\left(-\frac{n}{r}\right) + \sqrt{-2rn - n^2} ;$$

$$\therefore \quad m = r \text{ vers}^{-1}\left(-\frac{n}{r}\right) + \sqrt{-2rn - n^2} ; \qquad (4)$$

which is the same as equation (2). Hence we see that the equation of the evolute, OA (2), is the same as that of the cycloid, O'X (4). That is, *the evolute of a cycloid is an equal cycloid.*[*]

126. A Normal to an Involute is Tangent to the Evolute.—This was shown *geometrically* in Art. 124. It may also be shown as follows :

Let (x, y) be any point P of the involute (Fig. 40), from which the normal PQ is drawn, and let (m, n) be the point Q on the evolute through which the normal passes.

The equation of PQ is

$$y - n = - \frac{dx}{dy} (x - m) ; \qquad (1)$$

or $$\qquad x - m + \frac{dy}{dx} (y - n) = 0. \qquad (2)$$

Now when we pass from a point P to a consecutive point on the involute, Q also will change to a consecutive

---

This property was first discovered by Huygens.

point of the evolute, therefore we differentiate (2) with respect to x, regarding x, y, m, n, as variables, and get

$$1 - \frac{dm}{dx} + \frac{d^2y}{dx^2}(y - n) + \frac{dy^2}{dx^2} - \frac{dy}{dx} \cdot \frac{dn}{dx} = 0. \quad (3)$$

But, since (m, n) is the centre of curvature corresponding to P, we have, by (8) of Art. 119,

$$(y - n)\frac{d^2y}{dx^2} + 1 + \frac{dy^2}{dx^2} = 0;$$

which in (3) gives

$$-\frac{dm}{dx} - \frac{dy}{dx}\frac{dn}{dx} = 0, \quad \text{or} \quad -\frac{dx}{dy} = \frac{dn}{dm};$$

and this in (1) gives

$$y - n = \frac{dn}{dm}(x - m) ; \quad (4)$$

therefore (1) or (4), which is the equation of a normal to the involute at P (x, y), is also the equation of a tangent to the evolute at Q (m, n).

127. Envelopes of Curves.—Let us suppose that in the equation of any plane curve of the form

$$f(x, y, a) = 0, \quad (1)$$

we assign to the arbitrary constant a, a series of different values, then for each value of a we get a distinct curve, different from any of the others in form and position, and (1) may be regarded as representing an indefinite number of curves, each of which is determined when the corresponding value of a is known, and varies as a varies.

The quantity a is called *a variable parameter*, the name being applied to a quantity which is constant for any one curve of a series, but varies in changing from one curve to another, and the equation $f(x, y, a) = 0$, is said to represent a *family of curves*.

If we suppose a to change continuously, *i. e.*, by

infinitesimal increments, the curves of the series represented by (1) will differ in position by infinitesimal amounts ; and any two adjacent curves of the series will, in general, intersect; the intersections of these curves are points in the *envelope*. Hence an envelope may be defined as *the locus of the intersection of consecutive curves of a series.*

It can be easily seen that the envelope is tangent to each of the intersecting curves of the series; for, if we consider four consecutive curves, and suppose P_1 to be the point of intersection of the first and second, P_2 that of the second and third, and P_3 that of the third and fourth, the line $P_1 P_2$ joins two infinitely near points on the envelope and on the second of the four curves, and hence is a tangent both to the envelope and the second curve ; in the same way it may be shown that the line $P_2 P_3$ is a tangent to the envelope and the *third* consecutive curve, and so on.

128. To Find the Equation of the Envelope of a given Series of Curves.

Let
$$f(x, y, a) = 0, \qquad (1)$$
$$f(x, y, a+da) = 0, \qquad (2)$$

be the equations of two consecutive curves of the series; then the co-ordinates of the points of intersection of (1) and (2) will satisfy both (1) and (2), and therefore also will satisfy the equation

$$\frac{f(x, y, a) - f(x, y, a+da)}{da} = 0 \text{ (Anal. Geom., Art. 30),}$$

or
$$\frac{df(x, y, a)}{da} = 0, \qquad (3)$$

and therefore the points of intersection of two infinitely near curves of the series satisfy each of the equations (1) and (3). Hence, to find the equation of the envelope, we eliminate a between (1) and (3), *i. e.*, we eliminate the variable parameter between the equation of the locus and its first differential equation.

EXAMPLES.

1. Find the envelope of $y = ax + \dfrac{m}{a}$, when a varies.

Differentiating with respect to a, x, and y, being constant, we have

$$0 = x - \frac{m}{a^2}; \qquad\qquad \therefore\ a = \pm \left(\frac{m}{x}\right)^{\frac{1}{2}};$$

$$\therefore\ y = \pm \left[\sqrt{mx} + \sqrt{mx} \right] \quad \text{or} \quad y^2 = 4mx,$$

which is the equation of a parabola.

2. A right line of given length slides down between two rectangular axes; to find the envelope of the line in all positions.

Let c be the length of the line, a and b the intercepts OA and OB; then the equation of the line is

$$\frac{x}{a} + \frac{y}{b} = 1, \qquad\qquad (1)$$

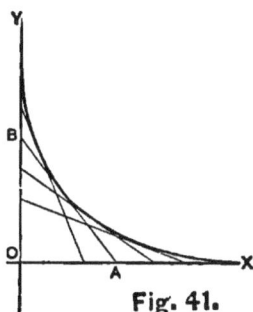

Fig. 41.

in which the variable parameters a and b are connected by the equation

$$a^2 + b^2 = c^2. \qquad\qquad (2)$$

Differentiating (1) and (2), regarding a and b as variable, we have

$$-\frac{x}{a^2} da - \frac{y}{b^2} db = 0, \quad \text{or} \quad -\frac{x}{a^2} da = \frac{y}{b^2} db. \qquad (3)$$

$$a\, da + b\, db = 0, \quad \text{or} \quad - a\, da = b\, db. \qquad (4)$$

Dividing (3) by (4), we have

$$\frac{x}{a^3} = \frac{y}{b^3}, \quad \text{or} \quad \frac{\dfrac{x}{a}}{a^2} = \frac{\dfrac{y}{b}}{b^2} = \frac{\dfrac{x}{a} + \dfrac{y}{b}}{a^2 + b^2} = \frac{1}{c^2};$$

$$\therefore\ a = (xc^2)^{\frac{1}{3}} \quad \text{and} \quad b = (yc^2)^{\frac{1}{3}},$$

which in (2) gives

$$(xc^2)^{\frac{2}{3}} + (yc^2)^{\frac{2}{3}} = c^2;$$

$$\therefore \quad x^{\frac{2}{3}} + y^{\frac{2}{3}} = c^{\frac{2}{3}},$$

which is the equation required. The form of the locus is given in Fig. 42, and is called a *hypo-cycloid*, which is a curve generated by a point in the circumference of a circle as it rolls on the concave arc of a fixed circle.

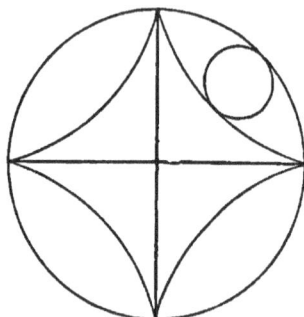

Fig. 42.

3. Find the envelope of a series of ellipses whose axes are coincident in direction, their product being constant.

Here
$$\frac{x^2}{a^2} + \frac{y^2}{b^2} = 1. \tag{1}$$

Let
$$a \cdot b = c ; \tag{2}$$

$$\therefore \quad \frac{x^2}{a^3} da + \frac{y^2}{b^3} db = 0, \quad \text{or} \quad \frac{x^2}{a^3} da = - \frac{y^2}{b^3} db. \tag{3}$$

$$\frac{da}{a} + \frac{db}{b} = 0, \quad \text{or} \quad \frac{da}{a} = - \frac{db}{b}. \tag{4}$$

Dividing (3) by (4), we have

$$\frac{x^2}{a^2} = \frac{y^2}{b^2} = \tfrac{1}{2}, \text{ by substituting in (1)}.$$

$$\therefore \quad a = \pm x\sqrt{2}, \quad \text{and} \quad b = \pm y\sqrt{2};$$

$$\therefore \quad \text{from (2),} \quad xy = \pm \frac{c}{2},$$

which is the equation of an hyperbola referred to its asymptotes as axes.

This example may also be solved as follows: Eliminating b from (1) and (2), we have

$$\frac{x^2}{a^2} + \frac{a^2}{c^2} y^2 = 1, \tag{5}$$

in which we have only the variable parameter a.

$$\therefore \quad -\frac{x^2}{a^3} + \frac{ay^2}{c^2} = 0; \quad \therefore \quad a^2 = \frac{cx}{y}; \tag{6}$$

which in (5) gives

$$\frac{xy}{c} + \frac{xy}{c} = 1; \quad \therefore \quad xy = \tfrac{1}{2}c.$$

4. Find the envelope of the right lines whose general equation is

$$y = mx + (a^2m^2 + b^2)^{\frac{1}{2}}, \tag{1}$$

where m is the variable parameter.

We find
$$m = -\frac{b}{a} \frac{x}{\sqrt{a^2 - x^2}},$$

which in (1) gives $\dfrac{x^2}{a^2} + \dfrac{y^2}{b^2} = 1$ for the required envelope.

Hence the envelope of (1) is an ellipse, as we might have inferred, since (1) is a tangent to an ellipse. (See Anal. Geom., Art. 74.)

EXAMPLES.

1. Find the radius of curvature of the logarithmic curve $x = \log y$.
$$r = \frac{(m^2 + y^2)^{\frac{3}{2}}}{my}.$$

2. Find the radius of curvature of the cubical parabola $y^3 = a^2x$.
$$r = \frac{(9y^4 + a^4)^{\frac{3}{2}}}{6a^4y}.$$

3. Find the radius of curvature of the curve
$$y = x^3 - x^2 + 1$$

where it cuts the axis of y, and also at the point of minimum ordinate.

At the first point, $r = -\frac{1}{2}$; at the second, $r = \frac{1}{2}$.

4. Find the radius of curvature of the curve
$$y^3 = 6x^2 + x^3.$$
$$r = \frac{[y^4 + (4x + x^2)^2]^{\frac{3}{2}}}{-8x^2y}.$$

5. Find the radius of curvature of the rectangular hyperbola $xy = m^2$.
$$r = \frac{(x^2 + y^2)^{\frac{3}{2}}}{2m^2}.$$

6. Find the radius of curvature of the Lemniscate of Bernouilli $r^2 = a^2 \cos 2\theta$.
$$R = \frac{a^2}{3r}.$$

7. Find the equation of the evolute of the ellipse
$$a^2y^2 + b^2x^2 = a^2b^2.$$
$$(am)^{\frac{2}{3}} + (bn)^{\frac{2}{3}} = (a^2 - b^2)^{\frac{2}{3}}.$$

8. Find the equation of the evolute of the hyperbola
$$a^2y^2 - b^2x^2 = -a^2b^2.$$
$$(am)^{\frac{2}{3}} - (bn)^{\frac{2}{3}} = (a^2 + b^2)^{\frac{2}{3}}.$$

9. Prove that, in Fig. 39, $OM = 4OA = 4p$, and $MP' = 2p\sqrt{2}$.

10. Find the length of the evolute AP' in Fig. 39.

Ans. $(3^{\frac{3}{2}} - 1)p.$

11. Find the length of the evolute of the ellipse. (See Art. 123, Ex. 1, and Art. 124.)

Ans. $4\dfrac{a^3 - b^3}{ab}.$

12. Find the length of the cycloidal arc $OO'X$, Fig. 40.

Ans. $8r.$

13. Find the envelope of the series of parabolas whose equation is $y^2 = m(x - m)$, m being the variable parameter.

$$y = \pm \frac{x}{2}.$$

14. Find the envelope of the series of parabolas expressed by the equation $y = ax - \dfrac{1 + a^2}{2p} x^2$, where a is the variable parameter.

The result is a parabola whose equation is

$$x^2 = 2p\left(\frac{p}{2} - y\right).$$

This is the equation of the curve touched by the parabolas described by projectiles discharged from a given point with a constant velocity, but at different inclinations to the horizon. The problem was the first of the kind proposed, and was solved by John Bernouilli, but not by any general method.

15. Find the envelope of the hypothenuse of a right-angled triangle of constant area c.

$$xy = \frac{c}{2}.$$

16. One angle of a triangle is fixed in position, find the envelope of the opposite side when the area is constant $= c$.

$$xy = \frac{c}{2}.$$

17. Find the envelope of $x \cos \alpha + y \sin \alpha = p$, in which α is the variable parameter. $\qquad x^2 + y^2 = p^2.$

18. Find the envelope of the consecutive normals to the parabola $y^2 = 2px$.

Ans. $y^2 = \dfrac{8}{27p} (x - p)^3$, which is the same as was found for the *evolute* in Ex. 1, Art. 125, as it clearly should be. (See Art. 124.)

19. Find the envelope of the consecutive normals to the ellipse $a^2y^2 + b^2x^2 = a^2b^2$.

Ans. $(ax)^{\frac{2}{3}} + (by)^{\frac{2}{3}} = (a^2 - b^2)^{\frac{2}{3}}$, which is the same as was found in (7) for the *evolute of the ellipse*.

PART II.

INTEGRAL CALCULUS.

————————•◆••————————

CHAPTER I.

ELEMENTARY FORMS OF INTEGRATION.

129. Definitions.—The Integral Calculus is the inverse of the Differential Calculus, its object being to find the relations between finite values of variables from given relations between the infinitesimal elements of those variables ; or, it may be defined as the process of finding the function from which any given differential may have been obtained. The function which being differentiated produces the given differential, is called the *integral of the differential*. The process by which we obtain the integral function from its differential is called *integration*.

The primary problem of the Integral Calculus is to effect the summation of a certain infinite series of infinitesimals, and hence the letter S was placed before the differential to show that its sum was to be taken. This was elongated into the symbol $\int$ (a long S), which is *the sign of integration*, and when placed before a differential, denotes that its integral is to be taken. Thus, $\int 3x^2dx$, which is read, "the integral of $3x^2dx$," denotes that the integral of $3x^2dx$ is to be taken. The signs of integration and differentiation are inverse operations, and when placed before a quantity, neutralize each other. Thus,

$$\int d\,(ax) = ax,$$

and $\qquad d\int axdx = axdx.$

130. Elementary Rules for Integration.—In the elementary forms of integration, the rules and methods are obtained by reversing the corresponding rules for differentiation. When a differential is given for integration, if we cannot see *by inspection* what function, being differentiated, produces it, or *if it cannot be integrated by known rules*, we proceed to transform the differential into an equivalent expression of known form, whose integral we can see *by inspection*, or can obtain *by known rules*. In every case, a sufficient reason that one function is the integral of another is that *the former, being differentiated, gives the latter.**

(*1.*) Since

$$d\,(v + y - z) = dv + dy - dz; \qquad \text{(Art. 14.)}$$

$$\therefore \int (dv + dy - dz) = \int d\,(v + y - z) = v + y - z$$

$$= \int dv + \int dy - \int dz.$$

Hence, *the integral of the algebraic sum of any number of differentials is equal to the algebraic sum of their integrals.*

(*2.*) Since

$$d\,(ax \pm b) = adx; \qquad \text{(Art. 15.)}$$

---

* While there is no quantity whose *differential* cannot be found, there is a large class of differentials whose *integrals* cannot be obtained ; either because there is no quantity which, being differentiated, will give them, or because the method for their integration has not yet been found.

$$\therefore \quad \int a dx = \int d\,(ax + b) = ax \pm b$$

$$= a \int dx \pm b.$$

Hence, *a constant factor can be moved from one side of the integral sign to the other without affecting the value of the integral.* Also, since constant terms, connected by the sign $\pm$, disappear in differentiation, therefore in returning from the differential to the integral, *an arbitrary constant, as* C, *must be added,* whose value must be determined afterwards by the data of the problem, as will be explained hereafter.

(3.)   Since

$$d\,\frac{a}{n}\,[f(x)]^n = a\,[f(x)]^{n-1}\,df(x); \qquad \text{(Arts. 15 and 19.)}$$

$$\therefore \quad \int a\,[f(x)]^{n-1}\,df(x) = \int d\,\frac{a}{n}\,[f(x)]^n = \frac{a}{n}\,[f(x)]^n + c.$$

Hence, *whenever a differential is the product of three factors, viz, a constant factor, a variable factor with any exponent except* -1, *and a differential factor which is the differential of the variable factor without its exponent, its integral is the product of the constant factor by the variable factor with its exponent increased by 1, divided by the new exponent.* *

It will be seen that the rule fails when $n = -1$, since if we divide by $1 - 1 = 0$, the result will be ∞.

(4.)   Since $\qquad d\,(a \log x) = \dfrac{a dx}{x};$ $\qquad$ (Art. 20, Cor.)

$$\therefore \quad \int \frac{a dx}{x} = \int d\,(a \log x) = a \log x.$$

---

* The arbitrary constant is not mentioned since its addition is always understood, and in the following integrals it will be omitted, as it can always be supplied when necessary.

Hence, *whenever a differential is a fraction whose numerator is the product of a constant by the differential of the denominator, its integral is the product of the constant by the Naperian logarithm of the denominator.*

EXAMPLES.

1. To integrate $dy = ax^5dx$.

$$y = \int ax^5dx = \int a \cdot x^5 \cdot dx = \frac{ax^6}{6}. \quad \text{[by (3)]}.$$

2. To integrate $dy = (a + 5x^3)^4 x^2 dx$.

The differential of the quantity within the parenthesis being $15x^2dx$, we write

$$y = \int \tfrac{1}{15}(a + 5x^3)^4 \, 15x^2dx = \frac{(a + 5x^3)^5}{75}. \quad \text{[by (3)]}.$$

This example might also be integrated by expanding the quantity within the parenthesis, and integrating each term separately by (1), but the process would be more lengthy than the one employed.

3. To integrate

$$dy = a (ax^2 + bx^3)^{\frac{1}{2}} 2xdx + 3bx^2 (ax^2 + bx^3)^{\frac{1}{2}} dx.$$

$$y = \int [a (ax^2 + bx^3)^{\frac{1}{2}} 2xdx + 3bx^2 (ax^2 + bx^3)^{\frac{1}{2}} dx]$$

$$= \int (ax^2 + bx^3)^{\frac{1}{2}} (2ax + 3bx^2) \, dx = \tfrac{2}{3} (ax^2 + bx^3)^{\frac{3}{2}} \text{ [by (3)]}.$$

4. To integrate $dy = \dfrac{adx}{a + bx}$.

Since the numerator must be bdx to be the differential of the denominator, we must multiply it by b, taking care to divide by b also ; hence,

$$y = \int \frac{adx}{a + bx} = \frac{a}{b} \int \frac{bdx}{a + bx} = \frac{a}{b} \log (a + bx). \quad \text{[by (4)]}.$$

11

131. Fundamental Forms.—On referring to the forms of differentials established in Chap. II, we may write down at once the following integrals from inspection, *the arbitrary constant being always understood.*

1. $\quad y = \int ax^n dx \qquad\qquad = \dfrac{ax^{n+1}}{n+1}.$

2. $\quad y = \int \dfrac{a\,dx}{x^n} \qquad\qquad = -\dfrac{a}{(n-1)\,x^{n-1}}.$

3. $\quad y = \int \dfrac{a\,dx}{x} \qquad\qquad = a\log x.*$

4. $\quad y = \int a^x \log a\; dx \qquad = ax.$

5. $\quad y = \int e^x dx \qquad\qquad = e^x.$

6. $\quad y = \int \cos x\; dx \qquad\qquad = \sin x.$

7. $\quad y = \int -\sin x\; dx \qquad\quad = \cos x.$

8. $\quad y = \int \sec^2 x\; dx \qquad\qquad = \tan x.$

9. $\quad y = \int -\mathrm{cosec}^2 x\; dx \qquad = \cot x.$

10. $\quad y = \int \sec x \tan x\; dx \qquad = \sec x.$

11. $\quad y = \int -\mathrm{cosec}\,x\,\mathrm{cotan}\,x\; dx = \mathrm{cosec}\,x.$

12. $\quad y = \int \sin x\; dx \qquad\qquad = \mathrm{vers}\,x.$

13. $\quad y = \int -\cos x\; dx \qquad\quad = \mathrm{covers}\,x.$

---

* Since the constant c to be added is arbitrary, $\log c$ is arbitrary, and we may write the integral in the form

$$\int a\,\frac{dx}{x} = a\log x + \log c = \log cx^a.$$

14. $\quad y = \displaystyle\int \frac{dx}{\sqrt{1-x^2}} \qquad\qquad = \sin^{-1} x.$

15. $\quad y = \displaystyle\int -\frac{dx}{\sqrt{1-x^2}} \qquad\quad = \cos^{-1} x.$

16. $\quad y = \displaystyle\int \frac{dx}{1+x^2} \qquad\qquad = \tan^{-1} x.$

17. $\quad y = \displaystyle\int -\frac{dx}{1+x^2} \qquad\quad = \cot^{-1} x.$

18. $\quad y = \displaystyle\int \frac{dx}{x\sqrt{x^2-1}} \qquad\quad = \sec^{-1} x.$

19. $\quad y = \displaystyle\int -\frac{dx}{x\sqrt{x^2-1}} \qquad = \csc^{-1} x.$

20. $\quad y = \displaystyle\int \frac{dx}{\sqrt{2x-x^2}} \qquad\quad = \mathrm{vers}^{-1} x.$

21. $\quad y = \displaystyle\int -\frac{dx}{\sqrt{2x-x^2}} \;. \qquad = \mathrm{covers}^{-1} x.$

These integrals are called the *fundamental* or *elementary forms*, to which all other forms, that admit of integration in a finite number of terms, can be ultimately reduced. It is in this *algebraic reduction* that the chief difficulty of the Integral Calculus is found ; and the processes of the whole subject are little else than a succession of transformations and artifices by which this reduction may be effected. The student must commit these fundamental forms to memory; they are as essential in integration as the multiplication table is in arithmetic.

132. Integration of other Circular and Trigonometric Functions by Transformation into the Fundamental Forms.

1. To integrate $dy = \dfrac{dx}{\sqrt{a^2 - b^2x^2}}.$

We see that this has the general form of the differential

of an arc in terms of its sine (see form 14 of Art. 131); hence we transform our expression into this form, as follows:

$$y = \int \frac{dx}{\sqrt{a^2 - b^2x^2}} = \int \frac{dx}{a\sqrt{1 - \frac{b^2x^2}{a^2}}} = \int \frac{\frac{1}{a}\,dx}{\sqrt{1 - \frac{b^2x^2}{a^2}}}.$$

To make this quantity the differential of an arc in terms of its sine, the numerator must be the differential of the square root of the second term in the denominator, which is $\frac{b}{a}\,dx$. Therefore we need to multiply the numerator by b, which can be done by multiplying also by the reciprocal of b, or putting the reciprocal of b outside the sign of integration. Hence,

$$y = \int \frac{dx}{\sqrt{a^2 - b^2x^2}} = \int \frac{\frac{1}{a}\,dx}{\sqrt{1 - \frac{b^2x^2}{a^2}}} = \frac{1}{b}\int \frac{\frac{b}{a}\,dx}{\sqrt{1 - \frac{b^2x^2}{a^2}}}$$

$$= \frac{1}{b} \sin^{-1} \frac{bx}{a}.$$

2. To integrate $\quad dy = -\dfrac{dx}{\sqrt{a^2 - b^2x^2}}.$

Here $\quad y = \int -\dfrac{dx}{\sqrt{a^2 - b^2x^2}} = \int -\dfrac{dx}{a\sqrt{1 - \frac{b^2x^2}{a^2}}}$

$$= \frac{1}{b}\int \frac{-\frac{b}{a}\,dx}{\sqrt{1 - \frac{b^2x^2}{a^2}}} = \frac{1}{b} \cos^{-1} \frac{bx}{a}.$$

3. $y = \displaystyle\int \frac{dx}{a^2 + b^2x^2} \quad = \frac{1}{ab} \tan^{-1} \frac{bx}{a}.$

4. $y = \displaystyle\int -\frac{dx}{a^2 + b^2x^2} = \frac{1}{ab} \cot^{-1} \frac{bx}{a}.$

5. $y = \displaystyle\int \frac{dx}{x\sqrt{b^2x^2 - a^2}} \qquad = \dfrac{1}{a} \sec^{-1} \dfrac{bx}{a}.$

6. $y = \displaystyle\int - \frac{dx}{x\sqrt{b^2x^2 - a^2}} \qquad = \dfrac{1}{a} \operatorname{cosec}^{-1} \dfrac{bx}{a}.$

7. $y = \displaystyle\int \frac{dx}{\sqrt{2abx - b^2x^2}} \qquad = \dfrac{1}{b} \operatorname{vers}^{-1} \dfrac{bx}{a}.$

8. $y = \displaystyle\int - \frac{dx}{\sqrt{2abx - b^2x^2}} \qquad = \dfrac{1}{b} \operatorname{covers}^{-1} \dfrac{bx}{a}.$

9. $y = \displaystyle\int \tan x \, dx = \int \frac{\sin x \, dx}{\cos x} = - \int \frac{d \cos x}{\cos x}$

$$= - \log \cos x = \log \sec x.$$

10. $y = \displaystyle\int \cot x \, dx = \int \frac{\cos x \, dx}{\sin x} = \log \sin x.$

11. $y = \displaystyle\int \frac{dx}{\sin x} = \int \frac{dx}{2 \sin \frac{1}{2}x \cos \frac{1}{2}x}$

$$= \int \frac{\frac{1}{2} \sec^2 (\frac{1}{2}x) \, dx}{\tan \frac{1}{2}x} = \log \tan \tfrac{1}{2}x.$$

12. $y = \displaystyle\int \frac{dx}{\cos x} = \int \frac{dx}{\sin \left(\dfrac{\pi}{2} + x\right)}$

$$= \log \tan \left(\frac{\pi}{4} + \tfrac{1}{2}x\right) \text{ [by (11)]}.$$

13. $y = \displaystyle\int \frac{dx}{\sin x \cos x} = \int \frac{\sec^2 x \, dx}{\tan x} = \log \tan x.$

14. $y = \displaystyle\int \frac{dx}{\sin^2 x \cos^2 x} = \int \frac{(\sin^2 x + \cos^2 x) \, dx}{\sin^2 x \cos^2 x}$

$$\text{(since } \sin^2 x + \cos^2 x = 1) = \int (\sec^2 x + \operatorname{cosec}^2 x) \, dx$$

$$= \tan x - \cot x.$$

15. $y = \displaystyle\int \tan^2 x \, dx = \int (\sec^2 x - 1) \, dx = \tan x - x.$

16. $y = \int \cot^2 x \, dx = \int (\operatorname{cosec}^2 x - 1) \, dx$

$$= -\cot x - x.$$

17. $y = \int \cos^2 x \, dx = \int (\tfrac{1}{2} + \tfrac{1}{2} \cos 2x) \, dx$ (by **Trig.**)

$$= \tfrac{1}{2}x + \tfrac{1}{4} \sin 2x.$$

(See Price's Calculus, Vol. II, p. 68.)

18. $y = \int \sin^2 x \, dx = \tfrac{1}{2}x - \tfrac{1}{4} \sin 2x.$

REMARK.—It will be observed that in every case we reduce the function to a known form, and then pass to the integral by simple inspection or by the elementary rules. *Whenever there is any doubt as to whether the integral found be correct or not, it is well to differentiate it, and see if it gives the proposed differential.* (See Art. 130.)

EXAMPLES.

1. $dy = bx^{\frac{1}{2}} \, dx.$

Here $\quad y = \int bx^{\frac{1}{2}} \, dx = \int b \cdot x^{\frac{1}{2}} \cdot dx$

$$= \tfrac{2}{3} b x^{\frac{3}{2}} \text{ [by (3) of Art. 130].}$$

2. $dy = \dfrac{x \, dx}{\sqrt{a^2 + x^2}}.$

Here $\quad y = \int \dfrac{x \, dx}{\sqrt{a^2 + x^2}} = \int \tfrac{1}{2} (a^2 + x^2)^{-\frac{1}{2}} 2x \, \boldsymbol{dx}$

$$= (a^2 + x^2)^{\frac{1}{2}}.$$

3. $dy = 2x^{\frac{5}{9}} \, dx.$ $\qquad\qquad y = \tfrac{9}{7} x^{\frac{14}{9}}.$

4. $dy = 2x^{\frac{2}{3}} \, dx.$ $\qquad\qquad y = \tfrac{6}{5} x^{\frac{5}{3}}.$

5. $dy = -x^{-\frac{6}{5}} \, dx.$ $\qquad\qquad y = 5x^{-\frac{1}{5}}.$

6. $dy = \dfrac{dx}{\sqrt{x}}.$ $\qquad\qquad y = 2\sqrt{x}.$

7. $dy = -5mx^{-\frac{1}{3}}\, dx.$ $y = -\tfrac{15}{2}mx^{\frac{2}{3}}.$

8. $dy = \left(\tfrac{5}{2}ax^{\frac{3}{2}} - \tfrac{3}{2}bx^{\frac{1}{2}}\right) dx.$ $y = ax^{\frac{5}{2}} - bx^{\frac{3}{2}}.$

9. $dy = \left(\dfrac{12}{x^2} - \dfrac{3}{x^4}\right) dx.$ $y = -\dfrac{12}{x} + \dfrac{1}{x^3}.$

10. $dy = -\dfrac{(2ax - x^2)}{(3ax^2 - x^3)^{\frac{1}{3}}}\, dx.$

Here $y = \displaystyle\int -(3ax^2 - x^3)^{-\frac{1}{3}}(2ax - x^2)\, dx$

$$= \int -\tfrac{1}{3}(3ax^2 - x^3)^{-\frac{1}{3}}(6ax - 3x^2)\, dx$$

$$= -\tfrac{1}{2}(3ax^2 - x^3)^{\frac{2}{3}}.$$

11. $dy = (2ax - x^2)^{\frac{5}{2}}(a - x)\, dx.$ $y = \tfrac{1}{7}(2ax - x^2)^{\frac{7}{2}}.$

133. Integrating Factor.—It has been easy, in the examples already given, to find the factor necessary to make the *differential factor* the differential of the *variable factor* [see (3) of Art. 130]; but sometimes this factor is not easily found, and often no such factor exists. There is a general method of ascertaining whether there is such a factor or not, and when there is, of finding it, which will be given in the two following examples:

12. $dy = 3(4bx^2 - 2cx^3)^{\frac{1}{3}}(4bx - 3cx^2)\, dx.$

Suppose A to be the constant factor required; then we have

$$y = \int \frac{1}{A}(4bx^2 - 2cx^3)^{\frac{1}{3}}(12Abx - 9Acx^2)\, dx.$$

If A be the required factor, we must have

$$d(4bx^2 - 2cx^3) = (12Abx - 9Acx^2)\, dx,$$

or $8bx - 6cx^2 = 12Abx - 9Acx^2,$

and since this result is to be true for every value of x, the coefficients of the like powers of x must be equal to each other, from the principle of indeterminate coefficients, giv-ing us two conditions,

$$12Ab = 8b, \tag{1}$$

and
$$-9Ac = -6c, \tag{2}$$

or $\quad A = \frac{2}{3}$, from (1), and $\quad A = \frac{2}{3}$, from (2) ;

hence $\frac{2}{3}$ is the factor required, and we have

$$y = \int \frac{2}{3} (4bx^2 - 2cx^3)^{\frac{1}{3}} (8bx - 6cx^2) \, dx$$

$$= \frac{2}{3} (4bx^2 - 2cx^3)^{\frac{4}{3}}.$$

13. $\quad dy = (2ax - x^2)^{\frac{3}{2}} (5a - x) \, dx.$

Let A be the required factor, then

$$y = \int \frac{1}{A} (2ax - x^2)^{\frac{3}{2}} (5Aa - Ax) \, dx.$$

$$\therefore \quad d (2ax - x^2) = (5Aa - Ax) \, dx ;$$

or $\qquad 2a - 2x = 5Aa - Ax.$

$$\therefore \quad 2a = 5Aa ; \tag{1}$$

and $\qquad -2 = -A ; \tag{2}$

or $\quad A = \frac{2}{5}$, from (1), and $\quad A = 2$, from (2).

These values of A being incompatible with each other, we infer that the differential cannot be integrated in this form.

14. $\quad dy = (2b + 3ax^2 - 5x^3)^{-\frac{1}{3}} (2ax - 5x^2) \, dx.$

$$y = \frac{1}{2} (2b + 3ax^2 - 5x^3)^{\frac{2}{3}}.$$

15. $\quad dy = (3ax^2 + 4bx^3)^{\frac{4}{3}} (2ax + 4bx^2) \, dx.$

$$y = \frac{1}{7} (3ax^2 + 4bx^3)^{\frac{7}{3}}.$$

16. $\quad dy = ax^2 dx + \dfrac{dx}{2\sqrt{x}} \cdot$ $\qquad y = \dfrac{ax^3}{3} + x^{\frac{1}{2}}.$

17. $dy = 5(5x^2 - 2x)^{\frac{1}{3}} x \, dx - (5x^2 - 2x)^{\frac{1}{3}} \, dx.$

Here $\qquad y = \int (5x^2 - 2x)^{\frac{1}{3}} (5x - 1) \, dx.$

$\qquad\qquad y = \frac{3}{8} (5x^2 - 2x)^{\frac{4}{3}}.$

134. A Differential may often be brought to the form required in (3) of Art. 130, by transposing a *variable factor* from the differential factor to the variable factor, or vice versa.

18. $dy = \dfrac{2a \, dx}{x \sqrt{2ax - x^2}}.$

Here $\quad y = \int (2ax - x^2)^{-\frac{1}{2}} 2ax^{-1} dx$

$\qquad\quad = \int - (2ax^{-1} - 1)^{-\frac{1}{2}} (-2ax^{-2}) \, dx;$

$\therefore \quad y = -2 (2ax^{-1} - 1)^{\frac{1}{2}}$ [by (3) of Art. 130]

$\qquad\quad = - \dfrac{2 \sqrt{2ax - x^2}}{x}.$

19. $dy = \dfrac{x \, dx}{(2ax - x^2)^{\frac{3}{2}}}.$ $\qquad y = \dfrac{x}{a \sqrt{2ax - x^2}}.$

20. $dy = \dfrac{ax \, dx}{(2bx + x^2)^{\frac{3}{2}}}.$ $\qquad y = \dfrac{ax}{b \sqrt{2bx + x^2}}$

21. $dy = \dfrac{a \, dx}{x \sqrt{3bx + 4c^2x^2}}.$ $\quad y = - \dfrac{2a \sqrt{3bx + 4c^2x^2}}{3bx}.$

22. $dy = \dfrac{(b + 2cx) \, dx}{a + bx + cx^2}.$ [See (4) of Art. 130.]

$\qquad y = \int \dfrac{d (a + bx + cx^2)}{a + bx + cx^2} = \log (a + bx + cx^2).$

23. $dy = \dfrac{8x^3 dx}{a + 2x^4}.$ $\qquad\qquad y = \log (a + 2x^4).$

24. $dy = \dfrac{5x^3 dx}{3x^4 + 7}.$

      $y = \tfrac{5}{12} \log (3x^4 + 7) = \log (3x^4 + 7)^{\frac{5}{12}}.$

25. $dy = \dfrac{\frac{1}{3}x dx}{x^2 + \frac{3}{4}}.$ $y = \log (x^2 + \tfrac{3}{4})^{\frac{1}{6}}.$

26. $dy = \dfrac{7x dx}{8a - 3x^2}.$ $y = \log \dfrac{1}{(8a - 3x^2)^{\frac{7}{6}}}.$

27. $dy = (b - x^2)^3 x^{\frac{1}{2}} dx.$

      $y = \tfrac{2}{3}b^3 x^{\frac{3}{2}} - \tfrac{6}{7}b^2 x^{\frac{7}{2}} + \tfrac{6}{11}b x^{\frac{11}{2}} - \tfrac{2}{15}x^{\frac{15}{2}}.$

28. $dy = \dfrac{x^{n-1} dx}{a + bx^n}.$ $y = \log (a + bx^n)^{\frac{1}{bn}}.$

29. $dy = 2 \log^3 x \dfrac{dx}{x}.$ $y = \tfrac{1}{2} \log^4 x.$

30. $dy = a^{2x} \log a dx.$ [See (4) of Art. 131.]

Here $y = \displaystyle\int a^{2x} \log a dx = \int \tfrac{1}{2} a^{2x} \log a d (2x) = \tfrac{1}{2} a^{2x}.$

31. $dy = 3a^{x^2} \log a x dx.$ $y = \tfrac{3}{2}a^{x^2}.$

32. $dy = b a^{3x} dx.$ $y = \dfrac{b a^{3x}}{3 \log a}.$

33. $dy = 5 e^x dx.$ $y = 5 e^x.$

34. $dy = \cos 3x dx.$ [See (6) of Art. 131.]

                $y = \tfrac{1}{3} \sin 3x.$

35. $dy = \cos (x^2) x dx.$ $y = \tfrac{1}{2} \sin (x^2).$

36. $dy = e^{\sin x} \cos x dx.$ $y = e^{\sin x}.$

37. $dy = - e^{\cos x} \sin x dx.$ $y = e^{\cos x}.$

38. $dy = \sin^2 (2x) \cos (2x) dx.$

Here $y = \displaystyle\int \tfrac{1}{2} \sin^2 (2x) \cos (2x) d (2x) = \tfrac{1}{6} \sin^3 (2x).$

39. $dy = \cos^3 (2x) \sin (2x) \, dx.$ $\quad y = - \frac{1}{8} \cos^4 (2x).$

40. $dy = \sec^2 (x)^3 \, x^2 dx.$ $\quad y = \frac{1}{3} \tan (x)^3.$

41. $dy = \sec (3x) \tan (3x) \, dx.$ $\quad y = \frac{1}{3} \sec (3x).$

42. $dy = \sin (ax) \, dx.$ $\quad y = \dfrac{1}{a} \text{vers} (ax).$

43. $dy = \tan x \, dx.$ $\quad y = \log \sec x.$

44. $dy = \sin \theta \sec^2 \theta \, d\theta.$ $\quad y = \sec \theta.$

45. $dy = \dfrac{2dx}{\sqrt{1 - 4x^2}}.$ [See (14) of Art. 131.]

$$y = \sin^{-1} (2x).$$

46. $dy = \dfrac{axdx}{\sqrt{1 - x^4}}.$ $\quad y = \dfrac{a}{2} \sin^{-1} (x^2).$

47. $dy = \dfrac{x^{\frac{1}{2}}dx}{\sqrt{2 - 4x^3}}.$

Here $\quad y = \displaystyle\int \dfrac{x^{\frac{1}{2}}dx}{\sqrt{2} \cdot \sqrt{1 - 2x^3}}$

$$= \dfrac{1}{\sqrt{2} \cdot \frac{3}{2}} \int \dfrac{\sqrt{2} \cdot \frac{3}{2} x^{\frac{1}{2}}dx}{\sqrt{2} \cdot \sqrt{1 - 2x^3}}$$

$$= \frac{1}{3} \int \dfrac{d \sqrt{2x^3}}{\sqrt{1 - 2x^3}}.$$

$$\therefore \quad y = \frac{1}{3} \sin^{-1} \sqrt{2x^3}.$$

48. $dy = \dfrac{dx}{\sqrt{2 - 9x^2}}.$ $\quad y = \frac{1}{3} \sin^{-1} \dfrac{3x}{\sqrt{2}}.$

49. $dy = \dfrac{- xdx}{\sqrt{2 - 5x^4}}.$ $\quad y = \dfrac{1}{2\sqrt{5}} \cos^{-1} \left(\dfrac{\sqrt{5}}{\sqrt{2}} x^2 \right)$

50. $dy = - \dfrac{dx}{\sqrt{ax - x^2}}.$ $\quad y = 2 \cos^{-1} \left(\sqrt{\dfrac{x}{a}} \right).$

51. $dy = \dfrac{3dx}{4 + 9x^2}.$ $\qquad$ $y = \frac{1}{2}\tan^{-1}\dfrac{3x}{2}.$

52. $dy = \dfrac{x^2dx}{1 + x^6}.$ $\qquad$ $y = \frac{1}{3}\tan^{-1}x^3.$

53. $dy = \dfrac{3dx}{\sqrt{6x - 9x^2}}.$ $\qquad$ $y = \text{vers}^{-1}\,3x.$

54. $dy = \dfrac{2dx}{x\sqrt{3x^2 - 5}}.$

Here $y = \displaystyle\int \dfrac{2dx}{x\sqrt{5}\sqrt{\frac{3}{5}x^2 - 1}} = \dfrac{2}{\sqrt{5}}\int \dfrac{\sqrt{\frac{3}{5}}dx}{\sqrt{\frac{3}{5}}x\sqrt{\frac{3}{5}x^2 - 1}}.$

$\therefore\ y = \dfrac{2}{\sqrt{5}}\sec^{-1}(x\sqrt{\tfrac{3}{5}}).$

55. $dy = -\dfrac{2dx}{\sqrt{5x^4 - 2x^2}}.$ $\quad$ $y = \sqrt{2}\,\text{cosec}^{-1}(x\sqrt{\tfrac{5}{2}}).$

56. $dy = \dfrac{-2x^{-1}dx}{\sqrt{14x^2 - 3}}.$ $\qquad$ $y = \dfrac{2}{\sqrt{3}}\text{cosec}^{-1}(x\sqrt{\tfrac{14}{3}}).$

57. $dy = \dfrac{3x^2dx}{\sqrt{4x^3 - 9x^6}}.$ $\qquad$ $y = \frac{1}{3}\text{vers}^{-1}\dfrac{9x^3}{2}.$

58. $dy = \dfrac{-6adx}{\sqrt{6 - 9x^2}}.$ $\qquad$ $y = a\,\text{covers}^{-1}\,3x^2.$

59. $dy = \dfrac{8x^{-\frac{2}{3}}dx}{\sqrt{2x^{\frac{1}{3}} - 6x^{\frac{2}{3}}}}.$ $\quad$ $y = 4\sqrt{6}\,\text{vers}^{-1}(6x^{\frac{1}{3}}).$

60. $dy = \dfrac{1 - \sin x}{x + \cos x}dx.$ $\quad$ $y = \log(x + \cos x).$

61. $dy = \dfrac{xdx}{1 + x^4}.$ $\qquad$ $y = \frac{1}{2}\tan^{-1}x^2.$

62. $dy = \dfrac{x^2 - 4x + 3}{x^3 - 6x^2 + 9x}dx.$ $\quad$ $y = \log(x^3 - 6x^2 + 9x)^{\frac{1}{3}}.$

63. $dy = \dfrac{3dx}{2 + 5x^2}.$ $\qquad$ $y = \dfrac{3}{\sqrt{10}} \tan^{-1} x \sqrt{\tfrac{5}{2}}.$

64. $dy = \dfrac{(a + \sqrt{x})^2\, dx}{\sqrt{x}}.$ $\quad$ $y = \tfrac{2}{3} (a + \sqrt{x})^3.$

65. $dy = \dfrac{(2a^2 + 4x^2)\, dx}{\sqrt{a^2 + x^2}}.$ $\quad$ $y = 2x \sqrt{a^2 + x^2}.$

66. $dy = \tan^2 x \sec^2 x dx.$ $\quad$ $y = \tfrac{1}{3} \tan^3 x.$

67. $dy = \dfrac{(1 - x^2)^2\, dx}{x}.$ $\quad$ $y = \log x - x^2 + \dfrac{x^4}{4}.$

68. $dy = \dfrac{(x - 2)\, dx}{x \sqrt{x}}.$ $\quad$ $y = 2 \sqrt{x} + \dfrac{4}{\sqrt{x}}.$

69. $dy = \dfrac{x^4 dx}{x^2 + 1}.$ $\quad$ $y = \tfrac{1}{3}x^3 - x + \tan^{-1} x.$

70. $dy = \dfrac{dx}{1 + x + x^2}.$

Here $y = \displaystyle\int \dfrac{dx}{\tfrac{3}{4} + (x + \tfrac{1}{2})^2}$

$$= \dfrac{2}{\sqrt{3}} \int \dfrac{\dfrac{2}{\sqrt{3}}\, dx}{1 + \left[(x + \tfrac{1}{2}) \dfrac{2}{\sqrt{3}} \right]^2}$$

$$= \dfrac{2}{\sqrt{3}} \tan^{-1} (x + \tfrac{1}{2}) \dfrac{2}{\sqrt{3}}.$$

71. $dy = \dfrac{dx}{2 - 2x + x^2}.$ $\quad$ $y = \tan^{-1} (x - 1).$

72. $dy = \dfrac{(m + nx)\, dx}{a^2 + x^2}.$

$$y = \dfrac{m}{a} \tan^{-1} \dfrac{x}{a} + \dfrac{n}{2} \log (a^2 + x^2).$$

135. Trigonometric Reduction. — A very slight acquaintance with Trigonometry will enable the student to solve the following examples easily. After a simple trigonometric reduction, the integrals are written out by inspection.

1. $dy = \tan^3 x\, dx.$

Here $y = \displaystyle\int \tan^3 x\, dx = \int (\sec^2 x - 1)\tan x\, dx$

$$= \int [\sec^2 x \tan x\, dx - \tan x\, dx]$$

$$= \tfrac{1}{2}\tan^2 x - \log \sec x. \quad \text{[See (9) of Art. 132.]}$$

2. $dy = \tan^4 x\, dx.$

$y = \tfrac{1}{3}\tan^3 x - \tan x + x. \quad \text{[(15) of Art. 132.]}$

3. $dy = \tan^5 x\, dx.$

$$y = \tfrac{1}{4}\tan^4 x - \tfrac{1}{2}\tan^2 x + \log \sec x.$$

4. $dy = \cot^3 x\, dx.$ $y = -\tfrac{1}{2}\cot^2 x - \log \sin x.$

5. $dy = \cot^4 x\, dx.$ $y = -\tfrac{1}{3}\cot^3 x + \cot x + x.$

6. $dy = \cot^5 x\, dx.$

$$y = -\tfrac{1}{4}\cot^4 x + \tfrac{1}{2}\cot^2 x + \log \sin x.$$

7. $dy = \sin^3 x\, dx.$

Here $y = \displaystyle\int \tfrac{1}{4}(3\sin x - \sin 3x)\, dx; \quad \text{(from Trig.),}$

$\therefore\ y = \tfrac{1}{12}\cos 3x - \tfrac{3}{4}\cos x;$

or $y = \tfrac{1}{3}\cos^3 x - \cos x.$*

8. $dy = \cos^3 x\, dx.$ $y = \sin x - \tfrac{1}{3}\sin^3 x.$

---

* By employing different methods, we often obtain integrals of the same expression which are different in form, and which sometimes appear at first sight not to agree. On examination, however, they will always be found to differ only by some constant; otherwise they could not have the same differential.

9. $dy = \cos^4 x\,dx.$

Here $y = \int (\cos^2 x)^2\, dx = \int \frac{1}{4}[1 + 2\cos 2x$

$$+ \cos^2 2x]\, dx \quad [\text{See (17) of Art. 132.}]$$

$$= \frac{1}{4}x + \frac{1}{4}\sin 2x + \frac{1}{8}\int \cos^2 2x\,d\,(2x)$$

$$= \frac{1}{4}x + \frac{1}{4}\sin 2x + \frac{1}{8}\left[x + \frac{1}{4}\sin 4x\right]; \ [\text{by (17) of Art. 132.}]$$

$$\therefore \ y = \frac{1}{32}\sin 4x + \frac{1}{4}\sin 2x + \frac{3}{8}x.$$

10. $dy = \sin^4 x\,dx.$ $\ y = \frac{1}{32}\sin 4x - \frac{1}{4}\sin 2x + \frac{3}{8}x.$

11. $dy = \sin^5 x\,dx.$ $\ y = -\cos x + \frac{2}{3}\cos^3 x - \frac{1}{5}\cos^5 x.$

12. $dy = \cos^5 x\,dx.$ $\ y = \sin x - \frac{2}{3}\sin^3 x + \frac{1}{5}\sin^5 x.$

13. $dy = \sin^3 x \cos^3 x\,dx.$

$$y = \frac{1}{192}\cos 6x - \frac{3}{64}\cos 2x. \quad (\text{From Ex. 7.})$$

14. $dy = \dfrac{dx}{\sin x \cos^3 x}.$ $\qquad y = \frac{1}{2}\sec^2 x + \log \tan x.$

15. $dy = \sin^3 x \cos^2 x\,dx.$ $\quad y = \frac{1}{5}\cos^5 x - \frac{1}{3}\cos^3 x.$

16. $dy = \cos^5 x \sin^5 x\,dx.$

$$y = -\frac{\cos^6 x}{2}\left(\frac{1}{3} - \frac{1}{2}\cos^2 x + \frac{1}{5}\cos^4 x\right).$$

17. $dy = \sin^6 x \cos^3 x\,dx = (\sin^6 x - \sin^8 x)\cos x\,dx.$

$$y = \sin^7 x \left(\frac{1}{7} - \frac{1}{9}\sin^2 x\right).$$

18. $dy = \dfrac{\sin^5 x}{\cos^2 x}\,dx.$

Here $y = \displaystyle\int \dfrac{(1 - 2\cos^2 x + \cos^4 x)\sin x\,dx}{\cos^2 x}$

$$= \sec x + 2\cos x - \frac{1}{3}\cos^3 x.$$

19. $dy = \cos^7 x\,dx.$

$$y = \sin x - \sin^3 x + \frac{3}{5}\sin^5 x - \frac{1}{7}\sin^7 x.$$

CHAPTER II.

136. Rational Fractions.—A fraction whose terms involve only positive and integral powers of the variable is called a rational fraction. Its general form is

$$\frac{ax^m + bx^{m-1} + cx^{m-2} + \text{etc.} \dots l}{a'x^n + b'x^{n-1} + c'x^{n-2} + \text{etc.} \dots l'}, \tag{1}$$

in which m and n are positive integers, and a, b, ..., a', b', ... are constants.

When m is $>$ or $= n$, (1) may, by common division, be reduced to the sum of an integral algebraic expression, and a fraction whose denominator will be the same as that of (1) and whose numerator will be at least one degree lower than the denominator. For example,

$$\frac{x^5}{x^3 - x^2 + x + 1} = x^2 + x - \frac{2x^2 + x}{x^3 - x^2 + x + 1}.$$

The former part can be integrated by the method of the preceding chapter; the fractional part may be integrated by decomposing it into a series of partial fractions, each of which can be integrated separately. There are three cases, which will be examined separately.

137. CASE I.—*When the denominator can be resolved into n real and unequal factors of the first degree*

For brevity, let $\dfrac{f(x)\,dx}{\phi(x)}$ denote the rational fraction whose integral is required, and let $(x-a)(x-b)\dots(x-l)$ be the n unequal factors of the denominator. Assume

$$\frac{f(x)}{\phi(x)} = \frac{A}{x-a} + \frac{B}{x-b} + \frac{C}{x-c} \cdots \frac{L}{x-l}, \quad (1)$$

where A, B, C, etc., are constants whose values are to be determined.

Clearing (1) of fractions, by multiplying each numerator by all the denominators except its own, we have

$$f(x) = A(x-b)(x-c)\ldots(x-l) + B(x-a)(x-c)\ldots(x-l)$$

$$+ \text{ etc. } + L(x-a)(x-b)\ldots(x-k), \quad (2)$$

which is an identical equation of the $(n-1)^{th}$ degree. To find A, B, C, etc., we may perform the operations indicated in (2), equate the coefficients of the like powers of x by the principle of indeterminate coefficients in Algebra, and solve the n resulting equations. The values of A, B, C, etc., thus determined, being substituted in (1) and the factor dx introduced, each term may be easily integrated by known methods.

In practice, however, in this first case, there is a simpler method of finding the values of A, B, etc., depending upon the fact that (2) is true for every value of x. If in (2) we make $x = a$, all the terms in the second member will reduce to 0, except the first, and we shall have

$$f(a) = A(a-b)(a-c)\ldots(a-l),$$

or $\qquad A = \dfrac{f(a)}{(a-b)(a-c)\ldots(a-l)} = \dfrac{f(a)}{\phi'(a)}.$

In the same way, making $x = b$, all the terms of (2) disappear except the second, giving us

$$f(b) = B(b-a)(b-c)\ldots(b-l),$$

or $\qquad B = \dfrac{f(b)}{(b-a)(b-c)\ldots(b-l)} = \dfrac{f(b)}{\phi'(b)}.$

Or, in general, the value of L is determined in any one of the terms, $\dfrac{L\,dx}{x-l}$, by substituting for x the corresponding root l of $\phi(x)$ in the expression $\dfrac{f(x)}{\phi'(x)}$; *i. e.*, $L = \dfrac{f(l)}{\phi'(l)}$.

EXAMPLES.

1. Integrate $dy = \dfrac{(x^2 + 1)\,dx}{x^3 + 6x^2 + 11x + 6}$.

In this example, the roots * of the denominator are found by Algebra to be -1, -2, -3.

$$\therefore\ x^3 + 6x^2 + 11x + 6 = (x+1)(x+2)(x+3).$$

Assume

$$\frac{x^2 + 1}{x^3 + 6x^2 + 11x + 6} = \frac{A}{x+1} + \frac{B}{x+2} + \frac{C}{x+3}. \quad (1)$$

$$\therefore\ x^2 + 1 = A(x+2)(x+3) + B(x+1)(x+3)$$
$$+ C(x+1)(x+2).$$

Making $x = -1$, we have $2 = 2A$, $\therefore A = 1$.

"　　　$x = -2$, " " $5 = -B$, $\therefore B = -5$.

"　　　$x = -3$, " " $10 = 2C$, $\therefore C = 5$.

Substituting these values of A, B, C, in (1), and multiplying by dx, we have

$$y = \int \frac{(x^2 + 1)\,dx}{x^3 + 6x^2 + 11x + 6}$$

$$= \int \frac{dx}{x+1} - 5\int \frac{dx}{x+2} + 5\int \frac{dx}{x+3}.$$

$$\therefore\ y = \log(x+1) - 5\log(x+2) + 5\log(x+3)$$

$$= \log \frac{(x+1)(x+3)^5}{(x+2)^5}.$$

---

* If the factors of the denominator are not easily seen, put it equal to 0, and solve the equation for x; the first root may be found by trial. x minus each of the several roots in turn will be the factors. (See Algebra.)

2. Integrate $dy = \dfrac{adx}{x^2 - a^2}.$

$$y = \tfrac{1}{2} \log \frac{x - a}{x + a} = \log \sqrt{\frac{x - a}{x + a}}.$$

3. Integrate $dy = \dfrac{(x^2 + 2)\,dx}{x^4 - 5x^2 + 4}.$

$$y = \tfrac{1}{2} \log \frac{x^2 - x - 2}{x^2 + x - 2}.$$

4. Integrate $dy = \dfrac{(5x + 1)\,dx}{x^2 + x - 2}.$

$$y = \log (x - 1)^2 (x + 2)^3.$$

5. Integrate $dy = \dfrac{dx}{a^2 - b^2 x^2}.$ $y = \dfrac{1}{2ab} \log \left(\dfrac{a + bx}{a - bx}\right).$

138. CASE II.—*When the denominator can be resolved into n real and equal factors of the first degree.*

Let the denominator of the rational fraction $\dfrac{f(x)}{\phi(x)}$ contain n factors, each equal to $x - a.$

Assume

$$\frac{f(x)}{\phi(x)} = \frac{A}{(x - a)^n} + \frac{B}{(x - a)^{n-1}} + \frac{C}{(x - a)^{n-2}}$$
$$+ \cdots \frac{L}{(x - a)} \qquad (1)$$

Clearing (1) of fractions by multiplying each term by the least common multiple of the denominators, we have

$$f(x) = A + B(x - a) + C(x - a)^2$$
$$+ \cdots L(x - a)^{n-1}, \qquad (2)$$

which is an identical equation of the $(n - 1)^{th}$ degree. To find the values of A, B, C, etc., we equate the coefficients of the like powers of x, as in the preceding Article, and solve the n resulting equations. The values of A, B, C, etc., thus determined, being substituted in (1), and the factor dx introduced, each term may be easily integrated by known methods.

In this case we cannot find the values of A, B, C, etc., by the second method used in Case I, but have to employ the first. When both equal and unequal factors, however, occur in the denominator, both methods may be combined to advantage.

E X A M P L E S.

1. Integrate $dy = \dfrac{(2 - 3x^2)\, dx}{(x + 2)^3}.$

Assume $\dfrac{2 - 3x^2}{(x + 2)^3} = \dfrac{A}{(x + 2)^3} + \dfrac{B}{(x + 2)^2} + \dfrac{C}{x + 2}.$ (1)

$\therefore \quad 2 - 3x^2 = A + Bx + 2B + Cx^2 + 4Cx + 4C.$

$$\therefore \quad A + 2B + 4C = 2. \tag{2}$$

$$B + 4C = 0. \tag{3}$$

$$C = -3. \tag{4}$$

Solving (2), (3), and (4), we get

$$A = -10, \qquad B = 12, \qquad C = -3.$$

Substituting these values of A, B, and C in (1), and multiplying by dx, we have

$$\frac{(2 - 3x^2)\, dx}{(x + 2)^3} = -\frac{10\, dx}{(x + 2)^3} + \frac{12\, dx}{(x + 2)^2} - \frac{3\, dx}{x + 2}.$$

$$\therefore \quad y = \int \frac{(2 - 3x^2)\, dx}{(x + 2)^3}$$

$$= \frac{5}{(x + 2)^2} - \frac{12}{x + 2} - 3 \log (x + 2).$$

2. Integrate $dy = \dfrac{(x^2 + x)\, dx}{(x - 2)^2 (x - 1)}.$

Assume

$$\frac{x^2 + x}{(x - 2)^2 (x - 1)} = \frac{A}{(x - 2)^2} + \frac{B}{(x - 2)} + \frac{C}{x - 1}. \quad (1)$$

$$\therefore \quad x^2 + x = A(x - 1) + B(x - 2)(x - 1)$$
$$+ C(x - 2)^2. \quad (2)$$

Here we may use the second method of Case I, as follows:

Making $x = 2$, we find $A = 6$.
" $x = 1$, " " $C = 2$.

Substituting in (2) for A and C their values, and making $x = 0$, we find

$$0 = -6 + 2B + 8; \quad \therefore \quad B = -1.$$

Substituting in (1), and multiplying by dx, we have

$$y = \int \frac{(x^2 + x)\, dx}{(x - 2)^2 (x - 1)}$$

$$= \int \frac{6dx}{(x - 2)^2} - \int \frac{dx}{x - 2} + \int \frac{2dx}{x - 1}$$

$$= -\frac{6}{x - 2} - \log(x - 2) + 2\log(x - 1).$$

$$\therefore \quad y = -\frac{6}{x - 2} + \log \frac{(x - 1)^2}{(x - 2)}.$$

3. Integrate $dy = \frac{(3x - 1)\, dx}{(x - 3)^2}.$

$$y = -\frac{8}{x - 3} + 3\log(x - 3).$$

4. Integrate $dy = \frac{x^2 - 4x + 3}{x^3 - 6x^2 + 9x}\, dx.$

$$y = \log\left[x(x - 3)^2\right]^{\frac{1}{3}}.$$

139. CASE III.—*When some of the simple factors of the denominator are imaginary.*

The methods given in Arts. 137 and 138 apply to the case of *imaginary*, as well as to real factors; but as the corresponding partial fractions appear in this case under an imaginary form, it is desirable to give an investigation in which the coefficients are all real. Since the denominator is real, if it contains imaginary factors, they must enter in pairs; that is, for every factor of the form $x \pm a + b\sqrt{-1}$, there must be another factor of the form $x \pm a - b\sqrt{-1}$,* otherwise the product of the factors would not be real. Every pair of conjugate imaginary factors of this form gives a real quadratic factor of the form $(x \pm a)^2 + b^2$.

Let the denominator contain n real and equal quadratic factors. Assume

$$\frac{f(x)}{\phi(x)} = \frac{Ax + B}{[(x \pm a)^2 + b^2]^n} + \frac{Cx + D}{[(x \pm a)^2 + b]^{n-1}}$$
$$+ \cdots \frac{Kx + L}{(x \pm a)^2 + b^2}. \tag{1}$$

If we clear (1) of fractions by multiplying each term by the least common multiple of the denominators, we shall have an identical equation of the $(2n - 1)^{th}$ degree. Equating the coefficients of the like powers of x, as in the two preceding Articles, and solving the $2n$ resulting equations, we find the values of A, B, C, etc. Substituting these values in (1), and introducing the factor dx, we have a series of partial fractions, the general form of each being

$$\frac{(Ax + B)\,dx}{[(x \pm a)^2 + b^2]^n},$$

in which n is an integer.

To integrate this expression, put $x \pm a = z$; $\therefore x = z \mp a$,

---

* Called *conjugate imaginary factors.*

$dx = dz$, $(x \pm a)^2 = z^2$. Substituting these values, we have,

$$\int \frac{(Ax + B)\, dx}{[(x \pm a)^2 + b^2]^n} = \int \frac{(Az \mp Aa + B)\, dz}{(z^2 + b^2)^n}$$

$$= \int \frac{Az\, dz}{(z^2 + b^2)^n} + \int \frac{(B \mp Aa)\, dz}{(z^2 + b^2)^n}$$

$$= -\frac{A}{2(n-1)(z^2 + b^2)^{n-1}} + \int \frac{A'\, dz}{(z^2 + b^2)^n},$$

(when $A' = B \mp Aa$);

so that the proposed integral is found to depend on the integral of this last expression; and it will be shown in Art. 151 that this integral may be made to depend finally upon $\int \frac{A\, dz}{z^2 + b^2}$, giving $\frac{A}{b} \tan^{-1} \frac{z}{b}$. (Art. 132, 3.)

EXAMPLES.

1. Integrate $dy = \dfrac{x\, dx}{x^3 - 1}$.

The factors of the denominator are

$$(x - 1) \quad \text{and} \quad (x^2 + x + 1).$$

therefore assume,

$$\frac{x}{x^3 - 1} = \frac{A}{x - 1} + \frac{Bx + C}{x^2 + x + 1}. \qquad (1)$$

$\therefore$ $x = Ax^2 + Ax + A + Bx^2 + Cx - Bx - C.$

$\therefore$ $A + B = 0$; $A - B + C = 1$; $A - C = 0.$

$\therefore$ $A = \frac{1}{3}$; $B = -\frac{1}{3}$; $C = \frac{1}{3}.$

$\therefore$ $y = \displaystyle\int \frac{x\, dx}{x^3 - 1} = \int \frac{1}{3}\frac{dx}{x - 1} - \int \frac{1}{3}\frac{(x - 1)\, dx}{x^2 + x + 1}$

$$= \tfrac{1}{3} \log (x - 1) - \tfrac{1}{3} \int \frac{(x - 1)\, dx}{(x + \tfrac{1}{2})^2 + \tfrac{3}{4}} ; \qquad (2)$$

(by changing the form of the denominator.)

Put $x + \tfrac{1}{2} = z$, then $x - 1 = z - \tfrac{3}{2}$, and $dx = dz$, and the second term of (2) becomes

$$- \tfrac{1}{3} \int \frac{(z - \tfrac{3}{2})\, dz}{z^2 + \tfrac{3}{4}} = - \tfrac{1}{3} \int \frac{z\, dz}{z^2 + \tfrac{3}{4}} + \tfrac{1}{2} \int \frac{dz}{z^2 + \tfrac{3}{4}}$$

$$= - \tfrac{1}{6} \log (z^2 + \tfrac{3}{4}) + \frac{1}{\sqrt{3}} \tan^{-1} \frac{2z}{\sqrt{3}} \quad \text{(Art. 132, 3.)}$$

$$= - \tfrac{1}{6} \log (x^2 + x + 1) + \frac{1}{\sqrt{3}} \tan^{-1} \frac{2x + 1}{\sqrt{3}},$$

(by restoring the value of z).

Substituting in (2), we have,

$$y = \tfrac{1}{3} \big[\log (x - 1) - \tfrac{1}{2} \log (x^2 + x + 1)$$
$$+ \sqrt{3} \tan^{-1} \frac{2x + 1}{\sqrt{3}} \big].$$

2. Integrate $dy = \dfrac{x^2 dx}{x^4 + x^2 - 2}.$

To find the factors of the denominator, put it $= 0$ and solve with respect to x^2; thus,

$$x^4 + x^2 - 2 = 0, \quad \text{or} \quad x^4 + x^2 = 2.$$

$$\therefore \quad x^2 = - \tfrac{1}{2} \pm \tfrac{3}{2} = 1 \text{ or } - 2.$$

$$\therefore \quad x^4 + x^2 - 2 = (x^2 - 1)(x^2 + 2).$$

Assume $\dfrac{x^2}{x^4 + x^2 - 2} = \dfrac{A}{x + 1} + \dfrac{B}{x - 1} + \dfrac{Cx + D}{x^2 + 2}.$ (1)

Hence $x^2 = A(x - 1)(x^2 + 2) + B(x + 1)(x^2 + 2)$
$$+ (Cx + D)(x - 1)(x + 1). \qquad (2)$$

We may equate the coefficients of the like powers of x, to find the values of A, B, C, D, or proceed as follows :

Making $x = -1$, we find $A = -\frac{1}{6}$.

" $x = 1$, " " $B = \frac{1}{6}$.

Substituting these values of A and B in (2) and equating the coefficients of x^3 and x^2, we have

$$6C = 0 \quad \text{and} \quad 6D = 4 ;$$

$$\therefore \quad C = 0 \quad \text{and} \quad D = \tfrac{2}{3}.$$

$$\therefore \quad y = -\tfrac{1}{6} \int \frac{dx}{x+1} + \tfrac{1}{6} \int \frac{dx}{x-1} + \tfrac{2}{3} \int \frac{dx}{x^2+2}$$

$$= \log \left(\frac{x-1}{x+1} \right)^{\frac{1}{6}} + \frac{\sqrt{2}}{3} \tan^{-1} \frac{x}{\sqrt{2}}.$$

EXAMPLES.

1. $dy = \dfrac{(x-1)\,dx}{x^2+6x+8}.$ $\qquad y = \log \dfrac{(x+4)^{\frac{5}{2}}}{(x+2)^{\frac{3}{2}}}.$

2. $dy = \dfrac{(2x+3)\,dx}{x^3+x^2-2x}.$ $\qquad y = \log \dfrac{(x-1)^{\frac{5}{6}}}{x^{\frac{3}{2}}(x+2)^{\frac{1}{6}}}.$

3. $dy = \dfrac{(3x^2-1)\,dx}{x(x-1)(x+1)}.$ $\quad y = \log(x^3-x).$

4. $dy = \dfrac{a\,dx}{x^2+bx}.$ $\qquad y = \log \left(\dfrac{x}{x+b} \right)^{\frac{a}{b}}.$

5. $dy = \dfrac{(2+3x-4x^2)\,dx}{4x-x^3}.$

$$y = \log \left[x^{\frac{1}{2}} (2+x)^{\frac{5}{3}} (2-x) \right].$$

6. $dy = \dfrac{(5x-2)\,dx}{x^3+6x^2+8x}.$ $\qquad y = \log \dfrac{(x+2)^3}{x^{\frac{1}{4}}(x+4)^{\frac{11}{4}}}.$

12

7. $dy = \dfrac{(a^3 + bx^2)\,dx}{a^2x - x^3}.$

$$y = a \log x - \frac{a + b}{2} \log (a^2 - x^2).$$

8. $dy = \dfrac{(3x - 5)\,dx}{x^2 - 6x + 8}.$ $\qquad y = \log \dfrac{(x - 4)^{\frac{7}{2}}}{(x - 2)^{\frac{1}{2}}}.$

9. $dy = \dfrac{x^2 dx}{(x - a)^2 (x + a)}.$

$$y = - \frac{a}{2(x - a)} + \log (x - a)^{\frac{3}{4}} (x + a)^{\frac{1}{4}}.$$

10. $dy = \dfrac{(x^3 + x^2 + 2)\,dx}{x\,(x - 1)^2 (x + 1)^2}.$

$$y = - \frac{3 + x}{2(x^2 - 1)} + \log \frac{x^2}{(x - 1)^{\frac{3}{4}} (x + 1)^{\frac{5}{4}}}.$$

11. $dy = \dfrac{x^2 dx}{(x + 2)^2 (x + 1)}.$

$$y = \frac{4}{x + 2} + \log (x + 1).$$

12. $dy = \dfrac{x dx}{(x + 2)(x + 3)^2}.$

$$y = - \frac{3}{x + 3} + \log \left(\frac{x + 3}{x + 2}\right)^2.$$

13. $dy = \dfrac{dx}{(x - 2)^2 (x + 3)^2}.$

$$y = - \frac{1}{25(x - 2)} - \tfrac{2}{125} \log (x - 2) - \frac{1}{25(x + 3)}$$
$$+ \tfrac{2}{125} \log (x + 3).$$

14. $dy = \dfrac{(x^2 - 4x + 3)\,dx}{x^3 - 6x^2 + 9x}.$

$$y = \log [x\,(x - 3)^2]^{\frac{1}{3}}.$$

15. $dy = \dfrac{x^2 dx}{(x+2)^2 (x+4)^2}.$

$$y = -\frac{5x+12}{x^2+6x+8} + \log\left(\frac{x+4}{x+2}\right)^2.$$

16. $dy = \dfrac{x\,dx}{(x+1)(x^2+1)}.$

$$y = \tfrac{1}{2}\tan^{-1}x + \log\frac{(x^2+1)^{\frac{1}{4}}}{(x+1)^{\frac{1}{2}}}.$$

17. $dy = \dfrac{dx}{x^3 - x^2 + 2x - 2}.$

$$y = \log\frac{(x-1)^{\frac{1}{3}}}{(x^2+2)^{\frac{1}{6}}} - \frac{1}{3\sqrt{2}}\tan^{-1}\frac{x}{\sqrt{2}}.$$

18. $dy = \dfrac{(x^2+x)\,dx}{x^3 - 2x^2 + x - 2}.$

$$y = \tfrac{3}{5}\tan^{-1}x + \log\frac{(x-2)^{\frac{6}{5}}}{(x^2+1)^{\frac{1}{10}}}.$$

19. $dy = \dfrac{(x^2 - x + 1)\,dx}{x^3 + x^2 + x + 1}.$

$$y = \log\frac{(x+1)^{\frac{3}{2}}}{(x^2+1)^{\frac{1}{4}}} - \tfrac{1}{2}\tan^{-1}x.$$

20. $dy = \dfrac{9x^2 + 9x - 128}{x^3 - 5x^2 + 3x + 9}dx.$

$$y = \log\frac{(x-3)^{17}}{(x+1)^8} + \frac{5}{x-3}.$$

21. $dy = \dfrac{2x\,dx}{(x^2+1)(x^2+3)}.$

$$y = \log\left(\frac{x^2+1}{x^2+3}\right)^{\frac{1}{2}}.$$

22. $dy = \dfrac{(x^3 - 1)\,dx}{x^2 - 4}.$

$$y = \frac{x^2}{2} + \log\left[(x + 2)^{\frac{3}{4}}\,(x - 2)^{\frac{7}{4}}\right].$$

23. $dy = \dfrac{x\,dx}{(x + 1)\,(x + 2)\,(x^2 + 1)}.$

$$y = \log\left[\frac{(x + 2)^{\frac{2}{5}}\,(x^2 + 1)^{\frac{1}{20}}}{(x + 1)^{\frac{1}{2}}}\right] + \tfrac{3}{10}\tan^{-1} x.$$

CHAPTER III.

INTEGRATION OF IRRATIONAL FUNCTIONS BY RATIONALIZATION.

140. Rationalization.—When an irrational function, which does not belong to one of the known elementary forms, is to be integrated, we endeavor to *rationalize* it ; that is, to transform it into an equivalent rational function of another variable, by suitable substitutions, and integrate the resulting functions by known methods.

141. Function containing only Monomial Surds.— When the function contains only monomial surds, it can be rationalized by substituting a new variable with an exponent equal to the least common multiple of all the denominators of the fractional exponents in the given function.

For example, let the expression be of the form,

$$dy = \frac{ax^{\frac{m'}{m}} + bx^{\frac{n'}{n}}}{a'x^{\frac{c'}{c}} + b'x^{\frac{e'}{e}}} dx.$$

Put
$$x = z^{mnce} ;$$

$$\therefore x^{\frac{m'}{m}} = z^{m'nce} : x^{\frac{n'}{n}} = z^{n'mce} : x^{\frac{c'}{c}} = z^{mnc'e} : x^{\frac{e'}{e}} = z^{mnce'}.$$

$$dx = mncez^{mnce-1}dz.$$

Hence
$$dy = \frac{az^{m'nce} + bz^{mn'ce}}{a'z^{mnc'e} + b'z^{mnce'}} mncez^{mnce-1}dz ;$$

which is evidently rational.

1. Integrate $dy = \dfrac{1 - x^{\frac{1}{2}}}{1 - x^{\frac{1}{3}}} dx.$ (1)

Put $\qquad\qquad x = z^6;$

then $\quad x^{\frac{1}{2}} = z^3, \quad x^{\frac{1}{3}} = z^2, \quad$ and $\quad dx = 6z^5 dz;$

$$\therefore \ dy = \frac{(1 - z^3)\, 6z^5 dz}{1 - z^2}$$

$$= 6\left(z^6 + z^4 - z^3 + z^2 - z + 1 - \frac{1}{1 + z}\right) dz.$$

Integrating by known methods, and replacing z by its value, we have

$$y = 6\left[\frac{x^{\frac{7}{6}}}{7} + \frac{x^{\frac{5}{6}}}{5} - \frac{x^{\frac{2}{3}}}{4} + \frac{x^{\frac{1}{2}}}{3} - \frac{x^{\frac{1}{3}}}{2} + x^{\frac{1}{6}} - \log\left(1 + x^{\frac{1}{6}}\right)\right].$$

2. Integrate $dy = \dfrac{3x^{\frac{2}{3}}\, dx}{2x^{\frac{1}{3}} - x^{\frac{2}{3}}}.$

$$y = -18\left[\frac{x^{\frac{5}{6}}}{5} + \frac{x^{\frac{2}{3}}}{2} + \frac{4x^{\frac{1}{2}}}{3} + 4x^{\frac{1}{3}} + 16x^{\frac{1}{6}} + 32\log\left(2 - x^{\frac{1}{6}}\right)\right].$$

142. Functions containing only Binomial Surds of the First Degree.— When the function involves no surd except one of the form $(a + bx)^{\frac{m}{n}}$, it can be rationalized as in the last Article, by treating $a + bx$ as the variable. And therefore can be integrated.

For example, let the expression be of the form,

$$dy = \frac{x^n dx}{\sqrt{a + bx}},$$

where n is a positive integer.

Put $\qquad\qquad a + bx = z^2;$

then $\quad dx = \dfrac{2z\,dz}{b}, \quad x = \dfrac{z^2 - a}{b}, \quad$ and $\quad x^n = \dfrac{(z^2 - a)^n}{b^n}.$

$$\therefore \quad dy = \frac{x^n dx}{\sqrt{a + bx}} = \frac{2(z^2 - a)^n dz}{b^{n+1}}.$$

This may be expanded by the Binomial Theorem, and each term integrated separately. It is also evident that the expression

$$\frac{x^n dx}{(a + bx)^{\frac{p}{q}}}$$

can be integrated by the same substitution.

1. Integrate $dy = \dfrac{dx}{x\sqrt{1 + x}}$.

Put $1 + x = z^2$; then $dx = 2z\,dz$ and $x = z^2 - 1$.

$$\therefore \quad dy = \frac{dx}{x\sqrt{1 + x}} = \frac{2dz}{z^2 - 1}$$

$$= \frac{dz}{z - 1} - \frac{dz}{z + 1} \quad \text{(Art. 137)};$$

or $\qquad y = \log \dfrac{z - 1}{z + 1} = \log \dfrac{\sqrt{1 + x} - 1}{\sqrt{1 + x} + 1}.$

2. Integrate $dy = \dfrac{x^3 dx}{(1 + 4x)^{\frac{5}{3}}}$.

Put $\qquad\qquad (1 + 4x) = z^2;$

then $\qquad dx = \dfrac{z\,dz}{2}, \quad x^3 = \dfrac{(z^2 - 1)^3}{64}, \quad (1 + 4x)^{\frac{5}{3}} = z^5.$

$$\therefore \quad dy = \frac{1}{128} \frac{(z^2 - 1)^3 \, dz}{z^4}$$

$$= \frac{1}{128}\left[z^2 - 3 + \frac{3}{z^2} - \frac{1}{z^4} \right] dz.$$

$$\therefore \quad y = \frac{1}{128}\left[\tfrac{1}{3}z^3 - 3z - \frac{3}{z} + \frac{1}{3z^3} \right],$$

or

$$y = \frac{1}{128}\left[\frac{(1+4x)^{\frac{3}{2}}}{3} - 3(1+4x)^{\frac{1}{2}} - \frac{3}{(1+4x)^{\frac{1}{2}}} + \frac{1}{3(1+4x)^{\frac{3}{2}}}\right].$$

143. Functions of the Form $\dfrac{x^{2n+1}\,dx}{(a + bx^2)^{\frac{1}{2}}}$, **where** n **is a Positive Integer.**

Put
$$a + bx^2 = z^2;$$

then
$$x\,dx = \frac{z\,dz}{b}, \quad x^2 = \frac{z^2 - a}{b}, \quad x^{2n} = \frac{(z^2 - a)^n}{b^n}.$$

$$\therefore \quad \frac{x^{2n+1}\,dx}{(a + bx^2)^{\frac{1}{2}}} = \frac{(z^2 - a)^n\,dz}{b^{n+1}},$$

which may be expanded by the Binomial Theorem, and each term integrated separately. It is also evident that the expression

$$\frac{x^{2n+1}\,dx}{(a + bx^2)^{\frac{r}{2}}}$$

can be integrated by the same substitution.

1. Integrate $dy = \dfrac{x^3 dx}{\sqrt{1 - x^2}}.$

Put $1 - x^2 = z^2$; then $x\,dx = -z\,dz,\ x^2 = 1 - z^2.$

$$\therefore \quad dy = \frac{x^3 dx}{\sqrt{1 - x^2}} = -(1 - z^2)\,dz.$$

$$\therefore \quad y = \tfrac{1}{3}z^3 - z = \tfrac{1}{3}(1 - x^2)^{\frac{3}{2}} - (1 - x^2)^{\frac{1}{2}}.$$

2. Integrate $dy = \dfrac{x^3 dx}{(a + cx^2)^{\frac{5}{2}}}.$ $y = -\dfrac{(2a + 3cx^2)}{3c^2(a + cx^2)^{\frac{3}{2}}}.$

144. Functions containing only Trinomial Surds of the Form $\sqrt{a + bx \pm x^2}.$

There are two cases, according as x^2 is $+$ or $-$.

CASE I.— *When x^2 is $+$.*

Assume
$$\sqrt{a + bx + x^2} = z - x\,;$$

then
$$a + bx = z^2 - 2zx\,; \qquad \therefore \quad x = \frac{z^2 - a}{b + 2z}.$$

$$dx = \frac{2(z^2 + bz + a)\,dz}{(b + 2z)^2}\,;$$

and
$$\sqrt{a + bx + x^2} = z - \frac{z^2 - a}{b + 2z} = \frac{z^2 + bz + a}{2z + b}.$$

The values of x, dx, and $\sqrt{a + bx + x^2}$ being expressed in rational terms of z, the transformed function will be rational, and may therefore be integrated and the z replaced by its value $\sqrt{a + bx + x^2} + x$.

CASE II.— *When x^2 is $-$.*

Let α and β be the two roots of the equation
$$x^2 - bx - a = 0\,;$$

then we have
$$x^2 - bx - a = (x - \alpha)(x - \beta)\,;$$

$$\therefore \quad a + bx - x^2 = -(x - \alpha)(x - \beta)$$
$$= (x - \alpha)(\beta - x).$$

Assume
$$\sqrt{a + bx - x^2} = \sqrt{(x - \alpha)(\beta - x)}$$
$$= (x - \alpha)z\,;$$

$$\therefore \quad (x - \alpha)(\beta - x) = (x - \alpha)^2 z^2,$$

or
$$(\beta - x) = (x - \alpha)z^2\,;$$

whence,
$$x = \frac{\alpha z^2 + \beta}{z^2 + 1},$$

$$\therefore \quad dx = 2\frac{(\alpha - \beta)z\,dz}{(z^2 + 1)^2},$$

and
$$\sqrt{a + bx - x^2} = \left(\frac{\alpha z^2 + \beta}{z^2 + 1} - \alpha\right)z$$

$$= \frac{(\beta - \alpha)z}{z^2 + 1}.$$

The values of x, dx, $\sqrt{a + bx - x^2}$, being expressed in rational terms of z, the transformed function will be rational.

1. Integrate $dy = \dfrac{dx}{\sqrt{a + bx + x^2}}.$

Assume $\sqrt{a + bx + x^2} = z - x;$

then, as in Case I, we have

$$a + bx = z^2 - 2zx; \qquad \therefore \quad x = \frac{z^2 - a}{b + 2z}.$$

$$dx = \frac{2\,(z^2 + bz + a)\,dz}{(b + 2z)^2}$$

$$\sqrt{a + bx + x^2} = \frac{z^2 + bz + a}{2z + b}.$$

$$\therefore \quad dy = \frac{2\,(z^2 + bz + a)\,dz \times (2z + b)}{(b + 2z)^2 \times (z^2 + bz + a)}$$

$$= \frac{2dz}{b + 2z} = \frac{dz}{\dfrac{b}{2} + z}.$$

$$y = \int \frac{dz}{\dfrac{b}{2} + z} = \log\left(\frac{b}{2} + z\right)$$

$$= \log\left[\frac{b}{2} + x + \sqrt{a + bx + x^2}\right].$$

If $b = 0$, we have

$$y = \int \frac{dx}{\sqrt{a + x^2}} = \log\left(x + \sqrt{a + x^2}\right);$$

and if $a = 1$, we have

$$y = \int \frac{dx}{\sqrt{1 + x^2}} = \log\left(x + \sqrt{1 + x^2}\right).$$

Had we integrated the expression $\dfrac{2dz}{b+2z}$ without dividing both terms of it by 2, we would have found for the integral the following : $y = \log(b+2z) = \log\left[b+2x+2\sqrt{a+bx+x^2}\right]$, which differs from the above integral only by the term, $\log 2$, which is a constant. (See Note to Art. 135.)

2. Integrate $dy = \dfrac{dx}{\sqrt{a+bx-x^2}}.$

Let α and β be the roots of $x^2 - bx - a = 0$; then, as in Case II, we have

$$\sqrt{a+bx-x^2} = \sqrt{(x-\alpha)(\beta-x)} = (x-\alpha)z.$$

$$(\beta-x) = (x-\alpha)z^2; \qquad x = \frac{\alpha z^2 + \beta}{z^2 + 1}.$$

$$dx = \frac{2(\alpha-\beta)zdz}{(z^2+1)^2}; \qquad \sqrt{a+bx-x^2} = \frac{(\beta-\alpha)z}{z^2+1}.$$

$$\therefore \quad dy = \frac{2(\alpha-\beta)zdz(z^2+1)}{(z^2+1)^2(\beta-\alpha)z} = -\frac{2dz}{1+z^2};$$

$$\therefore \quad y = \int \frac{dx}{\sqrt{a+bx-x^2}} = -2\int \frac{dz}{1+z^2}$$

$$= -2\tan^{-1}z = -2\tan^{-1}\sqrt{\frac{\beta-x}{x-\alpha}}.$$

3. Integrate $dy = \dfrac{dx}{x\sqrt{1+x+x^2}}.$

Assume $\sqrt{1+x+x^2} = z-x$, and we have, as in Case I,

$$x = \frac{z^2-1}{1+2z}; \qquad dx = \frac{2(z^2+z+1)dz}{(1+2z)^2}.$$

$$\sqrt{1+x+x^2} = \frac{z^2+z+1}{1+2z}.$$

$$\therefore \quad dy = \frac{2(z^2+z+1)dz(1+2z)(1+2z)}{(1+2z)^2(z^2+z+1)(z^2-1)} = \frac{2dz}{z^2-1}$$

$$= \frac{dz}{z-1} - \frac{dz}{z+1} \quad \text{(Art. 137)}.$$

$$\therefore \quad y = \int \frac{dx}{x\sqrt{1 + x + x^2}} = \int \frac{dz}{z - 1} - \int \frac{dz}{z + 1}$$

$$= \log \frac{z - 1}{z + 1} = \log \frac{x - 1 + \sqrt{1 + x + x^2}}{x + 1 + \sqrt{1 + x + x^2}}$$

$$= \log \frac{3x}{2 + x + 2\sqrt{1 + x + x^2}}.$$

145. Binomial Differentials.— Expressions of the form

$$dy = x^m (a + bx^n)^p\, dx,$$

in which m, n, p denote any numbers, positive, negative, or fractional, are called binomial differentials.

This expression can always be reduced to another, in which m and n are integers and n positive.

1st. For if m and n are fractional, and the binomial of the form

$$x^{-\frac{4}{3}} \left(a + bx^{\frac{1}{2}}\right)^p dx,$$

we may substitute for x another variable whose exponent is equal to the least common multiple of the denominators of the exponents of x, as in Art. 141. We shall then have an expression in which the exponents are whole numbers.

Thus, if we put $x = z^6$, we have

$$x^{-\frac{4}{3}} \left(a + bx^{\frac{1}{2}}\right)^p dx = 6z^{-3} (a + bz^3)^p\, dz,$$

in which the exponents of z are whole numbers, and the exponent of z within the parenthesis is positive.

2d. If n be negative, or the binomial of the form

$$x^m (a + bx^{-n})^p\, dx,$$

we may put $x = \dfrac{1}{z}$, and obtain

$$x^m (a + bx^{-n})^p\, dx = -z^{-m-2} (a + bz^n)^p\, dz,$$

in which the exponents of z are whole numbers, and the one within the parenthesis is positive.

3d. If x be in both terms, or the binomial is of the form

$$x^m (ax^t + bx^n)^p \, dx,$$

we may take x^t out of the parenthesis, and we shall have

$$x^{m+pt} (a + bx^{n-t})^p \, dx,$$

in which only one of the terms within the parenthesis contains the variable.

146. The Conditions under which the General Form

$$dy = x^m (a + bx^n)^{\frac{p}{q}} dx,$$

can be rationalized, any or all of the exponents being fractional.

(1.) Assume $\quad a + bx^n = z^q$.

Then $\qquad (a + bx^n)^{\frac{p}{q}} = z^p.$ \hfill (1)

Also $\qquad\qquad x = \left(\dfrac{z^q - a}{b} \right)^{\frac{1}{n}},$

and $\qquad\qquad x^m = \left(\dfrac{z^q - a}{b} \right)^{\frac{m}{n}}.$ \hfill (2)

$$dx = \frac{q}{nb} z^{q-1} \left(\frac{z^q - a}{b} \right)^{\frac{1}{n}-1} dz. \qquad (3)$$

Multiplying (1), (2), and (3) together, we have

$$dy = x^m (a + bx^n)^{\frac{p}{q}} dx = \frac{q}{nb} z^{p+q-1} \left(\frac{z^q - a}{b} \right)^{\frac{m+1}{n}-1} dz, \quad (4)$$

an expression which is rational when $\dfrac{m+1}{n}$ is an integer, or 0.

(2.) Assume $\quad a + bx^n = z^q x^n$.

Then $\qquad\qquad x^n = a (z^q - b)^{-1}.$ \hfill (1)

$$\therefore \quad x = a^{\frac{1}{n}} (z^q - b)^{-\frac{1}{n}}. \tag{2}$$

$$\therefore \quad x^m = a^{\frac{m}{n}} (z^q - b)^{-\frac{m}{n}}, \tag{3}$$

$$dx = - \frac{q}{n} a^{\frac{1}{n}} (z^q - b)^{-\frac{1}{n}-1} z^{q-1} dz. \tag{4}$$

Multiplying (1) by b, adding a, and taking $\frac{p}{q}$ power, we have

$$(a + bx^n)^{\frac{p}{q}} = a^{\frac{p}{q}} (z^q - b)^{-\frac{p}{q}} z^p. \tag{5}$$

Multiplying (3), (4), and (5) together, we have

$$x^m (a + bx^n)^{\frac{p}{q}} dx = - \frac{q}{n} a^{\left(\frac{m}{n}+\frac{p}{q}+\frac{1}{n}\right)} (z^q - b)^{-\left(\frac{m+1}{n}+\frac{p}{q}+1\right)} z^{p+q-1} dz,$$

an expression which is rational when $\dfrac{m+1}{n} + \dfrac{p}{q}$ is an integer, or 0.*

Therefore there are two cases in which the general binomial differential can be rationalized:

1st. *When the exponent of the variable without the parenthesis increased by unity, is exactly divisible by the exponent of the variable within the parenthesis.*

2d. *When the fraction thus formed, increased by the exponent of the parenthesis, is an integer.*

REM.—These two cases are called the *conditions of integrability* of binomial differentials,† and when either of them is fulfilled, the integration may be effected. If, in the former case, $\dfrac{m+1}{n} - 1$ is a positive integer or 0, or in the latter case, $\dfrac{m+1}{n} + \dfrac{p}{q} + 1$ is a negative integer

---

* The student will observe that Art. 143 is a particular case of this Article, resulting from making m an odd positive integer, and $n = 2$.

† These are the only cases of the general form which, in the present state of analysis, can be made rational. When neither of these conditions is satisfied, the expression, if $\dfrac{p}{q}$ be a fractional index, is, in general, incapable of integration in a finite number of terms.

or 0, the binomial $(z^q - a)$ or $(z^q - b)$ will have a positive integral exponent, and hence can be expanded by the Binomial Theorem, and each term integrated separately. But if, in the former case, $\dfrac{m+1}{n} - 1$ is a negative integer, or in the latter, $\dfrac{m+1}{n} + \dfrac{p}{q} + 1$ is a positive integer, the exponent of the binomial $(z^q - b)$ will be negative, and the form will be reduced to a rational fraction whose denominator is a binomial, and hence the integration may be performed by means of Chapter II. But as the integration by this method usually gives complicated results, it is expedient generally not to rationalize in such cases, but to integrate by the *reduction formulæ* given in the next Chapter.

1. Integrate $dy = x^5 (a + x^2)^{\frac{1}{3}} dx$.

Here $\dfrac{m+1}{n} - 1 = 2$, a positive integer, and therefore it can be integrated by the first method.

Let $\qquad\qquad (a + x^2) = z^3$.

Then $\qquad\qquad (a + x^2)^{\frac{1}{3}} = z$. $\hfill (1)$

$$x^6 = (z^3 - a)^3,$$

$$x^5 dx = \tfrac{3}{2}(z^3 - a)^2 z^2 dz. \hfill (2)$$

Multiplying (1) and (2) together, we have

$$dy = x^5 (a + x^2)^{\frac{1}{3}} dx = \tfrac{3}{2}(z^3 - a)^2 z^3 dz.$$

$$\therefore \; y = \frac{3}{2} \int (z^9 - 2z^6 a + a^2 z^3) \, dz = \frac{3}{2}\left(\frac{z^{10}}{10} - \frac{2az^7}{7} + \frac{a^2 z^4}{4}\right)$$

$$= \tfrac{3}{20}(a + x^2)^{\frac{10}{3}} - \frac{3a}{7}(a + x^2)^{\frac{7}{3}} + \frac{3a^2}{8}(a + x^2)^{\frac{4}{3}}.$$

2. Integrate $dy = \dfrac{dx}{x^4 (1 + x^2)^{\frac{1}{2}}} = x^{-4}(1 + x^2)^{-\frac{1}{2}} dx$.

Here $\dfrac{m+1}{n} + \dfrac{p}{q} + 1 = \dfrac{-4+1}{2} - \dfrac{1}{2} + 1 = -1$, a negative integer, and hence it can be integrated by the second method.

Let
$$(1 + x^2) = z^2x^2.$$

Then
$$x^2 = (z^2 - 1)^{-1}.$$
$$x = (z^2 - 1)^{-\frac{1}{2}};$$
$$x^{-4} = (z^2 - 1)^2. \tag{1}$$
$$(1 + x^2) = 1 + (z^2 - 1)^{-1}$$
$$= z^2 (z^2 - 1)^{-1}.$$
$$(1 + x^2)^{-\frac{1}{2}} = z^{-1} (z^2 - 1)^{\frac{1}{2}}. \tag{2}$$
$$dx = - (z^2 - 1)^{-\frac{3}{2}} z\, dz. \tag{3}$$

Multiplying (1), (2), (3) together, we have

$$dy = x^{-4} (1 + x^2)^{-\frac{1}{2}} dx = - (z^2 - 1)\, dz.$$

$$\therefore\ y = - \int (z^2 - 1)\, dz = z - \tfrac{1}{3} z^3$$

$$= \frac{(1 + x^2)^{\frac{1}{2}}}{x} - \frac{\tfrac{1}{3}(1 + x^2)^{\frac{3}{2}}}{x^3} = \frac{(1 + x^2)^{\frac{1}{2}}}{3x^3}(2x^2 - 1).$$

EXAMPLES.

1. $dy = \dfrac{\left(2x^{\frac{1}{2}} - 3x^{\frac{2}{3}}\right) dx}{5x^{\frac{1}{5}}}.$ (Art. 141.)

$y = \tfrac{3}{10}x^{\frac{4}{3}} - \tfrac{2}{5}x^{\frac{8}{5}}.$

2. $dy = \dfrac{x^{\frac{1}{2}} - 2x^{\frac{1}{3}}}{1 + x^{\frac{1}{6}}}\, dx.$

$y = \tfrac{6}{7}x^{\frac{7}{6}} - 2x - \tfrac{6}{5}x^{\frac{5}{6}} + 3x^{\frac{2}{3}} + 2x^{\frac{1}{2}} - 6x^{\frac{1}{3}} - 6x^{\frac{1}{6}}$
$\qquad + 6 \log \left(x^{\frac{1}{6}} + 1\right) + 6 \tan^{-1} x^{\frac{1}{6}}.$

3. $dy = \dfrac{2x^{\frac{1}{2}} - 3x^{\frac{1}{6}}}{3x^{\frac{2}{3}} + x^{\frac{3}{4}}}\, dx.$

$y = 12\left(\tfrac{2}{9}x^{\frac{3}{4}} - \tfrac{3}{4}x^{\frac{2}{3}} + \tfrac{18}{7}x^{\frac{7}{12}} - 9x^{\frac{1}{4}}\right)$
$+ 1908\left[\tfrac{1}{5}x^{\frac{5}{12}} - \tfrac{3}{4}x^{\frac{1}{3}} + 3x^{\frac{1}{4}} - \tfrac{27}{2}x^{\frac{1}{6}} + 81x^{\frac{1}{12}} - 243 \log\left(x^{\frac{1}{12}} + 3\right)\right].$

4. $dy = \dfrac{1 + x^{\frac{1}{4}}}{1 + x^{\frac{1}{3}}}\, dx.$

$$y = 12 \left\{ \frac{x^{\frac{11}{12}}}{11} + \frac{x^{\frac{2}{3}}}{8} - \frac{x^{\frac{7}{12}}}{7} - \frac{x^{\frac{1}{2}}}{4} + \frac{x^{\frac{1}{3}}}{3} + \tfrac{1}{4} \log\left(1 + x^{\frac{1}{3}}\right) \right.$$

$$\left. - \frac{1}{2^{\frac{3}{2}}} \left[\log\left(\frac{x^{\frac{1}{6}} - 2^{\frac{1}{2}} x^{\frac{1}{12}} + 1}{x^{\frac{1}{6}} + 2^{\frac{1}{2}} x^{\frac{1}{12}} + 1} \right) + 2 \tan^{-1}\left(\frac{2^{\frac{1}{2}} x^{\frac{1}{12}}}{1 - x^{\frac{1}{6}}} \right) \right] \right\}.$$

5. $dy = \dfrac{x\,dx}{(1 + x)^{\frac{3}{2}}}.$ (Art. 142.)

$$y = 2 \left[\sqrt{1 + x} + \frac{1}{\sqrt{1 + x}} \right].$$

6. $dy = \dfrac{dx}{x\sqrt{a + bx}}.$

$$y = \frac{2}{\sqrt{a}} \log \frac{\sqrt{a + bx} - \sqrt{a}}{\sqrt{x}}.$$

7. $dy = \dfrac{dx}{(1 + x)^{\frac{3}{4}} + (1 + x)^{\frac{1}{2}}}.$

$$y = 2 \tan^{-1} (1 + x)^{\frac{1}{4}}.$$

8. $dy = 4\left(x + \sqrt{x + 3} + \sqrt[3]{x + 3} \right) dx.$

$$y = 2\,(x+3)^2 - 12\,(x+3) + \tfrac{8}{3}(x+3)^{\frac{3}{2}} + 3\,(x+3)^{\frac{4}{3}}.$$

9. $dy = \dfrac{x^5 dx}{\sqrt{1 + x^2}}.$ (Art. 143.)

$$y = \tfrac{1}{5}(1 + x^2)^{\frac{5}{2}} - \tfrac{2}{3}(1 + x^2)^{\frac{3}{2}} + (1 + x^2)^{\frac{1}{2}}.$$

10. $dy = \dfrac{x^3 dx}{(1 + x^2)^{\frac{5}{2}}}.$ $\qquad y = - \dfrac{3x^2 + 2}{3\,(1 + x^2)^{\frac{3}{2}}}.$

11. $dy = \dfrac{x^3 dx}{(1 + x^2)^{\frac{3}{2}}}.$ $\qquad y = \dfrac{x^2 + 2}{\sqrt{1 + x^2}}.$

12. $dy = \dfrac{dx}{\sqrt{1 + x + x^2}}.$ (Art. 144.)

$y = \log\left(1 + 2x + 2\sqrt{1 + x + x^2}\right).$

(See Art. 144, Ex. 1.)

13. $dy = \dfrac{dx}{\sqrt{x^2 - x - 1}}.$

$y = \log\left(2x - 1 + 2\sqrt{x^2 - x - 1}\right).$

14. $dy = \dfrac{dx}{\sqrt{2 - x - x^2}}.$ $y = -2\tan^{-1}\sqrt{\dfrac{1 - x}{x + 2}}.$

15. $dy = \dfrac{dx}{\sqrt{1 + x - x^2}}.$

$y = -2\tan^{-1}\sqrt{\dfrac{\frac{1}{2} - \frac{1}{2}\sqrt{5} - x}{x - \frac{1}{2} - \frac{1}{2}\sqrt{5}}}.$

16. $dy = \dfrac{dx}{\sqrt{a^2 + b^2 x^2}}.$

$\left(\text{Assume } \sqrt{\dfrac{a^2}{b^2} + x^2} = z - x, \text{ etc.}\right)$

$y = \dfrac{1}{b}\log\left(bx + \sqrt{a^2 + b^2 x^2}\right).$

17. $dy = \dfrac{dx}{x\sqrt{a^2 + b^2 x^2}}.$ $y = \dfrac{1}{a}\log\left(\dfrac{\sqrt{a^2 + b^2 x^2} - a}{bx}\right)$

$\text{or}\quad -\dfrac{1}{a}\log\left(\dfrac{a + \sqrt{a^2 + b^2 x^2}}{x}\right)$

18. $dy = \dfrac{(2x + x^2)^{\frac{1}{2}}\,dx}{x^2}.$

$y = \log\left(x + 1 + \sqrt{2x + x^2}\right) - \dfrac{4}{x + \sqrt{2x + x^2}}.$

19. $dy = \dfrac{dx}{(1 + x^2)\sqrt{1 - x^2}}.$

$y = \dfrac{1}{\sqrt{2}}\tan^{-1}\left(\dfrac{x\sqrt{2}}{\sqrt{1 - x^2}}\right).$

20. $dy = \dfrac{a\,dx}{\sqrt{2ax + x^2}}.$

$y = a\log\left(x + a + \sqrt{2ax + x^2}\right).$

21. $dy = \dfrac{dx}{\sqrt{4x^2 - 7}}.$ $y = \tfrac{1}{2}\log\left(2x + \sqrt{4x^2 - 7}\right).$

(Compare with Ex. 16.)

22. $dy = \dfrac{2\,dx}{\sqrt{3x + 4x^2}}.$ $y = \log\left(x + \tfrac{3}{8} + \sqrt{\tfrac{3}{4}x + x^2}\right).$

(Compare with Ex. 20.)

23. $dy = x^5(2 + 3x^2)^{\frac{1}{3}}\,dx.$ (Art. 146.)

$y = \dfrac{1}{27}\left[\dfrac{(2 + 3x^2)^{\frac{7}{3}}}{7} - \tfrac{4}{5}(2 + 3x^2)^{\frac{5}{3}} + \tfrac{4}{3}(2 + 3x^2)^{\frac{3}{3}}\right].$

24. $dy = x^3(a + bx^2)^{\frac{3}{2}}\,dx.$

$y = (a + bx^2)^{\frac{5}{2}}\left(\dfrac{5bx^2 - 2a}{35b^2}\right).$

25. $dy = x^3(a + bx^2)^{\frac{1}{2}}\,dx.$

$y = \dfrac{1}{15b^2}(3bx^2 - 2a)(a + bx^2)^{\frac{3}{2}}.$

26. $dy = x^3(a - x^2)^{-\frac{1}{2}}\,dx.$

$y = -\tfrac{1}{3}(a - x^2)^{\frac{1}{2}}(2a + x^2).$

27. $dy = \dfrac{dx}{x^2(1 + x^2)^{\frac{3}{2}}}.$ $y = -\dfrac{1}{(1 + x^2)^{\frac{1}{2}}}\left(\dfrac{1}{x} + 2x\right).$

28. $dy = a(1 + x^2)^{-\frac{3}{2}}\,dx.$ $y = \dfrac{ax}{(1 + x^2)^{\frac{1}{2}}}.$

29. $dy = x^{-4} (1 - 2x^2)^{-\frac{1}{2}} dx.$

$$y = - \frac{(1 + 4x^2)}{3x^3} (1 - 2x^2)^{\frac{1}{2}}.$$

30. $dy = (1 + x^2)^{\frac{1}{2}} x^3 dx.$

$$y = \frac{(3x^2 - 2)(1 + x^2)^{\frac{3}{2}}}{15}.$$

31. $dy = x^{-2} (a + x^3)^{-\frac{5}{3}} dx.$

$$y = - \frac{(3x^3 + 2a)}{2a^2x (a + x^3)^{\frac{2}{3}}}.$$

32. $dy = x^3 (a^2 + x^2)^{\frac{1}{3}} dx.$

$$y = \tfrac{3}{56} (a^2 + x^2)^{\frac{4}{3}} (4x^2 - 3a^2).$$

33. $dy = x^5 (a + bx^2)^{\frac{2}{3}} dx.$

$$y = \frac{3z^5}{2b^3} \left(\frac{z^6}{11} - \frac{az^3}{4} + \frac{a^2}{5} \right);$$

in which $z = (a + bx^2)^{\frac{1}{3}}.$

34. $dy = \dfrac{(a + bx)^{\frac{3}{2}} dx}{x}.$

$$y = \tfrac{2}{3} (a + bx)^{\frac{3}{2}} + 2a (a + bx)^{\frac{1}{2}}$$
$$+ a^{\frac{3}{2}} \log \frac{\sqrt{a + bx} - \sqrt{a}}{\sqrt{a + bx} + \sqrt{a}}.$$

35. $dy = (a^2 + x^2)^{\frac{1}{2}} dx.$

$$y = \frac{x (a^2 + x^2)^{\frac{1}{2}}}{2} + \frac{a^2}{2} \log [x + (a^2 + x^2)^{\frac{1}{2}}].$$

CHAPTER IV.

147. Formulæ of Reduction.—When a binomial differential satisfies either of the conditions of integrability, it can be rationalized and integrated, as in the last chapter. But, instead of rationalizing the integral directly, it may be reduced to others of a simpler kind, and finally be made to depend upon forms whose integrals are fundamental, or have already been determined. This method is called *integration by successive reduction*, and is the process which in practice is generally the most convenient. It is effected by *formulæ of reduction*. These formulæ are obtained by applying another, known as *the formula for integration by parts*, and which is deduced directly as follows :

Since
$$d(uv) = udv + vdu, \quad (\text{Art. 16})$$

we have
$$uv = \int udv + \int vdu;$$

$$\therefore \int udv = uv - \int vdu;$$

a formula in which the integral of udv depends upon that of vdu.

148. To find a formula for diminishing the exponent of x without the parenthesis by the exponent of x within, in the general binomial form

$$\int x^m (a + bx^n)^p \, dx.$$

Let $\quad y = \int x^m \, (a + bx^n)^p \, dx = \int u \, dv$

$$= uv - \int v \, du \, ; \qquad (1)$$

and put $\quad dv = x^{n-1} \, (a + bx^n)^p \, dx \quad$ and $\quad u = x^{m-n+1}.$

Then $\qquad\qquad v = \dfrac{(a + bx^n)^{p+1}}{nb \, (p + 1)} \, ;$

and $\qquad\qquad du = (m - n + 1) \, x^{m-n} dx.$

Substituting these values of u, v, du, dv, in (1), we have

$$y = \int x^m \, (a + bx^n)^p \, dx = \frac{x^{m-n+1} \, (a + bx^n)^{p+1}}{nb \, (p + 1)}$$

$$- \frac{(m - n + 1)}{nb \, (p + 1)} \int x^{m-n} \, (a + bx^n)^{p+1} \, dx. \qquad (2)$$

This formula diminishes the exponent m by n as was desired, but it increases the exponent p by 1, which is generally an objection. We must therefore change the last term in (2) into an expression in which p shall not be increased.

Now $\quad x^{m-n} \, (a + bx^n)^{p+1} = x^{m-n} \, (a + bx^n)^p \, (a + bx^n)$

$$= ax^{m-n} \, (a + bx^n)^p + bx^m \, (a + bx^n)^p \, ;$$

which in (2) gives

$$y = \int x^m \, (a + bx^n)^p \, dx = \frac{x^{m-n+1} \, (a + bx^n)^{p+1}}{nb \, (p + 1)}$$

$$- \frac{(m - n + 1)}{nb \, (p + 1)} \, a \int x^{m-n} \, (a + bx^n)^p \, dx$$

$$- \frac{m - n + 1}{n \, (p + 1)} \int x^m \, (a + bx^n)^p \, dx.$$

Transposing the last term to the first member and reducing, we have

$$\left(\frac{np + m + 1}{n(p + 1)}\right) \int x^m (a + bx^n)^p \, dx = \frac{x^{m-n+1}(a + bx^n)^{p+1}}{nb(p + 1)}$$

$$- \frac{m - n + 1}{nb(p + 1)} \, a \int x^{m-n} (a + bx^n)^p \, dx.$$

Therefore we have

$$y = \int x^m (a + bx^n)^p \, dx$$

$$= \frac{x^{m-n+1}(a+bx^n)^{p+1} - (m - n + 1)a \int x^{m-n}(a+bx^n)^p dx}{b(np + m + 1)} \; ; (A)$$

which is the formula required.

149. To find a formula for increasing the exponent of x without the parenthesis by the exponent of x within, in the general binomial form

$$y = \int x^{-m} (a + bx^n)^p \, dx.$$

Clearing (A) of fractions, transposing the first member to the second, and the last term of the second to the first, and dividing by $(m - n + 1) \, a$, we have

$$\int x^{m-n} (a + bx^n)^p \, dx$$

$$= \frac{x^{m-n+1}(a+bx^n)^{p+1} - b(np + m + 1) \int x^m(a+bx^n)^p dx}{a(m - n + 1)} . \quad (1)$$

Writing $- m$ for $m - n$, and therefore $- m + n$ for m, **(1)** becomes

$$y = \int x^{-m} (a + bx^n)^p \, dx$$

$$= \frac{x^{-m+1}(a + bx^n)^{p+1} + b(m - np - n - 1) \int x^{-m+n}(a + bx^n)^p dx}{- a(m - 1)} \; ; (B)$$

which is the formula required.

**150. To find a formula for diminishing the expo-
nent of the parenthesis by 1, in the general binomial form**

$$y = \int x^m (a + bx^n)^p \, dx.$$

$$\int x^m (a + bx^n)^p \, dx = \int x^m (a + bx^n)^{p-1} (a + bx^n) \, dx$$

$$= a \int x^m (a + bx^n)^{p-1} \, dx + b \int x^{m+n} (a + bx^n)^{p-1} \, dx. \quad (1)$$

By formula (A), we have for the last term of (1), by writing $m + n$ for m and $p - 1$ for p,

$$\int x^{m+n} (a + bx^n)^{p-1} \, dx$$

$$= \frac{x^{m+1} (a + bx^n)^p - (m + 1) a \int x^m (a + bx^n)^{p-1} \, dx}{b \, [n \, (p - 1) + m + n + 1]} \, ;$$

which in (1) gives

$$y = \int x^m (a + bx^n)^p \, dx = a \int x^m (a + bx^n)^{p-1} \, dx$$

$$+ \frac{x^{m+1} (a + bx^n)^p - (m + 1) a \int x^m (a + bx^n)^{p-1} \, dx}{(np + m + 1)} \, .$$

Therefore, uniting the first and third terms of the second
member, we have

$$y = \int x^m (a + bx^n)^p \, dx$$

$$= \frac{x^{m+1} (a + bx^n)^p + anp \int x^m (a + bx^n)^{p-1} \, dx}{np + m + 1} \, ; \qquad (C)$$

which is the formula required.

151. To find a formula for increasing the exponent of the parenthesis by 1, in the general binomial form

$$y = \int x^m \, (a + bx^n)^{-p} \, dx.$$

By transposing and reducing (C), as we did (A) to find (B), we have

$$\int x^m \, (a + bx^n)^{p-1} \, dx$$

$$= \frac{x^{m+1} \, (a + bx^n)^p - (np + m + 1) \int x^m \, (a + bx^n)^p \, dx}{- anp} \cdot (1)$$

Writing $- p$ for $p - 1$, and therefore $- p + 1$ for p, (1) becomes

$$y = \int x^m \, (a + bx^n)^{-p} \, dx$$

$$= \frac{x^{m+1}(a + bx^n)^{-p+1} - (m + n + 1 - np) \int x^m(a + bx^n)^{-p+1}dx}{an \, (p - 1)} ;(D)$$

which is the required formula.

REMARK.—A careful examination of the process of reduction by these formulæ, will give a clearer insight into the method than can be given by any general rules. We therefore proceed at once to examples for illustration, and shall then leave it to the industry and ingenuity of the student to apply the method to the different cases that he may meet with.

EXAMPLES.

1. Integrate $\quad dy = \dfrac{x^m dx}{\sqrt{a^2 - x^2}}$.

Here $\quad y = \int x^m \, (a^2 - x^2)^{-\frac{1}{2}} \, dx,$

a form which corresponds to

$$\int x^m \, (a + bx^n)^p \, dx.$$

13

We see that by applying formula (A) we may diminish m by 2, and by continued applications of this formula, we can reduce m to 0 or 1 according as it is even or odd, so that the integral will finally depend upon

$$\int \frac{dx}{\sqrt{a^2 - x^2}} = \sin^{-1} \frac{x}{a}, \quad \text{when } m \text{ is even};$$

or

$$\int \frac{x\,dx}{\sqrt{a^2 - x^2}} = -(a^2 - x^2)^{\frac{1}{2}}, \quad \text{when } m \text{ is odd.}$$

Making $m = m$, $a = a^2$, $b = -1$, $n = 2$, $p = -\frac{1}{2}$, we have from formula (A),

$$y = \int x^m (a^2 - x^2)^{-\frac{1}{2}}\,dx$$

$$= \frac{x^{m-2+1}(a^2 - x^2)^{\frac{1}{2}} - a^2(m - 2 + 1)\int x^{m-2}(a^2 - x^2)^{-\frac{1}{2}}\,dx}{-[2(-\frac{1}{2}) + m + 1]}$$

$$= -\frac{x^{m-1}(a^2 - x^2)^{\frac{1}{2}}}{m}$$

$$+ \frac{(m-1)a^2 \int x^{m-2}(a^2 - x^2)^{-\frac{1}{2}}\,dx}{m}. \qquad (1)$$

When $m = 2$, (1) becomes

$$y = \int \frac{x^2\,dx}{\sqrt{a^2 - x^2}} = -\frac{x}{2}(a^2 - x^2)^{\frac{1}{2}} + \frac{a^2}{2}\sin^{-1}\frac{x}{a}.$$

When $m = 3$, (1) becomes

$$y = \int \frac{x^3\,dx}{\sqrt{a^2 - x^2}} = -\frac{1}{3}x^2(a^2 - x^2)^{\frac{1}{2}} - \frac{2}{3}a^2(a^2 - x^2)^{\frac{1}{2}}$$

$$= -\frac{1}{3}(a^2 - x^2)^{\frac{1}{2}}(x^2 + 2a^2).$$

When $m = 6$, (1) becomes, by applying (A) twice in succession,

$$y = \int \frac{x^6 dx}{\sqrt{a^2 - x^2}}$$

$$= -(a^2 - x^2)^{\frac{1}{2}} \left(\frac{x^5}{6} + \frac{5a^2 x^3}{6 \cdot 4} + \frac{5a^4 x}{4 \cdot 4} \right) + \frac{5a^6}{4 \cdot 4} \sin^{-1} \frac{x}{a};$$

(which the student may show.)

2. Integrate $dy = \frac{x^m dx}{\sqrt{a^2 + x^2}}.$

Here $y = \int x^m (a^2 + x^2)^{-\frac{1}{2}} dx.$

Making $m = m,\ a = a^2,\ b = 1,\ n = 2,\ p = -\frac{1}{2},$ we have from (A),

$$y = \int x^m (a^2 + x^2)^{-\frac{1}{2}} dx$$

$$= \frac{x^{m-1} (a^2 + x^2)^{\frac{1}{2}}}{m} - \frac{(m-1) a^2}{m} \int x^{n-2} (a^2 + x^2)^{-\frac{1}{2}} dx. \quad (1)$$

By continued applications of this formula, the integral will finally depend on

$$\int \frac{dx}{\sqrt{a^2 + x^2}} = \log (x + \sqrt{a^2 + x^2}), \quad \text{when } m \text{ is even,}$$

or $\int \frac{x dx}{\sqrt{a^2 + x^2}} = (a^2 + x^2)^{\frac{1}{2}}, \quad$ when m is odd.

3. Integrate $dy = \frac{dx}{x^m (a^2 - x^2)^{\frac{1}{2}}}.$

Here $y = \int x^{-m} (a^2 - x^2)^{-\frac{1}{2}} dx,$

from which we see that by applying (B) we may increase m by 2, and by continued applications of (B), we may reduce m to 0 or 1, according as it is even or odd, making the integral finally depend on a known form.

Making $m = m$, $a = a^2$, $b = -1$, $n = 2$, $p = -\frac{1}{2}$, (B) gives us

$$y = \int x^{-m} (a^2 - x^2)^{-\frac{1}{2}} dx$$

$$= \frac{x^{-m+1} (a^2 - x^2)^{\frac{1}{2}} - (m + 1 - 2 - 1) \int x^{-m+2}(a^2 - x^2)^{-\frac{1}{2}} dx}{- a^2 (m - 1)}$$

$$= -\frac{(a^2 - x^2)^{\frac{1}{2}}}{(m - 1) a^2 x^{m-1}} + \frac{(m - 2)}{(m - 1) a^2} \int \frac{dx}{x^{m-2} (a^2 - x^2)^{\frac{1}{2}}}. \quad (1)$$

When $m = 2$, (1) becomes

$$y = \int \frac{dx}{x^2 \sqrt{a^2 - x^2}} = -\frac{(a^2 - x^2)^{\frac{1}{2}}}{a^2 x};$$

(since the last term disappears.)

When $m = 3$, (1) becomes

$$y = \int \frac{dx}{x^3 \sqrt{a^2 - x^2}} = -\frac{(a^2 - x^2)^{\frac{1}{2}}}{2a^2 x^2} + \frac{1}{2a^2} \int \frac{dx}{x \sqrt{a^2 - x^2}}$$

$$= -\frac{\sqrt{a^2 - x^2}}{2a^2 x^2} + \frac{1}{2a^3} \log \frac{a - \sqrt{a^2 - x^2}}{x}.$$

(Ex. 17 of Art. 146.)

4. Integrate $dy = (a^2 - x^2)^{\frac{n}{2}} dx$, when n is odd.

Here we see that by applying (C) we may diminish $\frac{n}{2}$ by 1, and by continued applications of (C) we can reduce $\frac{n}{2}$ to $-\frac{1}{2}$, making the integral depend finally upon a known form.

Making $m = 0$, $a = a^2$, $b = -1$, $n = 2$, $p = \frac{n}{2}$, (C) gives us

$$y = \int (a^2 - x^2)^{\frac{n}{2}} dx$$

$$= \frac{x (a^2 - x^2)^{\frac{n}{2}} + na^2 \int (a^2 - x^2)^{\frac{n}{2}-1} dx}{n + 1}. \quad (1)$$

When $n = 1$, (1) becomes

$$y = \int (a^2 - x^2)^{\frac{1}{2}} dx = \frac{x (a^2 - x^2)^{\frac{1}{2}}}{2} + \frac{a^2}{2} \sin^{-1} \frac{x}{a}.$$

5. Integrate $dy = \dfrac{dx}{(a^2 - x^2)^{\frac{n}{2}}}$, when n is odd.

Here $y = \int (a^2 - x^2)^{-\frac{n}{2}} dx$,

from which we see that by applying (D) we may increase the exponent $\dfrac{n}{2}$ by 1, and by continued applications of (D) we can reduce $\dfrac{n}{2}$ to $-\frac{1}{2}$, making the integral depend finally on a known form.

Making $m = 0$, $a = a^2$, $b = -1$, $n = 2$, $p = \dfrac{n}{2}$, (D) gives us

$$y = \int (a^2 - x^2)^{-\frac{n}{2}} dx$$

$$= \frac{x (a^2 - x^2)^{-\frac{n}{2}+1} - (3 - n) \int (a^2 - x^2)^{-\frac{n}{2}+1} dx}{2a^2 \left(\dfrac{n}{2} - 1 \right)}$$

$$= \frac{x}{(n - 2) a^2 (a^2 - x^2)^{\frac{n}{2}-1}} + \frac{n - 3}{(n - 2) a^2} \int \frac{dx}{(a^2 - x^2)^{\frac{n}{2}-1}}. \quad (1)$$

When $n = 3$, (1) becomes

$$y = \int \frac{dx}{(a^2 - x^2)^{\frac{3}{2}}} = \frac{x}{a^2 (a^2 - x^2)^{\frac{1}{2}}}.$$

6. Integrate $dy = \dfrac{x^m dx}{\sqrt{2ax - x^2}}$.

Here $y = \int x^m (2ax - x^2)^{-\frac{1}{2}} dx = \int x^{m-\frac{1}{2}} (2a-x)^{-\frac{1}{2}} dx$,

which may be reduced by (A) to a known form.

Making $m = m - \frac{1}{2}$, $a = 2a$, $b = -1$, $n = 1$, $p = -\frac{1}{2}$, (A) gives us

$$y = \int \frac{x^m dx}{\sqrt{2ax - x^2}}$$

$$= \frac{x^{m-\frac{1}{2}} (2a - x)^{\frac{1}{2}} - 2a \left(m - \frac{1}{2}\right) \int x^{m-\frac{3}{2}} (2a - x)^{-\frac{1}{2}} dx}{-m}$$

$$= -\frac{x^{m-1}}{m} \sqrt{2ax - x^2} + \frac{(2m - 1) a}{m} \int \frac{x^{m-1} dx}{\sqrt{2ax - x^2}}. \qquad (1)$$

When $m = 2$, (1) becomes

$$y = \int \frac{x^2 dx}{\sqrt{2ax - x^2}} = -\frac{x + 3a}{2} \sqrt{2ax - x^2}$$

$$+ \tfrac{3}{2}a^2 \int \frac{dx}{\sqrt{2ax - x^2}}$$

$$= -\frac{x + 3a}{2} \sqrt{2ax - x^2} + \tfrac{3}{2}a^2 \, \text{vers}^{-1} \frac{x}{a}.$$

7. Integrate $dy = \dfrac{x^6 dx}{\sqrt{1 - x^2}}$.

$$y = -\left(\frac{x^5}{6} + \frac{1 \cdot 5}{4 \cdot 6} x^3 + \frac{1 \cdot 3 \cdot 5}{2 \cdot 4 \cdot 6} x\right)\sqrt{1 - x^2} + \frac{1 \cdot 3 \cdot 5}{2 \cdot 4 \cdot 6} \sin^{-1} x.$$

8. Integrate $dy = \dfrac{dx}{x^4 \sqrt{a + bx^2}}$.

$$y = \left(-\frac{1}{3ax^3} + \frac{2b}{3a^2 x}\right) \sqrt{a + bx^2}.$$

9. Integrate $dy = (1 - x^2)^{\frac{3}{2}} dx$.

$$y = \tfrac{1}{4}x (1 - x^2)^{\frac{3}{2}} + \tfrac{3}{8}x (1 - x^2)^{\frac{1}{2}} + \tfrac{3}{8} \sin^{-1} x.$$

10. Integrate $dy = \dfrac{dx}{(1 + x^2)^3}$.

$$y = \frac{x}{4 (1 + x^2)^2} + \frac{3}{8} \cdot \frac{x}{(1 + x^2)} + \frac{3}{8} \tan^{-1} x.$$

11. Integrate $dy = \dfrac{x^3 dx}{\sqrt{2ax - x^2}}.$

$y = -\left(\dfrac{x^2}{3} + \tfrac{5}{3}\cdot\dfrac{x}{2}\,a + \tfrac{5}{3}\cdot\tfrac{3}{2}a^2\right)\sqrt{2ax - x^2} + \tfrac{5}{3}\cdot\tfrac{3}{2}a^3\,\mathrm{vers}^{-1}\dfrac{x}{a}.$

12. Integrate $dy = \dfrac{dx}{x^5\sqrt{1 - x^2}}.$

$y = -\left(\dfrac{1}{4x^4} + \dfrac{1\cdot 3}{2\cdot 4x^2}\right)\sqrt{1 - x^2} - \dfrac{1\cdot 3}{2\cdot 4}\log\dfrac{1 + \sqrt{1 - x^2}}{x}. \; .$

These integrals might be determined by one or other of the methods of Chapter III, but the process of integration by reduction leads to a result more convenient and better suited in most cases for finding the definite integrals.*

LOGARITHMIC FUNCTIONS.

152. **Reduction of the Form** $\displaystyle\int X(\log x)^n\,dx$, **in which X is an Algebraic Function of x.**

Put $\qquad Xdx = dv \;\text{ and }\; \log^n x = u.$

$\therefore\quad v = \displaystyle\int Xdx \;\text{ and }\; du = n\log^{n-1}x\,\dfrac{dx}{x}.$

Substituting in $\quad \displaystyle\int udv = uv - \int vdu,\quad$ (Art. 147)

we have $\qquad y = \displaystyle\int X\log^n x\,dx$

$= \log^n x\displaystyle\int Xdx - \int n\log^{n-1}x\,\dfrac{dx}{x}\int(Xdx);$

or by making $\qquad \displaystyle\int(Xdx) = X_1,$

we have $\qquad y = \displaystyle\int X\log^n x\,dx$

$= X_1\log^n x - n\displaystyle\int\dfrac{X_1}{x}\log^{n-1}x\,dx;$

---

which diminishes the exponent of $\log x$ by 1, wherever it is possible to integrate the form $\int X dx$. By continued applications of this formula, when n is a positive integer, we can reduce n to 0 so that the integral will finally depend on

$$\int \frac{X_n}{x} dx.$$

Sch.—A useful case of this general form is that in which $X = x^m$, the form then being

$$y = \int x^m \log^n x dx;$$

and the formula of reduction becomes

$$y = \int x^m \log^n x dx = \frac{x^{m+1}}{m+1} \log^n x$$
$$- \frac{n}{m+1} \int x^m \log^{n-1} x dx.$$

by means of which the final integral, when n is a positive integer, becomes.

$$\int x^m dx = \frac{x^{m+1}}{m+1}.$$

EXAMPLES.

1. Integrate $dy = x^4 \log^2 x dx.$

Making $m = 4$, and $n = 2$,   we have

$$y = \int x^4 \log^2 x dx$$

$$= \frac{x^5 \log^2 x}{5} - \tfrac{2}{5} \int x^4 \log x dx. \qquad (1)$$

Making $m = 4$ and $n = 1$, we have

$$\int x^4 \log x \, dx = \frac{x^5 \log x}{5} - \frac{1}{5} \int x^4 dx \left(= \frac{x^5}{5} \right),$$

which substituted in (1) gives us

$$y = \int x^4 \log^2 x \, dx = \frac{x^5 \log^2 x}{5} - \frac{2}{5}\left(\frac{x^5 \log x}{5} - \frac{1}{5} \cdot \frac{x^5}{5}\right)$$

$$= \frac{x^5}{5}\left(\log^2 x - \tfrac{2}{5}\log x + \tfrac{2}{25}\right).$$

2. Integrate $dy = \dfrac{x \log x \, dx}{\sqrt{a^2 + x^2}}$.

Put $\qquad \dfrac{x dx}{\sqrt{a^2 + x^2}} = dv \qquad$ and $\quad \log x = u$;

then $\qquad\qquad v = \sqrt{a^2 + x^2} \quad$ and $\qquad du = \dfrac{dx}{x}$.

$$\therefore \;\; y = \int \frac{x \log x \, dx}{\sqrt{a^2 + x^2}} = (a^2 + x^2)^{\frac{1}{2}} \log x - \int \frac{\sqrt{a^2 + x^2}}{x} \, dx$$

$$= (a^2 + x^2)^{\frac{1}{2}} \log x - \int \frac{a^2 dx}{x\sqrt{a^2 + x^2}} - \int \frac{x dx}{\sqrt{a^2 + x^2}}$$

$$= (a^2 + x^2)^{\frac{1}{2}} \log x + a \log\left(\frac{a + \sqrt{a^2 + x^2}}{x}\right) - \sqrt{a^2 + x^2}.$$

(See Ex. 17, Art. 146.)

3. Integrate $dy = \dfrac{\log x \, dx}{(1 + x)^2}$.

$$y = \frac{x}{1 + x} \log x - \log(1 + x).$$

153. Reduction of $\displaystyle\int \frac{x^m \, dx}{\log^n x}$.

Put $\qquad x^{m+1} = u, \qquad\qquad \dfrac{1}{\log^n x}\dfrac{dx}{x} = dv$;

then $\qquad\qquad du = (m + 1) x^m \, dx$

and $\qquad\qquad v = \dfrac{1}{-(n - 1)\log^{n-1} x}$.

$$\therefore \quad y = \int \frac{x^m \, dx}{\log^n x} = -\frac{x^{m+1}}{(n-1)\log^{n-1} x} + \frac{m+1}{(n-1)}\int \frac{x^m \, dx}{\log^{n-1} x},$$

by means of which the final integral, when n is a positive integer, becomes

$$\int \frac{x^m \, dx}{\log x},$$

beyond which the reduction cannot be carried, for when $n = 1$ the formula ceases to apply. We may, however, express this final integral in a simpler form; thus,

Put $\qquad\qquad z = x^{m+1};$

then $dz = (m + 1) x^m \, dx$ and $\log z = (m + 1) \log x.$

$$\therefore \quad \int \frac{x^m \, dx}{\log x} = \int \frac{dz}{\log z},$$

an expression which, simple as it appears, has never yet been integrated, except by series, which gives only an approximate result.

Ex. 1. Integrate $dy = \dfrac{x^4 \, dx}{\log^2 x}.$

Here $m = 4$ and $n = 2$; therefore the formula gives us

$$y = \int \frac{x^4 \, dx}{\log^2 x} = -\frac{x^5}{\log x} + 5 \int \frac{x^4 \, dx}{\log x}.$$

Put $z = x^5$; then $dz = 5x^4 dx$ and $\log z = 5 \log x$;

therefore $\qquad \int \dfrac{x^4 \, dx}{\log x} = \int \dfrac{dz}{\log z}$

Now put $\log z = u$; then $z = e^u$ and $dz = e^u du.$

$$\therefore \int \frac{dz}{\log z} = \int \frac{e^u du}{u}.$$

$$= \int \left(1 + u + \frac{u^2}{2} + \frac{u^3}{2 \cdot 3} + \text{etc.}\right) \frac{du}{u} \quad \text{(Art. 62)}$$

$$= \log u + u + \frac{u^2}{2^2} + \frac{u^3}{2 \cdot 3^2} + \text{etc.}$$

$$= \log (\log x^5) + \log x^5 + \frac{1}{2^2} \log^2 x^5 + \frac{\log^3 x^5}{2 \cdot 3^2} + \text{etc.}$$

$$\therefore \ y = \int \frac{x^4 \, dx}{\log^2 x}$$

$$= -\frac{x^5}{\log x} + 5 \left[\log (\log x^5) + \log x^5 + \frac{\log^2 x^5}{2^2} \right.$$
$$\left. + \frac{\log^3 x^5}{2 \cdot 3^2} + \text{etc.} \right].$$

(See Strong's Calculus, p. 392; also, Young's Int. Cal., pp. 52 and 53.)

EXPONENTIAL FUNCTIONS.

154. Reduction of the Form $\int a^{mx} x^n \, dx$.

Put $\quad a^{mx} \, dx = dv \qquad$ and $\quad x^n = u$;

then $\qquad\qquad v = \dfrac{a^{mx}}{m \log a} \quad$ and $\quad du = nx^{n-1} \, dx.$

$$\therefore \ y = \int a^{mx} x^n \, dx = \frac{a^{mx} x^n}{m \log a} - \frac{n}{m \log a} \int a^{mx} x^{n-1} \, dx.$$

By successive applications of this formula, when n is a positive integer, it can be finally reduced to 0, and the integral made to depend on $\int a^{mx} \, dx = \dfrac{a^{mx}}{m \log a}.$

Only a very few of the logarithmic and exponential functions can be integrated by any general method at present known, except by the method of series, which furnishes only an approximation, and should therefore be resorted to only when exact methods fail.

EXAMPLES.

1. Integrate $dy = a^x x^3\, dx$.

Here $m = 1$ and $n = 3$; therefore, from the above formula, we have

$$y = \int a^x x^3 dx$$

$$= \frac{a^x x^3}{\log a} - \frac{3}{\log a} \int a^x x^2 dx; \text{ (by repeating the process)}$$

$$= \frac{a^x x^3}{\log a} - \frac{3}{\log a}\left(\frac{a^x x^2}{\log a} - \frac{2}{\log a}\int a^x x\, dx\right);$$

(by repeating the process)

$$= \frac{a^x x^3}{\log a} - \frac{3a^x x^2}{\log^2 a} + \frac{6}{\log^2 a}\left(\frac{a^x x}{\log a} - \frac{1}{\log^2 a} a^x\right)$$

$$= \frac{a^x}{\log a}\left(x^3 - \frac{3x^2}{\log a} + \frac{6x}{\log^2 a} - \frac{6}{\log^3 a}\right).$$

2. Integrate $dy = x^3 e^{ax}\, dx$.

$$y = e^{ax}\left(\frac{x^3}{a} - \frac{3x^2}{a^2} + \frac{6x}{a^3} - \frac{6}{a^4}\right).$$

155. Reduction of $\int \dfrac{a^x\, dx}{x^m}$, when m is a positive Integer.

Put $\qquad x^{-m}\, dx = dv, \qquad a^x = u$;

then $\qquad v = -\dfrac{x^{-m+1}}{m-1}$ and $du = a^x \log a\, dx$.

$$\therefore \quad y = \int \frac{a^x\, dx}{x^m} = -\frac{a^x}{(m-1)\, x^{m-1}} + \frac{\log a}{m-1}\int \frac{a^x dx}{x^{m-1}};$$

by means of which the final integral becomes

$$\int \frac{a^x dx}{x}.$$

which does not admit of integration in finite terms, but may be expressed in a series, and each term integrated separately. (See Lacroix, Calcul Intégral, Vol. II, p. 91.)

Ex. 1. Integrate $dy = \dfrac{e^x dx}{x^2}.$

By the formula just found, we have

$$y = \int \frac{e^x dx}{x^2} = -\frac{e^x}{x} + \int \frac{e^x dx}{x}$$

$$= -\frac{e^x}{x} + \int \left[1 + x + \frac{x^2}{2} + \frac{x^3}{2 \cdot 3} + \text{etc.} \right] \frac{dx}{x} \quad \text{(Art. 62)}$$

$$= -\frac{e^x}{x} + \log x + x + \frac{x^2}{2^2} + \frac{x^3}{2 \cdot 3^2} + \text{etc.}$$

TRIGONOMETRIC FUNCTIONS.

156. Cases in which $\sin^m \theta \cos^n \theta\, d\theta$ **is immediately Integrable.**—The value of this integral can be found immediately when *either m or n, or both, are odd positive integers; and also when m + n is an even negative integer.*

1st. Let $m = 2r + 1$; then

$$\int \sin^m \theta \cos^n \theta\, d\theta = \int (\sin \theta)^{2r+1} \cos^n \theta\, d\theta$$

$$= \int (1 - \cos^2 \theta)^r \cos^n \theta \sin \theta\, d\theta$$

$$= -\int (1 - \cos^2 \theta)^r \cos^n \theta\, d\cos \theta,$$

an expression in which the binomial $(1 - \cos^2 \theta)^r$ can be expanded, and each term integrated immediately. In like manner, if the exponent of $\cos \theta$ be an odd integer, we may assume $n = 2r + 1$, etc.

2d. Let $m + n = -2r$; then

$$\int \sin^m \theta \cos^n \theta \, d\theta = \int \tan^m \theta \, (\cos \theta)^{n+m} \, d\theta$$

$$= \int \tan^m \theta \, (\sec \theta)^{2r} \, d\theta$$

$$= \int \tan^m \theta \, (1 + \tan^2 \theta)^{r-1} \, d \cdot \tan \theta,$$

each term of which, after expansion, can be immediately integrated.

EXAMPLES.

1. $dy = \sin^2 \theta \cos^3 \theta \, d\theta.$

Here $y = \int \sin^2 \theta \cos^3 \theta \, d\theta$

$$= \int \sin^2 \theta \, (1 - \sin^2 \theta) \, d \cdot \sin \boldsymbol{\theta}$$

$$= \tfrac{1}{3} \sin^3 \theta - \tfrac{1}{5} \sin^5 \theta.$$

2. $dy = \dfrac{\sin^2 \theta}{\cos^4 \theta} \, d\theta.$

Here $y = \int \dfrac{\sin^2 \theta}{\cos^4 \theta} \, d\theta = \int \tan^2 \theta \sec^2 \theta \, d\theta$

$$= \tfrac{1}{3} \tan^3 \theta.$$

3. $dy = \sin^3 \theta \cos^4 \theta \, d\theta.$ $y = -\tfrac{1}{5} \cos^5 \theta + \tfrac{1}{7} \cos^7 \theta.$

4. $dy = \sin^5 \theta \cos^5 \theta \, d\theta.$
$$y = -\tfrac{1}{10} \cos^6 \theta \, (\sin^4 \theta + \tfrac{1}{2} \sin^2 \theta + \tfrac{1}{6}).$$

5. $dy = \sin^3 \theta \cos^7 \theta \, d\theta.$ $y = \tfrac{1}{10} \cos^{10} \theta - \tfrac{1}{8} \cos^8 \theta.$

6. $dy = \dfrac{\sin^2 \theta}{\cos^6 \theta} \, d\theta.$ $y = \tfrac{1}{3} \tan^3 \theta + \tfrac{1}{5} \tan^5 \theta.$

7. $dy = \dfrac{d\theta}{\sin \theta \cos^5 \theta}.$

Here $\quad y = \int \dfrac{\sec^4 \theta \, d\theta}{\tan \theta \cos^2 \theta} = \int \dfrac{(1 + \tan^2 \theta)^2 \sec^2 \theta \, d\theta}{\tan \theta}$

$\qquad = \log (\tan \theta) + \tan^2 \theta + \tfrac{1}{4} \tan^4 \theta.$

8. $\quad dy = \dfrac{d\theta}{\sin^{\frac{3}{2}} \theta \, \cos^{\frac{5}{2}} \theta}.$

Let $\qquad\qquad\qquad x = \tan \theta;$

then $\quad \cos \theta = \dfrac{1}{\sqrt{1 + x^2}}, \qquad \sin \theta = \dfrac{x}{\sqrt{1 + x^2}},$

and $\qquad\qquad\qquad d\theta = \dfrac{dx}{1 + x^2},$

$\therefore \quad y = \int \dfrac{d\theta}{\sin^{\frac{3}{2}} \theta \, \cos^{\frac{5}{2}} \theta} = \int \dfrac{(1 + x^2) \, dx}{x^{\frac{3}{2}}}$

$\qquad = \tfrac{2}{3} x^{\frac{3}{2}} - \dfrac{2}{x^{\frac{1}{2}}} = \tfrac{2}{3} \tan^{\frac{3}{2}} \theta - \dfrac{2}{\tan^{\frac{1}{2}} \theta}.$

9. $\quad dy = \dfrac{\sin^3 \theta}{\cos^5 \theta} \, d\theta. \qquad y = \tfrac{1}{4} \tan^4 \theta.$

10. $\quad dy = \dfrac{d\theta}{\cos^6 \theta}. \qquad y = \tan \theta + \tfrac{2}{3} \tan^3 \theta + \tfrac{1}{5} \tan^5 \theta.$

157. Formulæ of Reduction for

$$\int \sin^m \theta \, \cos^n \theta \, d\theta.$$

When neither of the above mentioned conditions as to m and n is fulfilled, the integration of this expression can be obtained only by aid of successive reduction.

We might produce formulæ for reducing the expression $\sin^m \theta \cos^n \theta$ directly;* but, as it would carry us beyond the limits of this book, we prefer to effect the integration by transforming the given expression into an equivalent algebraic form, and then reducing by one or more of the

---

* See Price, Lacroix, Williamson, Todhunter, Courtenay, etc.

formulæ (A), (B), $\{C\}$, (D). Thus, put $\sin\theta = x$, then $\sin^m\theta = x^m$, $\cos\theta = (1-x^2)^{\frac{1}{2}}$, $\cos^n\theta = (1-x^2)^{\frac{n}{2}}$, and $d\theta = (1-x^2)^{-\frac{1}{2}}\, dx$.

$$\therefore\ y = \int \sin^m\theta\,\cos^n\theta\, d\theta = \int x^m\,(1-x^2)^{\frac{n-1}{2}}\, dx.$$

or we may put $\cos\theta = x$, and get

$$y = \int \sin^m\theta\,\cos^n\theta\, d\theta = \int -x^n\,(1-x^2)^{\frac{m-1}{2}}\, dx\,;$$

either of which may be reduced by the above formulæ.

This process will always effect the integration when m and n are either positive or negative integers, and often when they are fractions. The method is exhibited by the following examples.

EXAMPLES.

1. $$dy = \sin^6\theta\, d\theta.$$

Put $$\sin\theta = x,$$

then $$d\theta = (1-x^2)^{-\frac{1}{2}}\, dx.$$

$$\therefore\ y = \int \sin^6\theta\, d\theta = \int x^6\,(1-x^2)^{-\frac{1}{2}}\, dx$$

$$= -\left(\frac{x^5}{6} + \frac{5x^3}{4\cdot6} + \frac{3\cdot5x}{2\cdot4\cdot6}\right)(1-x^2)^{\frac{1}{2}} + \frac{3\cdot5}{2\cdot4\cdot6}\sin^{-1}x$$

$$\text{(by Ex. 7, Art. 151);}$$

$$= -\frac{\cos\theta}{6}\left(\sin^5\theta + \frac{5}{4}\sin^3\theta + \frac{5\cdot3}{4\cdot2}\sin\theta\right) + \frac{5\cdot3}{2\cdot4\cdot6}\theta.$$

2. $$dy = \frac{d\theta}{\sin^5\theta}.$$

Put $$\sin\theta = x,$$

then $$d\theta = (1-x^2)^{-\frac{1}{2}}\, dx.$$

$$\therefore \quad y = \int \frac{d\theta}{\sin^5 \theta} = \int x^{-5} (1 - x^2)^{-\frac{1}{2}} \, dx.$$

$$= - \left(\frac{1}{4x^4} + \frac{1 \cdot 3}{2 \cdot 4x^2} \right) \sqrt{1 - x^2} - \frac{1 \cdot 3}{2 \cdot 4} \log \frac{1 + \sqrt{1 - x^2}}{x}$$

(by Ex. 12, Art. 151);

$$= - \frac{\cos \theta}{4} \left(\frac{1}{\sin^4 \theta} + \frac{3}{2 \sin^2 \theta} \right) + \frac{1 \cdot 3}{2 \cdot 4} \log \tan \tfrac{1}{2}\theta.$$

(since $- \log \dfrac{1 + \cos \theta}{\sin \theta} = \log \dfrac{\sin \theta}{1 + \cos \theta} = \log \tan \tfrac{1}{2}\theta.$)

3. $dy = \sin^4 \theta d\theta.$

$$y = - \frac{\cos \theta}{4} (\sin^3 \theta + \tfrac{3}{2} \sin \theta) + \tfrac{3}{8}\theta.$$

(See Ex. 10, Art. 135.)

4. $dy = \cos^4 \theta d\theta.$

$$y = \frac{\sin \theta \cos^3 \theta}{4} + \tfrac{3}{8} \sin \theta \cos \theta + \tfrac{3}{8}\theta.$$

(See Ex. 9, Art. 135.)

158. Integration of $\sin^m \theta \cos^n \theta \, d\theta$ in terms of the sines and cosines of the multiple arcs, when m and n are positive integers.

The above integrations have been effected in terms of the *powers* of the trigonometric functions. When m and n are positive integers, the integration may be effected without introducing any powers of the trigonometric functions by converting the powers of sines, cosines, etc., into the sines and cosines of *multiple arcs*, before the integration is performed. The numerical results obtained by this process are more easily calculated than from the powers.

Three transformations can always be made by the use of the three trigonometric formulæ.

(*1.*) $\sin a \sin b = \frac{1}{2} \cos (a - b) - \frac{1}{2} \cos (a + b)$.

(*2.*) $\sin a \cos b = \frac{1}{2} \sin (a + b) + \frac{1}{2} \sin (a - b)$.

(*3.*) $\cos a \cos b = \frac{1}{2} \cos (a + b) + \frac{1}{2} \cos (a - b)$.

EXAMPLES.

1. $dy = \sin^3 \theta \cos^2 \theta d\theta$.

Here $\sin^3 \theta \cos^2 \theta = \sin \theta (\sin \theta \cos \theta)^2$

$$= \sin \theta \left(\tfrac{1}{2} \sin 2\theta\right)^2 \qquad \text{[by (2)]}$$

$$= \tfrac{1}{4} \sin \theta (\sin^2 2\theta)$$

$$= \tfrac{1}{4} \sin \theta \left(\frac{1 - \cos 4\theta}{2}\right) \qquad \text{[by (1)]}$$

$$= \tfrac{1}{8} \sin \theta - \tfrac{1}{8} \sin \theta \cos 4\theta$$

$$= \tfrac{1}{8} \sin \theta - \tfrac{1}{8} \left(\tfrac{1}{2} \sin 5\theta - \tfrac{1}{2} \sin 3\theta\right)$$

$$\qquad \text{[by (2)]}$$

$$= \tfrac{1}{8} \sin \theta - \tfrac{1}{16} \sin 5\theta + \tfrac{1}{16} \sin 3\theta$$

$$\therefore \ y = \int \sin^3 \theta \cos^2 \theta d\theta$$

$$= \int \left(\tfrac{1}{8} \sin \theta d\theta - \tfrac{1}{16} \sin 5\theta d\theta + \tfrac{1}{16} \sin 3\theta d\theta\right)$$

$$= - \tfrac{1}{8} \cos \theta + \tfrac{1}{80} \cos 5\theta - \tfrac{1}{48} \cos 3\theta.$$

2. $dy = \sin^3 \theta \cos^3 \theta d\theta$.

$$y = - \tfrac{3}{64} \cos 2\theta + \tfrac{1}{192} \cos 6\theta.$$

3. $dy = \sin^3 \theta d\theta$.

$$y = \tfrac{1}{12} \cos 3\theta - \tfrac{3}{4} \cos \theta.$$

4. $dy = \cos^3 \theta d\theta$.

$$y = \tfrac{1}{12} \sin 3\theta + \tfrac{3}{4} \sin \theta.$$

159. Reduction of the Form

$$\int x^n \cos ax \, dx.$$

Put $\qquad\qquad\qquad u = x^n,$

and $\qquad\qquad\qquad dv = \cos ax \, dx;$

then $\qquad\qquad\qquad du = nx^{n-1} \, dx,$

and $\qquad\qquad\qquad v = \dfrac{1}{a} \sin ax.$

$$\therefore \; y = \int x^n \cos ax \, dx = \frac{1}{a} x^n \sin ax - \frac{n}{a} \int x^{n-1} \sin ax \, dx.$$

Again, put $\qquad\qquad u = x^{n-1},$

and $\qquad\qquad\qquad dv = \sin ax \, dx;$

then $\qquad\qquad\qquad du = (n-1) x^{n-2} \, dx,$

and $\qquad\qquad\qquad v = -\dfrac{1}{a} \cos ax.$

$$\therefore \; \int x^{n-1} \sin ax \, dx = -\frac{1}{a} x^{n-1} \cos ax$$

$$+ \frac{n-1}{a} \int x^{n-2} \cos ax \, dx.$$

$$\therefore \; y = \int x^n \cos ax \, dx = \frac{1}{a} x^n \sin ax$$

$$- \frac{n}{a} \left(-\frac{1}{a} x^{n-1} \cos ax + \frac{n-1}{a} \int x^{n-2} \cos ax \, dx \right)$$

$$= \frac{x^{n-1} (ax \sin ax + n \cos ax)}{a^2} - \frac{n(n-1)}{a^2} \int x^{n-2} \cos ax \, dx.$$

The formula of reduction for $\int x^n \sin ax \, dx$ can be obtained in like manner.

EXAMPLE.

1. $dy = x^3 \cos x \, dx.$

$\qquad y = x^3 \sin x + 3x^2 \cos x - 6x \sin x - 6 \cos x.$

160. Reduction of the Form

$$\int e^{ax} \cos^n x \, dx.$$

Put $\qquad u = \cos^n x,$

and $\qquad dv = e^{ax} \, dx \, ;$

then $\qquad du = -n \cos^{n-1} x \sin x \, dx,$

and $\qquad v = \dfrac{e^{ax}}{a}.$

$$\therefore \;\; y = \int e^{ax} \cos^n x \, dx = \frac{e^{ax} \cos^n x}{a}$$

$$+ \frac{n}{a} \int e^{ax} \cos^{n-1} x \sin x \, dx. \quad (1)$$

Again, put $\qquad u = \cos^{n-1} x \sin x,$

and $\qquad dv = e^{ax} \, dx \, ;$

then $\qquad du = -(n-1) \cos^{n-2} x \sin^2 x \, dx$
$$+ \cos^n x \, dx,$$

and $\qquad v = \dfrac{e^{ax}}{a}.$

$$\therefore \quad \int e^{ax} \cos^{n-1} x \sin x \, dx$$

$$= \frac{1}{a} e^{ax} \cos^{n-1} x \sin x - \frac{1}{a} \int e^{ax} [-(n-1) \cos^{n-2} x \sin^2 x$$
$$+ \cos^n x] \, dx$$

$$= \frac{1}{a} e^{ax} \cos^{n-1} x \sin x + \frac{(n-1)}{a} \int e^{ax} \cos^{n-2} x \, dx$$

$$- \frac{n}{a} \int e^{ax} \cos^n x \, dx. \quad \text{(Since } \sin^2 x = 1 - \cos^2 x.\text{)}$$

Substituting in (1), and transposing and solving for $\int e^{ax} \cos^n x \, dx$, we get

$$y = \int e^{ax} \cos^n x \, dx = \frac{e^{ax} \cos^{n-1} x \, (a \cos x + n \sin x)}{a^2 + n^2}$$

$$+ \frac{n(n-1)}{a^2 + n^2} \int e^{ax} \cos^{n-2} x \, dx \, ; \quad (2)$$

which diminishes the exponent of cos x by 2. By continued applications of this formula, we can reduce n to 0 or 1, so that the integral will depend finally on

$$\int e^{ax} dx = \frac{e^{ax}}{a}, \quad \text{when } n \text{ is even;}$$

or $\qquad \int e^{ax} \cos x\, dx, \quad$ when n is odd.

(2) gives the value of $\int e^{ax} \cos x\, dx$ without an integration, since the last term then contains the factor $n - 1 = 1 - 1 = 0$, and therefore that term disappears.

The reduction of $\int e^{ax} \sin^n x\, dx$ can be obtained in like manner.

EXAMPLES.

1. $dy = e^{ax} \cos x\, dx.$

$$y = \frac{e^{ax}}{a^2 + 1} (a \cos x + \sin x).$$

2. $dy = e^{ax} \cos^2 x\, dx.$

$$y = \frac{e^{ax} \cos x(a \cos x + 2 \sin x)}{4 + a^2} + \frac{2}{4 + a^2} \cdot \frac{e^{ax}}{a}.$$

161. Integration of the Forms

$$f(x) \sin^{-1} x\, dx, \quad f(x) \tan^{-1} x\, dx, \quad \text{etc.}$$

Integrals of these forms must be determined by the formula for integration by parts (Art. 147); the method is best explained by examples.

EXAMPLES.

1. $dy = \sin^{-1} x\, dx.$

Put $\qquad dv = dx, \qquad$ and $\qquad u = \sin^{-1} x;$

then $\qquad v = x, \qquad$ and $\qquad du = \dfrac{dx}{\sqrt{1 - x^2}}.$

$$\therefore \ y = \int \sin^{-1} x \, dx$$

$$= x \sin^{-1} x - \int \frac{x \, dx}{\sqrt{1 - x^2}}$$

$$= x \sin^{-1} x + (1 - x^2)^{\frac{1}{2}}.$$

2. $dy = \dfrac{x^2 \tan^{-1} x \, dx}{1 + x^2}.$

Put $dv = \dfrac{x^2 dx}{1 + x^2} = dx - \dfrac{dx}{1 + x^2},$

and $u = \tan^{-1} x \,;$

then $v = x - \tan^{-1} x,$

and $du = \dfrac{dx}{1 + x^2}.$

$$\therefore \ y = x \tan^{-1} x - (\tan^{-1} x)^2 - \int \left(\frac{x \, dx}{1 + x^2} - \frac{\tan^{-1} x \, dx}{1 + x^2} \right)$$

$$= x \tan^{-1} x - (\tan^{-1} x)^2 - \tfrac{1}{2} \log (1 + x^2) + \tfrac{1}{2} (\tan^{-1} x)^2$$

$$= x \tan^{-1} x - \tfrac{1}{2} (\tan^{-1} x)^2 - \tfrac{1}{2} \log (1 + x^2).$$

3. $dy = x^2 \sin^{-1} x \, dx.$

$$y = \frac{x^3}{3} \sin^{-1} x + \tfrac{1}{9} (x^2 + 2) \sqrt{1 - x^2}.$$

4. $dy = \sin^{-1} x \dfrac{dx}{(1 - x^2)^{\frac{1}{2}}}$ $y = \tfrac{1}{2} (\sin^{-1} x)^2.$

162. Integration of $dy = \dfrac{d\theta}{a + b \cos \theta}.$

$$y = \int \frac{d\theta}{a + b \cos \theta}$$

$$= \int \frac{d\theta}{a \left(\cos^2 \dfrac{\theta}{2} + \sin^2 \dfrac{\theta}{2} \right) + b \left(\cos^2 \dfrac{\theta}{2} - \sin^2 \dfrac{\theta}{2} \right)}$$

$$= \int \frac{d\theta}{(a + b) \cos^2 \frac{\theta}{2} + (a - b) \sin^2 \frac{\theta}{2}}$$

$$= \int \frac{\sec^2 \frac{\theta}{2} \, d\theta}{a + b + (a - b) \tan^2 \frac{\theta}{2}}$$

$$= 2 \int \frac{d \cdot \tan \frac{\theta}{2}}{a + b + (a - b) \tan^2 \frac{\theta}{2}}. \tag{1}$$

When $a > b$,

$$= \frac{2}{\sqrt{a^2 - b^2}} \tan^{-1} \left[\left(\frac{a - b}{a + b} \right)^{\frac{1}{2}} \tan \frac{\theta}{2} \right]$$

(by Ex. 3, Art. 132.)

When $a < b$, we have, from (1),

$$y = \int \frac{d\theta}{a + b \cos \theta} = 2 \int \frac{d \cdot \tan \frac{\theta}{2}}{b + a - (b - a) \tan^2 \frac{\theta}{2}}$$

$$= \frac{1}{\sqrt{b^2 - a^2}} \log \frac{\sqrt{b + a} + \sqrt{b - a} \tan \frac{\theta}{2}}{\sqrt{b + a} - \sqrt{b - a} \tan \frac{\theta}{2}}$$

(by Ex. 5, Art. 137).

The integral of $\dfrac{d\theta}{a + b \sin \theta}$ can be found in like manner to be

$$= \frac{2}{(a^2 - b^2)^{\frac{1}{2}}} \tan^{-1} \frac{a \sin \frac{\theta}{2} + b \cos \frac{\theta}{2}}{(a^2 - b^2)^{\frac{1}{2}} \cos \frac{\theta}{2}}, \text{ when } a > b;$$

$$\text{and} \quad = \frac{1}{(b^2 - a^2)^{\frac{1}{2}}} \log \frac{a \tan \dfrac{\theta}{2} + b - (b^2 - a^2)^{\frac{1}{2}}}{a \tan \dfrac{\theta}{2} + b + (b^2 - a^2)^{\frac{1}{2}}},$$

when $a < b$.

There are other forms which can be integrated by the application of the formula for integration by parts (Art. 147). Those which we have given are among the most important, and which occur the most frequently in the practical applications of the Calculus. The student who has studied the preceding formulæ carefully should find no difficulty in applying the methods to the solution of any expression that he may meet with, that is not too complicated.

The most suitable method of integration in every case can be arrived at only after considerable practice and familiarity with the processes of integration.

EXAMPLES.

1. $dy = \dfrac{x^3 dx}{\sqrt{1 - x^2}}.$ $\qquad y = -\frac{1}{3}(x^2 + 2)(1 - x^2)^{\frac{1}{2}}.$

2. $dy = \dfrac{x^4 dx}{\sqrt{1 - x^2}}.$

$$y = -\left(\frac{x^3}{4} + \frac{1 \cdot 3}{2 \cdot 4} x\right) \sqrt{1 - x^2} + \frac{1 \cdot 3}{2 \cdot 4} \sin^{-1} x.$$

3. $dy = \dfrac{x^7 dx}{\sqrt{1 - x^2}}.$

$$y = -\left(\frac{x^6}{7} + \frac{1 \cdot 6}{5 \cdot 7} x^4 + \frac{1 \cdot 4 \cdot 6}{3 \cdot 5 \cdot 7} x^2 + \frac{1 \cdot 2 \cdot 4 \cdot 6}{1 \cdot 3 \cdot 5 \cdot 7}\right) \sqrt{1 - x^2}.$$

4. $dy = \dfrac{x^5 dx}{\sqrt{a + bx^2}}.$

$$y = \frac{1}{15b}\left(3x^4 - \frac{4ax^2}{b} + \frac{8a^2}{b^2}\right) \sqrt{a + bx^2}.$$

5. $dy = \dfrac{dx}{x^7\sqrt{1-x^2}}.$

$$y = -\left(\frac{1}{6x^6} + \frac{1\cdot 5}{4\cdot 6x^4} + \frac{1\cdot 3\cdot 5}{2\cdot 4\cdot 6x^2}\right)\sqrt{1-x^2}$$
$$-\frac{1\cdot 3\cdot 5}{2\cdot 4\cdot 6}\log\frac{1+\sqrt{1-x^2}}{x}.$$

6. $dy = \dfrac{dx}{x^2\sqrt{1+x}}.$

$$y = -\frac{\sqrt{1+x}}{x} - \tfrac{1}{2}\log\left(\frac{\sqrt{1+x}-1}{\sqrt{1+x}+1}\right).$$

(See Ex. 1, Art. 142.)

7. $dy = (a^2 - x^2)^{\frac{5}{2}}\,dx.$

$$y = \tfrac{1}{6}x\,(a^2-x^2)^{\frac{5}{2}} + \frac{5}{6\cdot 4}a^2x\,(a^2-x^2)^{\frac{3}{2}}$$
$$+ \frac{5\cdot 3}{6\cdot 4\cdot 2}a^4x\,(a^2-x^2)^{\frac{1}{2}} + \frac{5\cdot 3}{6\cdot 4\cdot 2}a^6\sin^{-1}\frac{x}{a}.$$

8. $dy = x^3(1+x^2)^{\frac{3}{2}}\,dx.$ $\qquad y = \dfrac{5x^2 - 2}{3\cdot 5}(1+x^2)^{\frac{5}{2}}.$

9. $dy = (1 - x^2)^{\frac{3}{2}}\,dx.$

$$y = \tfrac{1}{4}x\,(1-x^2)^{\frac{3}{2}} + \tfrac{3}{8}x\,(1-x^2)^{\frac{1}{2}} + \tfrac{3}{8}\sin^{-1}x.$$

10. $dy = \dfrac{dx}{(a+bx^2)^{\frac{5}{2}}}.$

$$y = \left(\frac{1}{a+bx^2} + \frac{2}{a}\right)\frac{x}{3a\sqrt{a+bx^2}}.$$

11. $dy = \dfrac{dx}{(a+bx^2)^{\frac{7}{2}}}.$

$$y = \left[\frac{1}{(a+bx^2)^2} + \frac{4}{3a\,(a+bx^2)} + \frac{8}{3a^2}\right]\frac{x}{5a\sqrt{a+bx^2}}.$$

14

12. $dy = \dfrac{x^4 dx}{(1-x^2)^{\frac{3}{2}}}.$

$\qquad y = -\dfrac{x^3 - 3x}{2\sqrt{1-x^2}} - \tfrac{3}{2}\sin^{-1} \boldsymbol{x}.$

13. $dy = \dfrac{x^4 dx}{\sqrt{2ax - x^2}}.$

$\qquad y = -\left(\dfrac{x^3}{4} + \dfrac{7x^2}{4\cdot 3}\,a + \dfrac{7\cdot 5x}{4\cdot 3\cdot 2}\,a^2 + \dfrac{7\cdot 5\cdot 3}{4\cdot 3\cdot 2}\,a^3\right)\sqrt{2ax - x^2}$

$\qquad\qquad\qquad\qquad\qquad + \dfrac{7\cdot 5\cdot 3}{4\cdot 3\cdot 2}\,a^4 \operatorname{vers}^{-1}\dfrac{x}{a}.$

14. $dy = \log x\, dx.$ $y = x(\log x - 1).$

15. $dy = x^2 \log^2 x\, dx.$ $y = \tfrac{1}{3}x^3\left(\log^2 x - \tfrac{2}{3}\log x + \tfrac{2}{9}\right).$

16. $dy = \dfrac{dx}{x \log x}.$ $y = \log(\log x).$

17. $dy = x \log^3 x\, dx.$

$\qquad y = \dfrac{x^2}{2}\log^3 x - \dfrac{3x^2}{4}\log^2 x + \dfrac{3x^2}{4}\log x - \tfrac{3}{8}x^2.$

18. $dy = x^3 \log x\, dx.$ $y = \dfrac{x^4}{4}\log x - \dfrac{x^4}{16}.$

19. $dy = \dfrac{dx}{x \log^2 x}.$ $y = -\dfrac{1}{\log x}.$

20. $dy = \dfrac{\log^2 x\, dx}{x^{\frac{3}{2}}}.$ $y = -\dfrac{2}{x^{\frac{1}{2}}}\left(\log^2 x + 4\log x + 8\right).$

21. $dy = \dfrac{x^4 dx}{\log^3 x}.$

$\qquad y = -\dfrac{x^5}{2\log^2 x} - \dfrac{5x^5}{2\log x} + \dfrac{25}{2}\left[\log(\log x^5)\right.$

$\qquad\qquad\qquad \left. + \log x^5 + \dfrac{1}{2^2}\log^2 x^5 + \dfrac{1}{2\cdot 3^2}\log^3 x^5\right].$

22. $dy = e^x x^4 dx.$ $\qquad y = e^x (x^4 - 4x^3 + 12x^2 - 24x + 24).$

23. $dy = x a^x dx.$ $\qquad y = \dfrac{a^x}{\log a}\left(x - \dfrac{1}{\log a}\right).$

24. $dy = x^2 e^x dx.$ $\qquad y = e^x (x^2 - 2x + 2).$

25. $dy = \dfrac{x^2 dx}{e^x}.$ $\qquad y = -e^{-x}(x^2 + 2x + 2).$

26. $dy = \dfrac{a^x dx}{x^3}.$

$$y = -\frac{a^x}{2x^2}(1 + x \log a) + \frac{\log^2 a}{2}\left(\log x + \log a \cdot x\right.$$

$$\left. + \tfrac{1}{2}\log^2 a \cdot \frac{x^2}{2} + \text{etc.}\right).$$

27. $dy = \dfrac{(e^{2x} - 1)\, dx}{e^{2x} + 1}.$ $\qquad y = \log(e^x + e^{-x}).$

28. $dy = e^{e^x} e^x\, dx.$ $\qquad y = e^{e^x}.$

29. $dy = \dfrac{e^x x\, dx}{(1 + x)^2}.$ $\qquad y = \dfrac{e^x}{1 + x}.$

30. $dy = \dfrac{(1 + x^2)\, e^x dx}{(1 + x)^2}.$ $\qquad y = e^x \left(\dfrac{x - 1}{1 + x}\right).$

[Put $(1 + x) = z$; then $x = z - 1$, $dx = dz$, etc.]

31. $dy = \dfrac{\sin^5 \theta\, d\theta}{\cos^2 \theta}.$ (Art. 156.)

$\qquad y = \sec \theta + 2 \cos \theta - \tfrac{1}{3}\cos^3 \theta.$

32. $dy = \sin^{\frac{1}{3}} \theta \cos^3 \theta\, d\theta.$ $\qquad y = \tfrac{2}{3}\sin^{\frac{3}{2}} \theta - \tfrac{2}{7}\sin^{\frac{7}{2}} \theta.$

33. $dy = \dfrac{\sin^3 \theta\, d\theta}{\cos^{\frac{1}{3}} \theta}.$ $\qquad y = \tfrac{2}{5}\cos^{\frac{5}{3}} \theta - 2\cos^{\frac{1}{3}} \theta.$

34. $dy = \dfrac{\cos^3 \theta\, d\theta}{\sin^{\frac{2}{3}} \theta}.$ $\qquad y = 3\sin^{\frac{1}{3}} \theta - \tfrac{3}{7}\sin^{\frac{7}{3}} \theta.$

35. $dy = \dfrac{\sin^5\theta\, d\theta}{\cos^2\theta}$.

$$y = -\frac{1}{3\cos\theta}(\sin^4\theta + 4\sin^2\theta - 8).$$

36. $dy = \dfrac{d\theta}{\sin^4\theta\,\cos^2\theta}$.

$$y = \frac{1}{\cos\theta\,\sin^3\theta} - \frac{4\cos\theta}{3\sin^3\theta} - \frac{8\cos\theta}{3\sin\theta}.$$

37. $dy = \dfrac{d\theta}{\sin^{\frac{1}{2}}\theta\,\cos^{\frac{7}{2}}\theta}$. $\quad y = 2\tan^{\frac{1}{2}}\theta\left(1 + \frac{1}{5}\tan^2\theta\right)$.

38. $dy = \dfrac{\sin^{\frac{1}{2}}\theta\, d\theta}{\cos^{\frac{5}{2}}\theta}$. $\qquad y = \frac{2}{3}\tan^{\frac{3}{2}}\theta$.

39. $dy = \dfrac{d\theta}{\sin^4\theta\,\cos^4\theta}$. $\quad y = -8\cot 2\theta - \frac{8}{3}\cot^3 2\theta$.

40. $dy = \sin^4\theta\,\cos^4\theta\, d\theta$. (Art. 157.)

$$y = \frac{\sin^5\theta}{8}\left(\cos^3\theta + \frac{1}{2}\cos\theta\right) - \frac{\cos\theta}{64}\left(\sin^3\theta + \frac{3}{2}\sin\theta\right)$$
$$+ \frac{3}{128}\theta.$$

41. $dy = \dfrac{d\theta}{\sin\theta\,\cos^2\theta}$. $\qquad y = \sec\theta + \log\tan\dfrac{\theta}{2}$.

42. $dy = \dfrac{d\theta}{\sin\theta\,\cos^4\theta}$.

$$y = \frac{1}{3\cos^3\theta} + \frac{1}{\cos\theta} + \log\tan\frac{\theta}{2}.$$

43. $dy = \sin^8\theta\,\cos^6\theta\, d\theta$.

$$y = -\frac{\cos^7\theta}{14}\left(\sin^7\theta + \tfrac{7}{12}\sin^5\theta + \tfrac{7}{24}\sin^3\theta + \tfrac{7}{64}\sin\theta\right)$$
$$+ \frac{\sin\theta}{768}\left(\cos^5\theta + \tfrac{5}{4}\cos^3\theta + \tfrac{15}{8}\cos\theta\right) + \frac{15\theta}{6144}.$$

44. $dy = \sin^4 \theta \, d\theta$. (Art. 158.)

$y = \frac{1}{32} \sin 4\theta - \frac{1}{4} \sin 2\theta + \frac{3}{8}\theta$.

45. $dy = \cos^4 \theta \, d\theta$.

$y = \frac{1}{32} \sin 4\theta + \frac{1}{4} \sin 2\theta + \frac{3}{8}\theta$.

46. $dy = \sin^6 \theta \, d\theta$.

$y = \frac{1}{32} (-\frac{1}{6} \sin 6\theta + \frac{3}{2} \sin 4\theta - \frac{15}{2} \sin 2\theta + 10\theta)$.

47. $dy = x^4 \sin x \, dx$. (Art. 159.)

$y = -x^4 \cos x + 4x^3 \sin x + 12x^2 \cos x$
$$- 24x \sin x - 24 \cos x.$$

48. $dy = e^{ax} \sin^2 x \, dx$. (Art. 160.)

$$y = \frac{e^{ax} \sin x}{4 + a^2} (a \sin x - 2 \cos x) + \frac{2e^{ax}}{a(4 + a^2)}.$$

49. $dy = e^x \sin^3 x \, dx$.

$y = \frac{1}{10}e^x (\sin^3 x + 3 \cos^3 x + 3 \sin x - 6 \cos x)$.

50. $dy = e^{-ax} \sin kx \, dx$. $y = -\dfrac{a \sin kx + k \cos kx}{(a^2 + k^2) e^{ax}}$.

51. $dy = \dfrac{x^2 dx \, \sin^{-1} x}{\sqrt{1 - x^2}}$. (Art. 161.)

Put $dv = \dfrac{x^3 dx}{\sqrt{1 - x^2}}$ and $u = \sin^{-1} x$;   then

$v = -\frac{1}{3}(x^2 + 2)\sqrt{1 - x^2}$ (by Ex. 1), etc.

$y = -\frac{1}{3}(x^2 + 2)\sqrt{1 - x^2} \cdot \sin^{-1} x + \dfrac{x^3}{9} + \frac{2}{3}x$.

52. $dy = \dfrac{x^4 dx}{\sqrt{1 - x^2}} \sin^{-1} x$.

$y = [-\frac{1}{4}(x^3 + \frac{3}{2}x)\sqrt{1 - x^2} + \frac{3}{16} \sin^{-1} x] \sin^{-1} x + \dfrac{x^4}{16} + \frac{3}{16}x^2$.

53. $dy = \dfrac{dx}{2 + \cos x}\cdot$ (Art. 162.)

$$y = \frac{2}{\sqrt{3}} \tan^{-1}\left[\frac{1}{\sqrt{3}} \tan \frac{x}{2}\right]\cdot$$

54. $dy = \dfrac{x^2 dx}{\sqrt{2 - 2x + x^2}}\cdot$ (See Formula 43, p. 345.)

$$y = \frac{x + 3}{2}(2 - 2x + x^2)^{\frac{1}{2}}$$

$$+ \tfrac{1}{2}\log\left[x - 1 + (2 - 2x + x^2)^{\frac{1}{2}}\right].$$

CHAPTER V

163. Integration by Series.—The number of differential expressions which can be integrated in finite terms is
very small ; the great majority of differentials can be integrated only by the aid of infinite series. When a differential can be developed into an infinite series, each term may
be integrated separately. If the result is a converging
series, the value of the integral may be found with sufficient
accuracy for practical purposes by summing a finite number
of terms ; and sometimes the law of the series is such that
its exact value can be found, even though the series is infinite. This method is not only a last resort when the
methods of exact integration fail, but it may often be em
ployed with advantage when an exact integration would
lead to a function of complicated form ; and the two methods
may be used together to discover the form of the developed
integral.

EXAMPLES.

1. Integrate $dy = \dfrac{dx}{a + x}$ in a series.

By division,

$$\frac{1}{a + x} = \frac{1}{a} - \frac{x}{a^2} + \frac{x^2}{a^3} - \frac{x^3}{a^4} + \text{etc.}$$

$$\therefore \ y = \int \frac{dx}{a + x} = \int \left(\frac{1}{a} - \frac{x}{a^2} + \frac{x^2}{a^3} - \frac{x^3}{a^4} + \text{etc.} \right) dx$$

$$= \frac{x}{a} - \frac{x^2}{2a^2} + \frac{x^3}{3a^3} - \frac{x^4}{4a^4} + \text{etc.}$$

But $\displaystyle\int \frac{dx}{a+x} = \log(a+x)$.　[Art. 130, (4).]

$$\therefore \quad \log(a+x) = \frac{x}{a} - \frac{x^2}{2a^2} + \frac{x^3}{3a^3} - \frac{x^4}{4a^4} + \text{etc.}$$

2.　$dy = x^{\frac{1}{3}}(1 - x^2)^{\frac{1}{2}}\, dx.$

Expanding $(1 - x^2)^{\frac{1}{3}}$ by the Binomial Theorem, we have

$$(1 - x^2)^{\frac{1}{2}} = 1 - \frac{x^2}{2} - \frac{x^4}{8} - \frac{x^6}{16} - \frac{5x^8}{128} - \text{etc.}$$

$$\therefore \quad y = \int x^{\frac{1}{2}}\left(1 - \frac{x^2}{2} - \frac{x^4}{8} - \frac{x^6}{16} - \frac{5x^8}{128} - \text{etc.}\right) dx.$$

$$= \tfrac{2}{3}x^{\frac{3}{2}} - \tfrac{1}{7}x^{\frac{7}{2}} - \tfrac{1}{44}x^{\frac{11}{2}} - \tfrac{1}{120}x^{\frac{15}{2}} - \tfrac{5}{1216}x^{\frac{19}{2}} - \text{etc.}$$

3.　$dy = \dfrac{dx}{1 + x^2}.$

$$y = \int \frac{dx}{1 + x^2} = \tan^{-1} x \qquad \text{[Art. 131, (16)]}$$

$$= x - \frac{x^3}{3} + \frac{x^5}{5} - \frac{x^7}{7} + \frac{x^9}{9} - \text{etc.}$$

4.　$dy = \dfrac{dx}{\sqrt{1 + x^2}}.$

$$y = \int \frac{dx}{\sqrt{1 + x^2}} = \log(x + \sqrt{1 + x^2})$$

$$\text{[Art. 144, (1)]}$$

$$= x - \frac{x^3}{2 \cdot 3} + \frac{3x^5}{2 \cdot 4 \cdot 5} - \frac{3 \cdot 5x^7}{2 \cdot 4 \cdot 6 \cdot 7} + \text{etc.}$$

5.　$dy = x^{-\frac{1}{2}}(x - 1)^{\frac{2}{3}}\, dx.$

$$y = \tfrac{6}{7}x^{\frac{7}{6}} - 4x^{\frac{1}{6}} + \tfrac{2}{15}x^{-\frac{5}{6}} + \tfrac{8}{261}x^{-\frac{11}{6}} - \text{etc.}$$

164. Successive Integration. — By applying the rules previously demonstrated for integration, we may obtain the original function from which second, third or n^{th} differentials, containing a single variable, may have been derived.

If the second derivative $\dfrac{d^2y}{dx^2} = X$ be given, when X is any function of x, two successive integrations will be required to determine the original function y in terms of x. Thus, multiplying by dx, we have

$$\frac{d^2y}{dx} = Xdx;$$

or
$$d\left(\frac{dy}{dx}\right) = Xdx.$$

Integrating, we get

$$\frac{dy}{dx} = \int Xdx = X_1 + C_1.$$

Multiplying again by dx and integrating, we get

$$y = \int X_1 dx + \int C_1 dx = X_2 + C_1 x + C_2.$$

Similarly, if we had $\dfrac{d^3y}{dx^3} = X$, three successive integrations would give

$$y = X_3 + C_1 \frac{x^2}{2} + C_2 x + C_3, \quad \text{and so on.}$$

Generally, let there be the nth derivative

$$\frac{d^n y}{dx^n} = X.$$

$$\therefore \quad d\left(\frac{d^{n-1}y}{dx^{n-1}}\right) = Xdx;$$

hence, by integrating we have

$$\frac{d^{n-1}y}{dx^{n-1}} = \int Xdx = X_1 + C_1.$$

Again, we get from this last equation,

$$d\left(\frac{d^{n-2}y}{dx^{n-2}}\right) = X_1 dx + C_1 dx\,;$$

and by integrating,

$$\frac{d^{n-2}y}{dx^{n-2}} = X_2 + C_1 x + C_2.$$

Also from this we obtain

$$d\left(\frac{d^{n-3}y}{dx^{n-3}}\right) = X_2 dx + C_1 x\,dx + C_2 dx,$$

and integrating,

$$\frac{d^{n-3}y}{dx^{n-3}} = X_3 + C_1\frac{x^2}{2} + C_2 x + C_3.$$

And continuing the process we get, after n integrations,

$$\int^n d^n y = \int^n Xdx^n$$

$$= X_n + C_1 \frac{x^{n-1}}{1\cdot 2\cdot 3\ldots(n-1)} + C_2 \frac{x^{n-2}}{1\cdot 2\cdot 3\ldots(n-2)}$$

$$+ \ldots\, C_{n-1}x + C_n. \qquad (1)$$

The symbol $\int^n Xdx^n$ is called the nth integral of Xdx^n, and denotes that n successive integrations are required. The first term X_n of the second member is the nth integral of Xdx^n, without the arbitrary constants; the remaining part of the series is the result of introducing at each integration, an arbitrary constant.

165. To Develop the nth Integral $\int^n X dx^n$ into a Series.—By Maclaurin's theorem, we have

$$\int^n X dx^n = \left(\int^n X dx^n\right) + \left(\int^{i-1} X dx^{n-1}\right)^* \frac{x}{1}$$

$$+ \left(\int^{n-2} X dx^{n-2}\right) \frac{x^2}{1 \cdot 2} + \text{etc.}$$

$$+ \left(\int X dx\right) \frac{x^{n-1}}{1 \cdot 2 \cdot 3 \ldots (n-1)} + (X) \frac{x^n}{1 \cdot 2 \cdot 3 \ldots n}$$

$$+ \left(\frac{dX}{dx}\right) \frac{x^{n+1}}{1 \cdot 2 \ldots (n+1)}$$

$$+ \left(\frac{d^2X}{dx^2}\right) \frac{x^{n+2}}{1 \cdot 2 \ldots (n+2)} + \text{etc.} \qquad (1)$$

in which the brackets

$$\left(\int^n X dx^n\right), \quad \left(\int^{n-1} X dx^{n-1}\right) \ldots \left(\int X dx\right),$$

are the arbitrary constants

$$C_n, \quad C_{n-1}, \ldots \ldots C_1,$$

for that is what these expressions become respectively, when $x = 0$.

By Maclaurin's theorem, we have

$$X = (X) + \left(\frac{dX}{dx}\right) \frac{x}{1} + \left(\frac{d^2X}{dx^2}\right) \frac{x^2}{2} + \left(\frac{d^3X}{dx^3}\right) \frac{x^3}{2 \cdot 3} + \text{etc.} \quad (2)$$

which may be converted into (1) by substituting for x^0, x^1, x^2, x^3, etc., in (2), the quantities

$$\frac{x^n}{1 \cdot 2 \ldots n}, \quad \frac{x^{n+1}}{2 \cdot 3 \ldots (n+1)}, \quad \frac{x^{n+2}}{3 \cdot 4 \ldots (n+2)}, \quad \text{etc.,}$$

---

* Since $\dfrac{d}{dx} \displaystyle\int^n X dx^n = \int^{n-1} X dx^{n-1}$.

and prefixing the terms containing the arbitrary constants as above shown, viz.,

$$C_n, \quad C_{n-1}\frac{x}{1}, \quad C_{n-2}\frac{x^2}{1\cdot 2}, \quad \cdots \cdots \quad C_1\frac{x^{n-1}}{1\cdot 2\cdot 3 \ldots (n-1)}.$$

(See Lacroix, Calcul Intégral, Vol. II, pp. 154 and 155.)

EXAMPLES.

1. Develop $\displaystyle\int^4 \frac{dx^4}{\sqrt{1-x^2}}.$

Here $\quad X = (1-x^2)^{-\frac{1}{2}}$

$$= 1 + \tfrac{1}{2}x^2 + \frac{1\cdot 3}{2\cdot 4}x^4 + \frac{1\cdot 3\cdot 5}{2\cdot 4\cdot 6}x^6 + \text{etc.}$$

Substituting in this series for x^0, x^2, x^4, x^6, etc., the quantities

$$\frac{x^4}{1\cdot 2\cdot 3\cdot 4}, \quad \frac{x^6}{3\cdot 4\cdot 5\cdot 6}, \quad \frac{x^8}{5\cdot 6\cdot 7\cdot 8}, \quad \frac{x^{10}}{7\cdot 8\cdot 9\cdot 10}, \quad \text{etc.};$$

and prefixing

$$C_4, \quad C_3\frac{x}{1}, \quad C_2\frac{x^2}{1\cdot 2}, \quad C_1\frac{x^3}{1\cdot 2\cdot 3},$$

we get

$$\int^4\frac{dx^4}{\sqrt{1-x^2}} = C_4 + C_3\frac{x}{1} + C_2\frac{x^2}{1\cdot 2} + C_1\frac{x^3}{1\cdot 2\cdot 3}$$

$$+ \frac{x^4}{1\cdot 2\cdot 3\cdot 4} + \frac{x^6}{2\cdot 3\cdot 4\cdot 5\cdot 6} + \frac{1\cdot 3x^8}{2\cdot 4\cdot 5\cdot 6\cdot 7\cdot 8}$$

$$+ \frac{1\cdot 3\cdot 5x^{10}}{2\cdot 4\cdot 6\cdot 7\cdot 8\cdot 9\cdot 10} + \text{etc.}$$

2. Integrate $\quad d^3y = 6a\,dx^3.$

Dividing by dx^2 we have

$$\frac{d^3y}{dx^2} = 6a\,dx, \quad \text{or} \quad d\left(\frac{d^2y}{dx^2}\right) = 6a\,dx.$$

$$\therefore \int d\left(\frac{d^2y}{dx^2}\right) = \int 6a\,dx;$$

or
$$\frac{d^2y}{dx^2} = 6ax + C_1.$$

Multiplying by dx and integrating again, we have

$$\frac{dy}{dx} = 3ax^2 + C_1x + C_2.$$

Multiplying again by dx and integrating, we have

$$y = ax^3 + C_1\frac{x^2}{2} + C_2x + C_3.$$

3. Integrate $\quad d^2y = \sin x \cos^2 x\, dx^2.$

Put $\quad\quad\quad \sin x = z ;$

$$\therefore \quad dz = \cos x\, dx,$$

and $\quad\quad\quad dz^2 = \cos^2 x\, dx^2 ;$

$$\therefore \quad d^2y = z\, dz^2 ;$$

from which we get

$$y = \frac{z^3}{6} + C_1z + C_2 ;$$

$$\therefore \quad y = \frac{\sin^3 x}{6} + C_1 \sin x + C_2.$$

4. Integrate $\quad d^3y = ax^2 dx^3.$

$$y = \frac{ax^5}{60} + \frac{C_1x^2}{2} + C_2x + C_3.$$

5. Integrate $\quad d^3y = 2x^{-3}dx^3.$

$$y = \log x + \frac{C_1x^2}{2} + C_2x + C_3.$$

6. Integrate $\quad d^4y = \cos x\, dx^4.$

$$y = \cos x + \frac{C_1x^3}{6} + \frac{C_2x^2}{2} + C_3x + C_4.$$

166. Integration of Functions of Two or More Variables.—Differential functions of two or more variables are either partial or total (Art. 80). When partial, they are obtained from the original function either by differentiating with respect to one variable only, or by differentiating first with respect to one variable, regarding the others as constant; then the result differentiated with respect to a second variable, regarding the rest as constant, and so on (Art. 83). For example,

$$\frac{d^2u}{dx^2} = f(x, y),$$

and

$$\frac{d^2u}{dx\,dy} = f(x, y)$$

are differential functions of the first and second kinds respectively, in which u is a function of the independent variables x and y. From the manner in which the expression

$$\frac{d^2u}{dx^2} = f(x, y)$$

was obtained (Art. 83), it is evident that the value of u may be found by integrating twice with respect to x, as in Art. 165, regarding y as constant; care being taken, at each integration, to add an arbitrary function of y, instead of a constant.

167. Integration of $\dfrac{d^2u}{dy\,dx} = f(x, y).$

This equation may be written

$$\frac{d}{dy}\left(\frac{du}{dx}\right) = f(x, y).$$

It is evident that $\dfrac{du}{dx}$ must be a function such that if we differentiate it with respect to y, regarding x as constant, the result will be $f(x, y)$.

Therefore we may write

$$\frac{du}{dx} = \int f(x, y)\,dy.$$

Here, also, it is evident that u must be such a function that if we differentiate it with respect to x, regarding y as constant, the result will be the function

$$\int f(x, y)\, dy.$$

Hence, $\qquad u = \int \left[\int f(x, y)\, dy \right] dx.$

Therefore, we first integrate with respect to y, regarding x as constant,* and then integrate the result with respect to x, regarding y as constant,* which is exactly reversing the process of differentiation. (Art. 83.)

The above expression for u may be abbreviated into

$$\int \int f(x, y)\, dy\, dx \qquad \text{or} \qquad \int \int f(x, y)\, dx\, dy.$$

We shall use the latter form;† that is, when we perform the y-integration before the x-integration, we shall write dy to the right of dx.

It is immaterial whether we first integrate with respect to y and then with respect to x, or first with respect to x and then with respect to y. (See Art. 84.)

In integrating with respect to y, care must be taken to add an arbitrary function of x, and in integrating with respect to x to add an arbitrary function of y.

In a similar manner, it may be shown that to find the value of u in the equation

$$\frac{d^3 u}{dx\, dy\, dz} = f(x, y, z),$$

we may write it

$$u = \int \int \int f(x, y, z)\, dx\, dy\, dz,$$

---

* Called the y-integration and x-integration, respectively.

† On this point of notation writers are not quite uniform. See Todhunter's Ca'., p. 78; also Price's Cal., Vol. II, p. 281.

which means that we first integrate with respect to z, regarding x and y as constant; then this result with respect to y, regarding x and z as constant; then this last result with respect to x, regarding y and z as constant, adding with the z-integration arbitrary functions of x and y, with the y-integration arbitrary functions of x and z, and with the x-integration arbitrary functions of y and z. (See Lacroix, Calcul Intégral, Vol. II, p. 206.)

EXAMPLES.

1. Integrate $d^2u = bx^2y\,dx^2$

Here $$d\left(\frac{du}{dx}\right) = bx^2y\,dx.$$

$$\therefore \quad \frac{du}{dx} = \int bx^2y\,dx = \tfrac{1}{3}bx^3y + f(y).$$

$$du = \tfrac{1}{3}bx^3y\,dx + f(y)\,dx.$$

$$\therefore \quad u = \tfrac{1}{12}bx^4y + f(y)\,x + \phi(y).$$

2. Integrate $d^2u = 2x^2y\,dx\,dy$.

Here $$d\left(\frac{du}{dx}\right) = 2x^2y\,dy.$$

$$\therefore \quad \frac{du}{dx} = \int 2x^2y\,dy = x^2y^2 + \phi(x).$$

$$du = x^2y^2dx + \phi(x)\,dx.$$

$$\therefore \quad u = \tfrac{1}{3}x^3y^2 + \int \phi(x)\,dx + f(y).$$

3. Integrate $d^2u = 3xy^3\,dx\,dy$.

$$u = \tfrac{3}{8}x^2y^4 + \int \phi(x)\,dx + f(y).$$

4. Integrate $d^2u = ax^3y^2\,dx\,dy$.

$$u = \frac{a}{12}\,x^4y^3 + \int \phi(x)\,dx + f(y).$$

168. Integration of Total Differentials of the First Order.

If
$$u = f(x, y),$$

we have (Art. 81),

$$du = \frac{du}{dx} dx + \frac{du}{dy} dy,$$

in which $\frac{du}{dx} dx$ and $\frac{du}{dy} dy$ are the partial differentials of u; also, we have (Art. 84),

$$\frac{d^2u}{dx\, dy} = \frac{d^2u}{dy\, dx},$$

or
$$\frac{d}{dy}\left(\frac{du}{dx}\right) = \frac{d}{dx}\left(\frac{du}{dy}\right). \tag{1}$$

Therefore, if an expression of the form

$$du = Pdx + Qdy \tag{2}$$

be a total differential of u, we must have

$$\frac{du}{dx} = P, \qquad \frac{du}{dy} = Q\,;$$

and hence, from (1), we must have the condition

$$\frac{dP}{dy} = \frac{dQ}{dx}, \tag{3}$$

which is called *Euler's Criterion of Integrability*. When this is satisfied, (2) is the differential of a function of x and y, and we shall obtain the function itself by integrating either term; thus,

$$u = \int Pdx + f(y), \tag{4}$$

in which $f(y)$ must be determined so as to satisfy the condition

$$\frac{du}{dy} = Q.$$

REMARK.—Since the differential with respect to x of every term of u which involves x must contain dx, therefore the integral of Pdx will give all the terms of u which involve x. The differential with respect to y of those terms of u which involve y and not x, will be found only in the expression Qdy. Hence, if we integrate those terms of Qdy which do not involve x, we shall have the terms of u which involve y only. This will be the value of $f(y)$, which added with an arbitrary constant to $\int Pdx$ will give the entire integral. Of course, if every term of the given differential contain x or dx, $f(y)$ will be constant. (See Church's Calculus, p. 274.)

EXAMPLES.

1. $du = 4x^3y^3dx + 3x^4y^2dy.$

Here $\qquad P = 4x^3y^3, \qquad\qquad Q = 3x^4y^2.$

$$\therefore \quad \frac{dP}{dy} = 12x^3y^2 \quad \text{and} \quad \frac{dQ}{dx} = 12x^3y^2.$$

Therefore (3) is satisfied, and since each term contains x or dx, we have from (4),

$$u = \int 4x^3y^3dx = x^4y^3 + C.$$

2. $du = \dfrac{dx}{y} + \left(2y - \dfrac{x}{y^2}\right)dy.$

(3) is satisfied, therefore from (4) we have

$$u = \int \frac{dx}{y} + f(y) = \frac{x}{y} + f(y).$$

Since the term $2ydy$ does not contain x, we must have, from the above Remark, $f(y) = \int 2ydy = y^2$, which must be added to $\dfrac{x}{y}$, giving for the entire integral,

$$u = \frac{x}{y} + y^2 + C.$$

3. $du = ydx + xdy.$ $\qquad\qquad\qquad u = xy + C.$

4. $du = (6xy - y^2)\,dx + (3x^2 - 2xy)\,dy.$
$$u = 3x^2y - y^2x + C.$$

5. $\quad du = (2axy - 3bx^2y)\, dx + (ax^2 - bx^3)\, dy.$

$$u = ax^2y - byx^3 + C.$$

The limits of this work preclude us from going further in this most interesting branch of the Calculus. The student who wishes to pursue the subject further is referred to Gregory's Examples; Price's Calculus, Vol. II; Lacroix's Calcul Intégral, Vol. II; and Boole's Differential Equations, where the subject is specially investigated.

169. Definite Integrals.—It was shown in Art. 130 that, to complete each integral, an arbitrary constant C must be added. While the value of this constant C remains unknown, the integral expression is called an *indefinite integral;* such are all the integrals that have been found by the methods hitherto explained.

When two different values of the variable have been substituted in the indefinite integral, and the difference between the two results is taken, the integral is said to be taken between *limits.*

In the application of the Calculus to the solution of real problems, the nature of the question will always require that the integral be taken between given limits. When an integral is taken between limits, it is called a *definite integral.**

The symbol for a definite integral is

$$\int_a^b f(x)\, dx,$$

which means that the expression $f(x)\, dx$ is first to be integrated; then in this result b and a are to be substituted successively for x, and the latter result is to be subtracted from the former; b and a are called the *limits of integration*, the former being the *superior*, and the latter the *inferior* limit. Whatever may be the value of the integral

---

* In the Integral Calculus, it is often the most difficult part of the work to pass from the indefinite to the definite integral.

at the inferior limit, that value is included in the value of the integral up to the superior limit. Hence, to find the integral between the limits, take the difference between the values of the integral at the limits.

In the preceding we assume that the function is continuous between the limits a and b, *i. e.*, that it does not become imaginary or infinite for any value of x between a and b.

Suppose u to be a function of x represented by the equation

$$u = f(x);$$

then

$$du = f'(x)\, dx.$$

Now if we wish the integral between the limits a and b, we have

$$u = \int_a^b f'(x)\, dx = f(b) - f(a).$$

If there is anything in the nature of the problem under consideration from which we can know the value of the integral for a particular value of the variable, the constant C can be found by substituting this value in the indefinite integral. Thus, if we have

$$du = (abx - bx^2)^{\frac{1}{2}}(ab - 2bx)\, dx,$$

and know that the integral must reduce to m when $x = a$, we can find the definite integral as follows:

Integrating by known rules, we have

$$u = \tfrac{2}{3}(abx - bx^2)^{\frac{3}{2}} + C,$$

which is the indefinite integral; and since $u = m$ when $x = a$, we have

$$m = 0 + C; \qquad \therefore \quad C = m,$$

which substituted in the value of u gives

$$u = \tfrac{2}{3}(abx - bx^2)^{\frac{3}{2}} + m.$$

E X A M P L E S.

1. Find the definite integral of $du = (1 + \frac{9}{4}ax)^{\frac{1}{2}} dx$, on the hypothesis that $u = 0$ when $x = 0$.

The indefinite integral is

$$u = \frac{8}{27a}(1 + \tfrac{9}{4}ax)^{\frac{3}{2}} + C.$$

Since when $x = 0$, $u = 0$, we have

$$0 = \frac{8}{27a} + C; \quad \therefore\ C = -\frac{8}{27a},$$

which substituted in the indefinite integral, gives

$$u = \frac{8}{27a}(1 + \tfrac{9}{4}ax)^{\frac{3}{2}} - \frac{8}{27a}$$

for the definite integral required.

2. Integrate $du = 6x^2 dx$ between the limits 3 and 0.

Here $\quad u = \int_0^3 6x^2 dx = \left[\, 2x^3 \,\right]_0^{3\,*} = 54.$

3. $u = \int_0^1 x^n dx = \left[\dfrac{x^{n+1}}{n+1}\right]_0^1 = \dfrac{1}{n+1}.$

4. $u = \int_0^\infty e^{-x}\, dx = \left[\, -e^{-x} \,\right]_0^\infty = -(0-1) = 1.$

5. $u = \int_0^\infty \dfrac{dx}{a^2 + x^2} = \dfrac{1}{a}\left[\tan^{-1}\dfrac{x}{a}\right]_0^\infty = \dfrac{\pi}{2a}.$

6. $u = \int_0^a \dfrac{dx}{a^2 + x^2} = \dfrac{1}{a}\left[\tan^{-1}\dfrac{x}{a}\right]_0^a = \dfrac{\pi}{4a}.$

---

* This notation signifies that the integral is to be taken between the limits 3 and 0.

$$= \frac{1}{a} \left[\tan^{-1} \infty - \tan^{-1} (-\infty) \right] = \frac{\pi}{a}.$$

8. $u = \displaystyle\int_0^a \frac{dx}{\sqrt{a^2 - x^2}} = \left[\sin^{-1} \frac{x}{a} \right]_0^a = \frac{\pi}{2}.$

REMARK.—It should be observed here that the value of the infinitesimal element corresponding to the superior limit is *excluded*, while that corresponding to the inferior limit is *included* in the definite integral ; for, were this not the case, as $\dfrac{1}{\sqrt{a^2 - x^2}}$ becomes equal to ∞ when $x = a$, the integral of Ex. 8 between the limits a and 0 would not be correct ; but as the limit a, being the *superior* limit in Ex. 8, and that which renders infinite the infinitesimal element, is *not included*, the definite integral is correct. (See Price's Calculus, Vol. II, p. 89.)

9. $u = \displaystyle\int_0^a (a^2 - x^2)^{\frac{1}{2}} dx.$ (See Ex. 4, Art. 151.]

$$= \left[\frac{x}{2} (a^2 - x^2)^{\frac{1}{2}} + \frac{a^2}{2} \sin^{-1} \frac{x}{a} \right]_0^a = \frac{a^2 \pi}{4}.$$

10. $u = \displaystyle\int_0^1 \frac{x^6 dx}{\sqrt{1 - x^2}}.$ (See Ex. 7, Art. 151.)

$$= \frac{1 \cdot 3 \cdot 5 \cdot \pi}{2 \cdot 4 \cdot 6 \cdot 2}.$$

11. $u = \displaystyle\int_0^{\frac{\pi}{2}} \sin^7 x \cos^4 x \, dx = \frac{4^2}{3 \cdot 5 \cdot 7 \cdot 11}.$

170. Change of Limits.—It is not necessary that the increment dx should be regarded as positive, for we may consider x as decreasing by infinitesimal elements, as well as increasing. Therefore, we have

$$\int_b^a \phi'(x)\,dx = \phi(a) - \phi(b) = -[\phi(b) - \phi(a)]$$

$$= -\int_a^b \phi'(x)\,dx.$$

That is, *if we interchange the limits, we change the sign of the definite integral.*

Also, it is obvious from the nature of integration (Art. 129), that

$$\int_a^c \phi(x)\,dx = \int_a^b \phi(x)\,dx + \int_b^c \phi(x)\,dx,$$

and so on. Hence,

$$\int_0^\pi \cos x\,dx = \int_0^{\frac14\pi} \cos x\,dx + \int_{\frac14\pi}^{\frac12\pi} \cos x\,dx + \int_{\frac12\pi}^{\frac34\pi} \cos x\,dx$$

$$+ \int_{\frac34\pi}^\pi \cos x\,dx$$

$$= \int_0^{\frac12\pi} \cos x\,dx + \int_{\frac12\pi}^\pi \cos x\,dx = 0.$$

Also, $\quad \displaystyle\int_{-a}^a f'(x)\,dx = \int_{-a}^0 f'(x)\,dx + \int_0^a f'(x)\,dx.$ \hfill (1)

Let $x = -x$; then $dx = -dx$, and the limits 0 and $-a$ become 0 and $+a$; therefore we have

$$\int_{-a}^0 f'(x)\,dx = -\int_a^0 f'(-x)\,dx = \int_0^a f'(-x)\,dx,$$

which in (1) gives

$$\int_{-a}^a f'(x)\,dx = \int_0^a f'(-x)\,dx + \int_0^a f'(x)\,dx$$

$$= \int_0^a [f'(-x)\,dx + f'(x)\,dx]. \hfill (2)$$

Now if $f'(-x) = -f'(x)$, (2) becomes,

$$\int_{-a}^a f'(x)\,dx = 0.$$

But if $f'(-x) = f'(x)$, we have

$$\int_{-a}^{0} f'(x)\, dx = \int_{0}^{a} f'(x)\, dx, \tag{3}$$

which in (1) gives $\quad \int_{-a}^{a} f'(x)\, dx = 2\int_{0}^{a} f'(x)\, dx. \tag{4}$

The following are examples of these principles.

1. $u = \int_{-\frac{1}{2}\pi}^{\frac{1}{2}\pi} \cos x\, dx.$

Here $\qquad\qquad f'(-x) = f'(x) = \cos x.$

$$\therefore \quad u = \int_{-\frac{1}{2}\pi}^{\frac{1}{2}\pi} \cos x\, dx = 2\int_{0}^{\frac{1}{2}\pi} \cos x\, dx = 2.$$

2. $u = \int_{-\frac{1}{2}\pi}^{\frac{1}{2}\pi} \sin x\, dx = 0.$ [Since $f'(-x) = -f'(x)$.]

3. $u = \int_{-a}^{a} (a^2 - x^2)^{\frac{1}{2}}\, dx = 2\int_{0}^{a} (a^2 - x^2)^{\frac{1}{2}}\, dx = \dfrac{\pi a^2}{2}.$

4. $u = \int_{0}^{\pi} \sin x\, dx = \int_{0}^{\frac{1}{2}\pi} \sin x\, dx + \int_{\frac{1}{2}\pi}^{\pi} \sin x\, dx$

$$= 2\int_{0}^{\frac{1}{2}\pi} \sin x\, dx.$$

$$\left[\text{Since } \int_{\frac{1}{2}\pi}^{\pi} \sin x\, dx = \int_{0}^{\frac{1}{2}\pi} \sin x\, dx \right] = 2.$$

5. $u = \int_{0}^{\pi} \cos x\, dx = \int_{0}^{\frac{1}{2}\pi} \cos x\, dx + \int_{\frac{1}{2}\pi}^{\pi} \cos x\, dx = 0.$

$$\left[\text{Since } \int_{\frac{1}{2}\pi}^{\pi} \cos x\, dx = -\int_{0}^{\frac{1}{2}\pi} \cos x\, dx. \right]$$

6. $u = \int_{-1}^{1} \dfrac{x^4 dx}{\sqrt{1 - x^2}} = 2\int_{0}^{1} \dfrac{x^4 dx}{\sqrt{1 - x^2}} = \frac{1}{2}\cdot\frac{3}{4}\pi.$

(See Ex. 2, Art. 162.)

EXAMPLES.

1. Integrate $du = \dfrac{dx}{\sqrt{a^2 - x^2}}$ by series.

$$u = \frac{x}{a} + \frac{x^3}{1 \cdot 2 \cdot 3 a^3} + \frac{1 \cdot 3 x^5}{2 \cdot 4 \cdot 5 a^5} + \frac{1 \cdot 3 \cdot 5 x^7}{2 \cdot 4 \cdot 6 \cdot 7 a^7} + \text{etc.}$$

But $\displaystyle\int \frac{dx}{\sqrt{a^2 - x^2}} = \sin^{-1} \frac{x}{a}$ (by Ex. 14 of Art. 131);

therefore,

$$\sin^{-1} \frac{x}{a} = \frac{x}{a} + \frac{x^3}{2 \cdot 3 a^3} + \frac{1 \cdot 3 x^5}{2 \cdot 4 \cdot 5 a^5} + \frac{1 \cdot 3 \cdot 5 x^7}{2 \cdot 4 \cdot 6 \cdot 7 a^7} + \text{etc.}$$

2. Integrate $du = \dfrac{dx}{\sqrt{1 + x^4}}$.

$$u = \frac{x}{1} - \frac{1 \cdot x^5}{2 \cdot 5} + \frac{1 \cdot 3 x^9}{2 \cdot 4 \cdot 9} - \frac{1 \cdot 3 \cdot 5 x^{13}}{2 \cdot 4 \cdot 6 \cdot 13} + \text{etc.}$$

3. Integrate $du = \dfrac{dx}{\sqrt{1 - x^2}} (1 - e^2 x^2)^{\frac{1}{2}}$.

By the Binomial Theorem,

$$\sqrt{1 - e^2 x^2} = 1 - \tfrac{1}{2} e^2 x^2 - \frac{1 \cdot e^4 x^4}{2 \cdot 4} - \frac{1 \cdot 3 e^6 x^6}{2 \cdot 4 \cdot 6} - \text{etc.}$$

Multiplying by $\dfrac{dx}{\sqrt{1 - x^2}}$, and integrating each term separately (see Ex. 1, Art. 151), we have

$$u = \int \frac{dx}{\sqrt{1 - x^2}} (1 - e^2 x^2)^{\frac{1}{2}}$$

$$= \sin^{-1} x + \tfrac{1}{2} e^2 \left[\frac{x}{2} \sqrt{1 - x^2} - \tfrac{1}{2} \sin^{-1} x \right]$$

$$+ \frac{e^4}{2 \cdot 4} \left[\left(\frac{x^3}{4} + \frac{1 \cdot 3 \cdot x}{2 \cdot 4} \right) \sqrt{1 - x^2} - \frac{1 \cdot 3}{2 \cdot 4} \sin^{-1} x \right]$$

15

$$+ \frac{1\cdot 3\cdot e^6}{2\cdot 4\cdot 6}\left[\left(\frac{x^5}{6} + \frac{1\cdot 5x^3}{4\cdot 6} + \frac{1\cdot 3\cdot 5x}{2\cdot 4\cdot 6}\right)\sqrt{1-x^2}\right.$$

$$\left. - \frac{1\cdot 3\cdot 5}{2\cdot 4\cdot 6}\sin^{-1}x\right] + \text{etc.}$$

4. Given $d^3y = \dfrac{2}{x^3}dx^3$, to find y.   (Art. 164.)

$$y = \log x + \tfrac{1}{2}C_1x^2 + C_2x + C_3.$$

5. $d^4y = -x^{-4}dx^4.$

$$y = \tfrac{1}{6}\log x + \tfrac{1}{6}C_1x^3 + \tfrac{1}{2}C_2x^2 + C_3x + C_4.$$

6. $d^3y = x^{\frac{1}{2}}dx^3.$ $y = \tfrac{8}{105}x^{\frac{7}{2}} + \tfrac{1}{2}C_1x^2 + C_2x + C_3.$

7. $d^2y = Sx^2dx^2.$ $y = \tfrac{1}{12}Sx^4 + C_1x + C_2.$

8. $d^2y = \cos x \sin^2 x\, dx^2.$

$$y = \tfrac{1}{6}\cos^3 x + C_1\cos x + C_2.$$

9. $d^4y = \cos x\, dx^4.$

$$y = \cos x + \tfrac{1}{6}C_1x^3 + \tfrac{1}{2}C_2x^2 + C_3x + C_4.$$

10. $d^3y = e^x dx^3.$ $y = e^x + \tfrac{1}{2}C_1x^2 + C_2x + C_3.$

11. $d^4y = (1+x^2)^{-\frac{1}{2}}dx^4.$ (Art. 165.)

$$y = C_4 + C_3x + C_2\frac{x^2}{2} + C_1\frac{x^3}{2\cdot 3} + \frac{x^4}{2\cdot 3\cdot 4}$$

$$- \frac{x^6}{2\cdot 3\cdot 4\cdot 5\cdot 6} + \frac{1\cdot 3x^8}{2\cdot 4\cdot 5\cdot 6\cdot 7\cdot 8}$$

$$- \frac{1\cdot 3\cdot 5x^{10}}{2\cdot 4\cdot 6\cdot 7\cdot 8\cdot 9\cdot 10} + \text{etc.}$$

12. $d^2u = ax^4y^5dxdy.$

$$u = \frac{a}{30}x^5y^6 + \int\phi(x)\,dx + f(y).$$

13. $du = (2xy^2 + 9x^2y + 8x^3)\,dx + (2x^2y + 3x^3)\,dy.$
$$u = x^2y^2 + 3x^3y + 2x^4 + C.$$

14. $u = \int_{-a}^{a} (a^2 - x^2)^{\frac{5}{2}} dx = 2 \int_{0}^{a} (a^2 - x^2)^{\frac{5}{2}} dx.$

(See Art. 170.)

$u = \dfrac{3 \cdot 5}{2 \cdot 4 \cdot 6} a^6 \pi.$ (See Ex. 7 of Art. 162.)

15. $u = \int_{0}^{2a} \dfrac{x^4 dx}{\sqrt{2ax - x^2}}.$

$u = \dfrac{7 \cdot 5 \cdot 3}{4 \cdot 3 \cdot 2} a^4 \pi.$

16. $u = \int_{0}^{1} x^3 (1 - x)^{\frac{5}{2}} dx = \dfrac{2^5}{3 \cdot 7 \cdot 11 \cdot 13}.$

17. $u = \int_{0}^{} \dfrac{dx}{\sqrt{1 - x^2}} (1 - e^2 x^2)^{\frac{1}{2}}.$ (See Ex. 3.)

$u = \dfrac{\pi}{2} - \dfrac{1}{4} e^2 \dfrac{\pi}{2} - \dfrac{1 \cdot 3}{2^2 \cdot 4^2} e^4 \dfrac{\pi}{2} - \dfrac{1 \cdot 3^2 \cdot 5 e^6}{2^2 \cdot 4^2 \cdot 6^2} \dfrac{\pi}{2} -\!-\ \text{etc.}$

18. $u = \int_{0}^{\frac{1}{4}\pi} \int_{0}^{x} \dfrac{x \, dx \, dy}{x^2 + y^2}.$ (Art. 167.)

We first perform the y-integration, regarding x as constant, and then the x-integration.

$\therefore \ u = \int_{0}^{\frac{1}{4}\pi} \left[\dfrac{1}{x} \tan^{-1} \dfrac{y}{x} \right]_{0}^{x} x \, dx.$ (Art. 132, Ex. 3.)

$= \int_{0}^{\frac{1}{4}\pi} \dfrac{\pi}{4} dx = \dfrac{\pi^2}{16}.$

19. $u = \int_{0}^{a} \int_{0}^{x} \int_{0}^{y} xyz \, dx \, dy \, dz$

$= \int_{0}^{a} \int_{0}^{x} \dfrac{x y^3}{2} dx \, dy = \int_{0}^{a} \dfrac{x^5}{8} dx = \dfrac{a^6}{48}.$

20. $u = \int_{0}^{a} \int_{0}^{a-x} \int_{0}^{a-x-y} dx \, dy \, dz = \dfrac{a^3}{6}.$

21. $u = \int_0^{2\pi} \int_0^a r^3 d\theta \, dr = \dfrac{a^4\pi}{2}.$

22. $u = \int_{-\frac{1}{2}\pi}^{\frac{1}{2}\pi} \int_0^{2a\cos\theta} r^3 d\theta \, dr = \frac{3}{2}a^4\pi.$

For the convenience of the student, the preceding formulæ are summed up in the following table.

TABLE OF INTEGRALS.

CHAPTER I.

ELEMENTARY FORMS. (Page 238.)

1. $\displaystyle\int (dv + dy - dz) = v + y - z.$ $\hspace{2em}$ (130)*

2. $\displaystyle\int ax^n dx = \dfrac{ax^{n+1}}{n+1}.$ $\hspace{2em}$ (131)

3. $\displaystyle\int \dfrac{adx}{x^n} = -\dfrac{a}{(n-1)\,x^{n-1}}.$

4. $\displaystyle\int \dfrac{adx}{x} = a \log x.$

5. $\displaystyle\int a^x \log a \, dx = a^x.$

6. $\displaystyle\int e^x dx = e^x.$

7. $\displaystyle\int \cos x \, dx = \sin x.$

8. $\displaystyle\int \sec^2 x \, dx = \tan x.$

9. $\displaystyle\int \sec x \tan x \, dx = \sec x.$

---

* The arbitrary constant is understood. (See Art. 131.)

10. $\int \dfrac{dx}{\sqrt{a^2 - b^2x^2}} = \dfrac{1}{b} \sin^{-1} \dfrac{bx}{a}.$ (132)

11. $\int \dfrac{dx}{a^2 + b^2x^2} = \dfrac{1}{ab} \tan^{-1} \dfrac{bx}{a}.$

12. $\int \dfrac{dx}{x \sqrt{b^2x^2 - a^2}} = \dfrac{1}{a} \sec^{-1} \dfrac{bx}{a}.$

13. $\int \dfrac{dx}{\sqrt{2abx - b^2x^2}} = \dfrac{1}{b} \text{vers}^{-1} \dfrac{bx}{a}.$

14. $\int \tan x\, dx = \log \sec x.$

15. $\int \dfrac{dx}{\sin x} = \log \tan \tfrac{1}{2}x.$

CHAPTER II.

RATIONAL FRACTIONS. (Page 256.)

16. $\int \dfrac{f(x)\, dx}{\phi(x)} = \int \dfrac{A\,dx}{x - a} + \dfrac{B\,dx}{x - b} + \cdots \int \dfrac{L\,da}{l - x}.$ (137)

17. $\int \dfrac{f(x)\, dx}{\phi(x)}$

$= \int \dfrac{A\,dx}{(x-a)^n} + \int \dfrac{B\,dx}{(x-a)^{n-1}} + \cdots \int \dfrac{L\,dx}{(x-a)}.$ (138)

18. $\int \dfrac{f(x)\, dx}{\phi(x)} = \int \dfrac{(Ax + B)\, dx}{[(x \pm a)^2 + b^2]^n}$

$+ \int \dfrac{(Cx + D)\, dx}{[(x \pm a)^2 + b^2]^{n-1}} + \cdots \int \dfrac{(Kx + L)\, dx}{(x \pm a)^2 + b^2}.$ (139)

19. $\int \dfrac{dx}{a^2 - b^2x^2} = \dfrac{1}{2ab} \log \left(\dfrac{a + bx}{a - bx} \right).$

20. $\int \dfrac{a\,dx}{x^2 - a^2} = \log \sqrt{\dfrac{x - a}{x + a}}.$

CHAPTER III.

IRRATIONAL FUNCTIONS. (Page 269.)

21. $\displaystyle \int \frac{x^n dx}{\sqrt{a + bx}} = \frac{2}{b^{n+1}} \int (z^2 - a^2)^n \, dz,$

[where $z = (a + bx)^{\frac{1}{2}}$]. $\hspace{2cm}$ (142,

22. $\displaystyle \int \frac{x^{2n+1} dx}{(a + bx^2)^{\frac{1}{2}}} = \int \frac{(z^2 - a)^n \, dz}{b^{n+1}},$

[where $z = (a + bx^2)^{\frac{1}{2}}$]. $\hspace{2cm}$ (143)

23. $\displaystyle \int \frac{dx}{\sqrt{a + bx + x^2}} = \log\!\left(\frac{b}{2} + x + \sqrt{a + bx + x^2}\right).$ (144)

24. $\displaystyle \int \frac{dx}{\sqrt{a + bx - x^2}} = -2 \tan^{-1} \sqrt{\frac{\beta - x}{x - \alpha}},$

[where $a + bx - x^2 = (x - \alpha)(\beta - x)$.

25. $\displaystyle \int x^m (a + bx^n)^{\frac{p}{q}} dx = \frac{q}{nb} \int z^{(p+q-1)} \left(\frac{z^q - a}{b}\right)^{\left(\frac{m+1}{n} - 1\right)} dz,$ (146)

(where $z^q = a + bx^n$);

or $\displaystyle \quad = -\frac{q}{n} a^{\left(\frac{m}{n} + \frac{p}{q} + \frac{1}{n}\right)} \int (z^q - b)^{\left(\frac{m+1}{n} + \frac{p}{q} + 1\right)} z^{(p+q-1)} \, dz,$

(where $x^n z^q = a + bx^n$).

26. $\displaystyle \int \frac{dx}{x\sqrt{a + bx}} = \frac{2}{\sqrt{a}} \log\!\left(\frac{\sqrt{a+bx} - \sqrt{a}}{x}\right).$ Ex. 6, (146)

27. $\displaystyle \int \frac{dx}{\sqrt{a^2 + b^2 x^2}} = \frac{1}{b} \log\left(bx + \sqrt{a^2 + b^2 x^2}\right).$ $\hspace{1cm}$ Ex. 16

28. $\displaystyle \int \frac{dx}{x\sqrt{a^2 + b^2 x^2}} = \frac{1}{a} \log\!\left(\frac{\sqrt{a^2 + b^2 x^2} - a}{bx}\right),$ $\hspace{1cm}$ Ex. 17

or $\displaystyle \hspace{2cm} = -\frac{1}{a} \log\!\left(\frac{a + \sqrt{a^2 + b^2 x^2}}{bx}\right).$

29. $\int \dfrac{a\,dx}{\sqrt{2ax+x^2}} = a \log (x+a+\sqrt{2ax+x^2}).$ Ex. 20.

30. $\int (a^2+x^2)^{\frac{1}{2}}\, dx$

$= \dfrac{x}{2} (a^2+x^2)^{\frac{1}{2}} + \dfrac{a^2}{2} \log [x+(a^2+x^2)^{\frac{1}{2}}].$ Ex. 35.

CHAPTER IV.

Successive Reduction. (Page 285.)

31. $\int u\,dv = uv - \int v\,du.$ $\qquad\qquad$ (147)

32. $\int x^m (a + bx^n)^p\, dx$ $\qquad\qquad$ (148)

$= \dfrac{x^{m-n+1}(a+bx^n)^{p+1} - (m-n+1)\, a \int x^{m-n}(a+bx^n)^p dx}{b\,(np + m + 1)} \cdot (A)$

33. $\int x^{-m} (a + bx^n)^p\, dx$ $\qquad\qquad$ (149)

$= \dfrac{x^{-m+1}(a+bx^n)^{p+1} + b(m-np-n-1)\int x^{-m+n}(a+bx^n)^p dx}{-a\,(m-1)} \cdot (B)$

34. $\int x^m (a + bx^n)^p\, dx$ $\qquad\qquad$ (150)

$= \dfrac{x^{m+1} (a + bx^n)^p + anp \int x^m (a + bx^n)^{p-1}\, dx}{np + m + 1} \cdot$ $\qquad$ (C)

35. $\int x^m (a + bx^n)^{-p}\, dx$ $\qquad\qquad$ (151)

$= \dfrac{x^{m+1}(a+bx^n)^{-p+1} - (m+n+1-np)\int x^m(a+bx^n)^{-p+1}dx}{an\,(p - 1)} \cdot (D)$

36. $\int x^m (a^2 - x^2)^{-\frac{1}{2}}\, dx$

$= -\dfrac{x^{m-1}}{m} (a^2-x^2)^{\frac{1}{2}} + \dfrac{(m-1)}{m}\, a^2 \int x^{m-2} (a^2 - x^2)^{-\frac{1}{2}}\, dx.$

37. $\displaystyle\int x^m \left(a^2 + x^2\right)^{-\frac{1}{2}} dx$

$$= \frac{x^{m-1}}{m} \left(a^2 + x^2\right)^{\frac{1}{2}} - \frac{(m-1)}{m} a^2 \int x^{m-2} \left(a^2 + x^2\right)^{-\frac{1}{2}} dx.$$

38. $\displaystyle\int \frac{dx}{x^m \left(a^2 - x^2\right)^{\frac{1}{2}}}$

$$= \frac{-\left(a^2 - x^2\right)^{\frac{1}{2}}}{(m-1)a^2 x^{m-1}} + \frac{m-2}{(m-1)\,a^2} \int \frac{dx}{x^{m-2}\left(a^2 - x^2\right)^{\frac{1}{2}}}.$$

39. $\displaystyle\int \left(a^2 - x^2\right)^{\frac{n}{2}} dx$

$$= \frac{x\left(a^2 - x^2\right)^{\frac{n}{2}} + na^2 \int \left(a^2 - x^2\right)^{\frac{n}{2}-1} dx}{n + 1}.$$

40. $\displaystyle\int \frac{dx}{\left(a^2 - x^2\right)^{\frac{n}{2}}}$

$$= \frac{x}{(n-2)a^2\left(a^2 - x^2\right)^{\frac{n}{2}-1}} + \frac{n-3}{(n-2)a^2} \int \frac{dx}{\left(a^2 - x^2\right)^{\frac{n}{2}-1}}.$$

41. $\displaystyle\int \frac{x^m dx}{\sqrt{2ax - x^2}}$

$$= -\frac{x^{m-1}}{m} \sqrt{2ax - x^2} + \frac{(2m-1)\,a}{m} \int \frac{x^{m-1} dx}{\sqrt{2ax - x^2}}$$

42. $\displaystyle\int \frac{x^n dx}{\sqrt{a + bx + cx^2}}$

$$= \frac{x^{n-1} \sqrt{a + bx + cx^2}}{nc} - \frac{n-1}{n} \cdot \frac{a}{c} \int \frac{x^{n-2} dx}{\sqrt{a + bx + cx^2}}$$

$$- \frac{2n-1}{2n} \cdot \frac{b}{c} \int \frac{x^{n-1} dx}{\sqrt{a + bx + cx^2}}.\,*$$

---

* See Price's Calculus, Vol. II, p. 63.

43. $\displaystyle\int\frac{x^2 dx}{\sqrt{a+bx+cx^2}}$

$$=\sqrt{a+bx+cx^2}\left(\frac{x}{2c}-\frac{3b}{4c^2}\right)+\left(\frac{3b^2}{8c^2}-\frac{a}{2c}\right)\int\frac{dx}{\sqrt{a+bx+cx^2}}\quad *.$$

44. $\displaystyle\int\frac{x\,dx}{\sqrt{a+bx+cx^2}}$

$$=\frac{\sqrt{a+bx+cx^2}}{c}-\frac{b}{2c}\int\frac{dx}{\sqrt{a+bx+cx^2}}$$

$$=\frac{\sqrt{a+bx+cx^2}}{c}-\frac{b}{2c}\left[\frac{1}{\sqrt{c}}\log\left(x+\frac{b}{2c}+\sqrt{x^2+\frac{bx}{c}+\frac{a}{c}}\right)^*\right].$$

45. $\displaystyle\int x^m \log^n x\,dx$

$$=\frac{x^{m+1}}{m+1}\log^n x-\frac{n}{m+1}\int x^m \log^{n-1} x\,dx.\quad(152)$$

46. $\displaystyle\int\frac{x^m\,dx}{\log^n x}$

$$=-\frac{x^{m+1}}{(n-1)\log^{n-1}x}+\frac{m+1}{n-1}\int\frac{x^m\,dx}{\log^{n-1}x}.\quad(153)$$

47. $\displaystyle\int\frac{dz}{\log z}$

$$=\log(\log z)+\log z+\frac{1}{2^2}\log^2 z+\frac{1}{2\cdot3^2}\log^3 z+\text{etc.}$$

48. $\displaystyle\int a^{mx}x^n\,dx=\frac{a^{mx}x^n}{m\log a}-\frac{n}{m\log a}\int a^{mx}x^{n-1}\,dx.\quad(154)$

49. $\displaystyle\int\frac{a^x\,dx}{x^m}=-\frac{a^x}{(m-1)\,x^{m-1}}+\frac{\log a}{m-1}\int\frac{a^x dx}{x^{m-1}}.\quad(155)$

50. $\displaystyle\int\sin^m\theta\,\cos^n\theta\,d\theta$

$$=-\int(1-\cos^2\theta)^r\cos^n\theta\,d\cos\theta,\quad(156)$$

---

* See Price's Calculus, Vol. II, p. 63.

$$\text{(when } m = 2r + 1) ;$$

or
$$= \int \tan^m \theta \, (1 + \tan^2 \theta)^{r-1} d \tan \theta,$$

$$\text{(when } m + n = - 2r) ;$$

or
$$= \int x^m (1 - x^2)^{\frac{n-1}{2}} dx, \text{ (when } x = \sin \theta). \quad (157)$$

51. $\displaystyle \int x^n \cos ax \, dx$

$$= \frac{x^{n-1}}{a^2}(ax \sin ax + n \cos ax) - \frac{n(n-1)}{a^2} \int x^{n-2} \cos ax \, dx. \quad (159)$$

52. $\displaystyle \int e^{ax} \cos^n x \, dx = \frac{e^{ax} \cos^{n-1} x \, (a \cos x + n \sin x)}{a^2 + n^2}$

$$+ \frac{n(n-1)}{a^2 + n^2} \int e^{ax} \cos^{n-2} x \, dx. \quad (160)$$

53. $\displaystyle \int \sin^{-1} x \, dx = x \sin^{-1} x + (1 - x^2)^{\frac{1}{2}}. \quad (161)$

54. $\displaystyle \int \frac{d\theta}{a + b \cos \theta}$

$$= \frac{2}{\sqrt{a^2-b^2}} \tan^{-1}\left[\left(\frac{a-b}{a+b}\right)^{\frac{1}{2}} \frac{\theta}{2}\right], \text{ (when } a>b). \quad (162)$$

$$= \frac{1}{\sqrt{b^2-a^2}} \log\left[\frac{\sqrt{b+a}+\sqrt{b-a} \tan \frac{\theta}{2}}{\sqrt{b+a}-\sqrt{b-a} \tan \frac{\theta}{2}}\right], \text{ (when } a<b).$$

55. $\displaystyle \int^n X dx^n = C_n + C_{n-1}\frac{x}{1} + C_{n-2}\frac{x^2}{1\cdot2} + \cdots C_1\frac{x^{n-1}}{1\cdot2\cdot3\cdot(n-1)}$

$$+ (X)\frac{x^n}{1\cdot2\cdot3 \ldots n} + \left(\frac{dX}{dx}\right)\frac{x^{n+1}}{1\cdot2\cdot3 \ldots (n+1)}$$

$$+ \left(\frac{d^2X}{dx^2}\right)\frac{x^{n+2}}{1\cdot2\cdot3 \ldots (n+2)} + \text{etc.} \quad (165)$$

CHAPTER VI.

LENGTHS OF CURVES.

171. Length of Plane Curves referred to Rectangular Axes.—Let P and Q be two consecutive points on the curve AB, and let (x, y) be the point P ; let s denote the length of the curve AP measured from a fixed point A up to P. Then

$$PQ = ds, \quad PR = dx, \quad RQ = dy,$$

Therefore, from the right-angled triangle PRQ we have

$$ds = \sqrt{dx^2 + dy^2};$$

Fig. 43.

hence, $\quad s = \int \sqrt{dx^2 + dy^2} = \int \left(1 + \frac{dy^2}{dx^2}\right)^{\frac{1}{2}} dx.$

To apply this formula to any particular curve, we find the value of $\frac{dy}{dx}$ in terms of x from the equation of the curve, and then by integration between proper limits s becomes known.

The process of finding the length of an arc of a curve is called the *rectification of the curve*.

It is evident that if y be considered the independent variable, we shall have

$$s = \int \left(1 + \frac{dx^2}{dy^2}\right)^{\frac{1}{2}} dy.$$

The curves whose lengths can be obtained in finite terms are very limited in number. We proceed to consider some of the simplest applications:

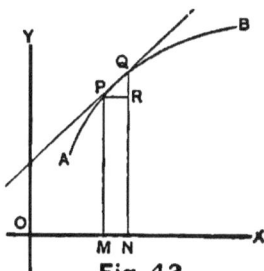

172. The Parabola.—The equation of the parabola is

$$y^2 = 2px;$$

hence, $$\frac{dy}{dx} = \frac{p}{y}.$$

$$\therefore \quad s = \int \left(1 + \frac{p^2}{y^2}\right)^{\frac{1}{2}} dx; \qquad dx = \frac{ydy}{p},$$

or $$s = \frac{1}{p} \int (p^2 + y^2)^{\frac{1}{2}} dy, \text{ (which, by Ex. 35 Art. 146)}$$

$$= \frac{y\sqrt{p^2 + y^2}}{2p} + \frac{p}{2} \log (y + \sqrt{p^2 + y^2}) + C. \quad (1)$$

If we estimate the arc from the vertex, then $s = 0$, $y = 0$, and we have

$$0 = \frac{p}{2} \log p + C; \qquad \therefore \quad C = -\frac{p}{2} \log p,$$

which in (1) gives

$$s = \frac{y\sqrt{p^2 + y^2}}{2p} + \frac{p}{2} \log \left(\frac{y + \sqrt{p^2 + y^2}}{p}\right), \quad (2)$$

which is the length of the curve from the vertex to the point which has any ordinate y. If, for example, we wish to find the length of the curve between the vertex and one extremity of the latus-rectum, $y = p$, we substitute p for y in (2), and get

$$s = \tfrac{1}{2}p\sqrt{2} + \frac{p}{2} \log \left(1 + \sqrt{2}\right)$$

for the required length.

We have here found the value of the constant C by the second method given in Art. 169. We might have found the definite integral at once by integrating between the limits 0 and p, as explained in the first method of Art. 169, and as illustrated in the examples of that Article. Hence, we need not take any notice of the constant C, but write our result

$$s = \frac{1}{p} \int_0^p (p^2 + y^2)^{\frac{1}{2}}\, dy, \quad \text{(see Art. 169)}$$

and integrate between these limits.

173. Semi-Cubical Parabola.*—The equation of this curve is of the form $y^2 = ax^3$. (See Fig. 39.)

Hence, $\quad \dfrac{dy}{dx} = \tfrac{3}{2}\sqrt{ax} \quad$ and $\quad \dfrac{dy^2}{dx^2} = \tfrac{9}{4}ax.$

$$\therefore \quad s = \int (1 + \tfrac{9}{4}ax)^{\frac{1}{2}}\, dx$$

$$= \frac{8}{27a} (1 + \tfrac{9}{4}ax)^{\frac{3}{2}} + C.$$

If we wish to find the length of the curve from A to P, we must integrate between the limits 0 and $3p$ (see Art. 128, Ex. 9) ; hence,

$$s = \int_0^{3p} (1 + \tfrac{9}{4}ax)^{\frac{1}{2}}\, dx = \left[\frac{8}{27a} (1 + \tfrac{9}{4}ax)^{\frac{3}{2}} \right]_0^{3p}$$

$$= \frac{8}{27a} (1 + \tfrac{27}{4}ap)^{\frac{3}{2}} - \frac{8}{27a}$$

$$= \frac{8}{27a} [(1 + \tfrac{27}{4}ap)^{\frac{3}{2}} - 1] = p\,(3^{\frac{3}{2}} - 1),$$

by substituting $\dfrac{8}{27p}$ for a. (See Art. 125, Ex. 1. Compare Ex. 10, Art. 128.)

174. The Circle.—From $x^2 + y^2 = r^2$, we have

$$\frac{dy}{dx} = -\frac{x}{y}.$$

---

* This was the first curve which was rectified. The author was William Neil, who was led to the discovery, about 1660, by a remark of Wallis, in his Arithmetica Infinitorum. See Gregory's Examples, p. 420.

Hence, for the length of a quadrant, we have (since the limits are 0 and r),

$$s = \int_0^r \left(1 + \frac{x^2}{y^2}\right)^{\frac{1}{2}} dx = \int_0^r \frac{r\,dx}{\sqrt{r^2 - x^2}}$$

$$= \left[\, r\sin^{-1}\frac{x}{r}\,\right]_0^r = \tfrac{1}{2}r\pi,$$

which involves a circular arc, the very quantity we wish to determine. The circle is therefore not a rectifiable curve; but the above integral may be developed into a series, and an approximate result obtained.

By Ex. 1, Art. 170, we have

$$s = \left[\, r\left(\frac{x}{r} + \frac{x^3}{2\cdot 3r^3} + \frac{1\cdot 3x^5}{2\cdot 4\cdot 5r^5} + \frac{1\cdot 3\cdot 5x^7}{2\cdot 4\cdot 6\cdot 7r^7} + \text{etc.}\right)\right]_0^r$$

$$= r\left(1 + \frac{1}{2\cdot 3} + \frac{1\cdot 3}{2\cdot 4\cdot 5} + \frac{3\cdot 5}{2\cdot 4\cdot 6\cdot 7} + \text{etc.}\right);$$

$$\therefore\ \ \tfrac{1}{2}\pi = 1 + \frac{1}{2\cdot 3} + \frac{1\cdot 3}{2\cdot 4\cdot 5} + \frac{3\cdot 5}{2\cdot 4\cdot 6\cdot 7} + \text{etc.}$$

By taking a sufficient number of terms, reducing each to a decimal, and adding, we have $\pi = 3.141592653589793+$. For the approximation usually employed in practice, π is taken as 3.1416, and for still ruder approximations as $3\tfrac{1}{7}$.

175. The Ellipse.—From $y^2 = (1-e^2)(a^2 - x^2)$, we have

$$\frac{dy}{dx} = -(1-e^2)\frac{x}{y} = -\frac{x\sqrt{1-e^2}}{\sqrt{a^2-x^2}}.$$

To find the length of a quadrant, we must integrate between the limits 0 and a; hence,

$$s = \int_0^a \sqrt{1 + \frac{x^2(1-e^2)}{a^2 - x^2}}\, dx = \int_0^a \sqrt{\frac{a^2 - e^2x^2}{a^2 - x^2}}\, dx$$

$$= \int_0^a \left(1 - \frac{e^2x^2}{a^2}\right)^{\frac{1}{2}} \frac{dx}{\sqrt{1 - \dfrac{x^2}{a^2}}}.$$

This integration cannot be effected in finite terms, but may be obtained by series.

Put $\dfrac{x}{a} = z$; then $dx = adz$. When $x = a$, $z = 1$, and when $x = 0$, $z = 0$; therefore the above integral becomes

$$s = a \int_0^1 (1 - e^2 z^2)^{\frac{1}{2}} \frac{dz}{\sqrt{1 - z^2}}$$

$$= a \frac{\pi}{2} \left(1 - \tfrac{1}{4}e^2 - \frac{1 \cdot 3}{2^2 \cdot 4^2} e^4 - \frac{1 \cdot 3^2 \cdot 5}{2^2 \cdot 4^2 \cdot 6^2} e^6 - \text{etc.} \right),$$

(by Ex. 17, Art. 170), which is the length of a quadrant of the ellipse whose semi-major axis is a and eccentricity e.

176. The Cycloid.—From $x = r \, \mathrm{vers}^{-1} \dfrac{y}{r} - \sqrt{2ry - y^2}$, we have

$$\frac{dx}{dy} = \frac{y}{\sqrt{2ry - y^2}}.$$

$$\therefore \quad s = \sqrt{2r} \int_0^{2r} (2r - y)^{-\frac{1}{2}} \, dy$$

$$= \left[-2 (2r)^{\frac{1}{2}} (2r - y)^{\frac{1}{2}} \right]_0^{2r} = 4r,$$

which is $\frac{1}{2}$ the cycloidal arc; hence the whole arc of the cycloid is $8r$ or 4 times the diameter of the generating circle.

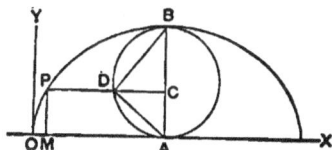

Fig. 44.

If we integrate the above expression between y and $2r$, we get

$$s = \sqrt{2r} \int_y^{2r} (2r - y)^{-\frac{1}{2}} \, dy = 2 (2r)^{\frac{1}{2}} (2r - y)^{\frac{1}{2}}$$

$$= 2\sqrt{2r (2r - y)} = \text{arc BP.}$$

But $BD = \sqrt{BA \times BC} = \sqrt{2r (2r - y)}$;

$$\therefore \quad \text{arc BP} = 2 \text{ times chord BD.*}$$

---

* This rectification was discovered by Wren. See Gregory's Examples, p. 421.

177. The Catenary.— A catenary is the curve assumed by a perfectly flexible string, when its ends are fastened at two points, A and B, nearer together than the length of the string.   Its equation is

$$y = \frac{a}{2}\left(e^{\frac{x}{a}} + e^{-\frac{x}{a}}\right).$$

Hence,

Fig. 45.

$$\frac{dy}{dx} = \frac{1}{2}\left(e^{\frac{x}{a}} - e^{-\frac{x}{a}}\right); \qquad \therefore \; ds = \frac{1}{2}\left(e^{\frac{x}{a}} + e^{-\frac{x}{a}}\right)dx.$$

If s be measured from the lowest point V, to any point P (x, y), we have

$$s = \frac{1}{2}\int_0^x \left(e^{\frac{x}{a}} + e^{-\frac{x}{a}}\right)dx = \frac{a}{2}\left(e^{\frac{x}{a}} - e^{-\frac{x}{a}}\right).$$

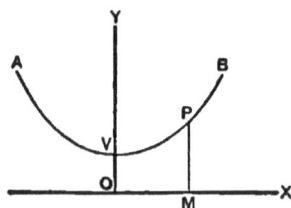

178. The Involute of a Circle.—(See Art. 124.)   Let C be the centre of the circle, whose radius is r; APR is a portion of the involute, T and T' are two consecutive points of the circle, P and Q two consecutive points of the involute, and ϕ the angle ACT. Then TCT' = PTQ = $d\phi$, and PT = AT = $r\phi$.

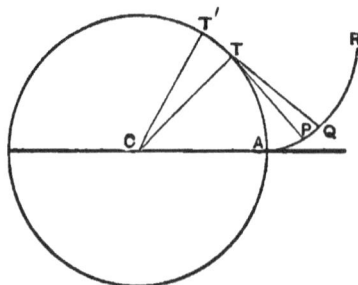

$$\therefore \; ds = \text{PQ} = r\phi d\phi;$$

Fig. 46.

$$\therefore \; s = r\int \phi d\phi = \tfrac{1}{2}r\phi^2 + C.$$

If the curve be estimated from A, $C = 0$, and we have

$$s = \tfrac{1}{2}r\phi^2.$$

For one circumference, $\phi = 2\pi$; $\therefore s = \tfrac{1}{2}r\,(2\pi)^2 = 2r\pi^2$.

For n circumferences, $\phi = 2n\pi$; $\therefore s = \tfrac{1}{2}r\,(2n\pi)^2 = 2rn^2\pi^2$.

179. Rectification in Polar Co-ordinates.—If the curve be referred to polar co-ordinates, we have (Art. 102),

$$ds^2 = r^2 d\theta^2 + dr^2;$$

hence we get
$$s = \int \left(r^2 + \frac{dr^2}{d\theta^2} \right)^{\frac{1}{2}} d\theta,$$

or
$$s = \int \left(1 + \frac{r^2 d\theta^2}{dr^2} \right)^{\frac{1}{2}} dr.$$

180. The Spiral of Archimedes.—From $r = a\theta$, we have

$$\frac{d\theta}{dr} = \frac{1}{a}.$$

$$\therefore \quad s = \frac{1}{a} \int_0^r (r^2 + a^2)^{\frac{1}{2}} dr$$

$$= \frac{r (a^2 + r^2)^{\frac{1}{2}}}{2a} + \frac{a}{2} \log \left(\frac{r + \sqrt{a^2 + r^2}}{a} \right),$$

(see Art. 172), from which it follows that the length of any arc of the Spiral of Archimedes, measured from the pole, is equal to that of a parabola measured from its vertex, r and a having the same numerical values as y and p.

181. The Cardioide.—The equation of this curve is

$$r = a(1 + \cos \theta).$$

Here
$$\frac{dr}{d\theta} = -a \sin \theta,$$

and hence
$$s = \int [a^2 (1 + \cos \theta)^2 + a^2 \sin^2 \theta]^{\frac{1}{2}} d\theta$$

$$= a \int (2 + 2 \cos \theta)^{\frac{1}{2}} d\theta$$

$$= 2a \int \cos \frac{\theta}{2} d\theta = 4a \sin \frac{\theta}{2} + C.$$

If we estimate the arc s from the point A, for which $\theta = 0$, we have

$$s = 0 ; \quad \therefore \ C = 0.$$

Making $\theta = \pi$ for the superior limit, we have

$$s = 4a \sin \frac{\pi}{2} = 4a,$$

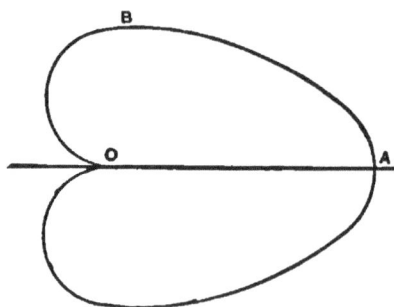

Fig. 47.

which is the length of the arc ABO; hence the whole perimeter is $8a$.

182. Lengths of Curves in Space.—The length of an infinitesimal element of a curve in space, whether plane or of double curvature, from the principles of Solid Geometry (see Anal. Geom., Art. 169) is easily seen to be

$$\sqrt{dx^2 + dy^2 + dz^2}.^*$$

Hence, if s denote the length of the curve, measured from some fixed point up to any point P (x, y, z), we have

$$s = \int \sqrt{dx^2 + dy^2 + dz^2}$$

$$= \int \left[1 + \left(\frac{dy}{dx} \right)^2 + \left(\frac{dz}{dx} \right)^2 \right]^{\frac{1}{2}} dx.$$

If the equations of the curve are given in the form

$$y = f(x) \qquad \text{and} \qquad z = \phi(x),$$

we may find the values of $\dfrac{dy}{dx}$ and $\dfrac{dz}{dx}$ in terms of x, and then by integration s is known in terms of x.

---

* The student who wants further demonstration of this, is referred to Price's Cal., Vol. I, Art. 341, and Vol. II, Art. 164; De Morgan's Dif. and Integral Cal., p. 444; and Homersham Cox's Integral Cal., p. 95.

183. The Intersection of Cycloidal and Parabolic Cylinders.—To find the length of the curve formed by the intersection of two right cylinders, of which one has its generating lines parallel to the axis of z and stands on a parabola in the plane of xy, and the other has its generating lines parallel to the axis of y and stands on a cycloid in the plane of xz, the equations of the curve of intersection being

$$y^2 = 4px, \qquad z = a \operatorname{vers}^{-1} \frac{x}{a} + \sqrt{2ax - x^2}.$$

Here $\quad \dfrac{dy}{dx} = \sqrt{\dfrac{p}{x}} \quad$ and $\quad \dfrac{dz}{dx} = \sqrt{\dfrac{2a - x}{x}};$

$$\therefore \quad ds = \left(1 + \frac{p}{x} + \frac{2a}{x} - 1\right)^{\frac{1}{2}} dx = (p + 2a)^{\frac{1}{2}} \frac{dx}{\sqrt{x}}.$$

Estimating the curve from the origin to any point P, we have

$$s = \int_0^x (p + 2a)^{\frac{1}{2}} \frac{dx}{x^{\frac{1}{2}}} = 2 (p + 2a)^{\frac{1}{2}} \sqrt{x}.$$

EXAMPLES.

1. Rectify the hypocycloid whose equation is

$$x^{\frac{2}{3}} + y^{\frac{2}{3}} = a^{\frac{2}{3}}.$$

   Ans. The whole length of the curve is $6a$.

2. Rectify the logarithmic curve $y = be^{\frac{z}{a}}$.

   Ans. $s = a \log \dfrac{y}{a + \sqrt{a^2 + y^2}} + \sqrt{a^2 + y^2} + C.$

3. Rectify the curve $e^y = \dfrac{e^z + 1}{e^z - 1}$ between the limits $x = 1$ and $x = 2$.

   Ans. $s = \log (e + e^{-1}).$

4. Rectify the evolute of the ellipse, its equation being

$$\left(\frac{x}{a}\right)^{\frac{2}{3}} + \left(\frac{y}{\beta}\right)^{\frac{2}{3}} = 1.$$

Put $\qquad x = a \cos^3 \theta, \qquad y = \beta \sin^3 \theta;$

then $\qquad dx = -3a \cos^2 \theta \sin \theta \, d\theta,$

$$dy = 3\beta \sin^2 \theta \cos \theta \, d\theta ;$$

$$\therefore \quad s = 3 \int_0^{\frac{1}{2}\pi} (a^2 \cos^2 \theta + \beta^2 \sin^2 \theta)^{\frac{1}{2}} \sin \theta \cos \theta \, d\theta$$

$$= -\frac{3}{4} \int_0^{\frac{1}{2}\pi} \left(\frac{a^2 + \beta^2}{2} + \frac{a^2 - \beta^2}{2} \cos 2\theta\right)^{\frac{1}{2}} d \cos 2\theta$$

$$= \frac{a^3 - \beta^3}{a^2 - \beta^2};$$

therefore the whole length is $4 \dfrac{a^3 - \beta^3}{a^2 - \beta^2}.$

If $\beta = a$, this result becomes $6a$, which agrees with that given in Ex. 1. (See Price's Calculus, Vol. II, p. 203.)

5. Find the length of the arc of the parabola $x^{\frac{1}{2}} + y^{\frac{1}{2}} = a^{\frac{1}{2}}$ between the co-ordinate axes.

Put $\qquad x = a \cos^4 \theta, \qquad y = a \sin^4 \theta;$

$$\therefore \quad s = 4a \int_0^{\frac{1}{4}\pi} (\cos^4 \theta + \sin^4 \theta)^{\frac{1}{2}} \sin \theta \cos \theta \, d\theta$$

$$= -\frac{a}{\sqrt{2}} \int_0^{\frac{1}{4}\pi} (1 + \cos^2 2\theta)^{\frac{1}{2}} d \cos 2\theta$$

$$= a + \frac{a}{\sqrt{2}} \log (\sqrt{2} + 1).$$

6. Find the length, measured from the origin, of the curve

$$x^2 = a^2 \left(1 - e^{\frac{y}{a}}\right).$$

$$Ans. \ \ s = a \log \left(\frac{a + x}{a - x}\right) - x.$$

7. Rectify the logarithmic spiral $\log r = \theta$ between the limits r_0 and r_1.   *Ans.* $s = (1 + m^2)^{\frac{1}{2}} (r_1 - r_0)$.

8. If 100 yards of cord be wound in a single coil upon an upright post an inch in diameter, what time will it take a man to unwind it, by holding one end in his hand and traveling around the post so as to keep the cord continually tight, supposing he walks 4 miles per hour; and what is the length of the path that the man walks over?

Ans. Time $= 51\frac{3}{22}$ hours; distance $= 204\frac{6}{11}$ miles.

9. Find the length of the tractrix or equitangential curve.

If AB is a curve such that PT, the length of the intercepted tangent between the point of contact and the axis of x, is always equal to OA, then the locus of P is the equitangential curve.

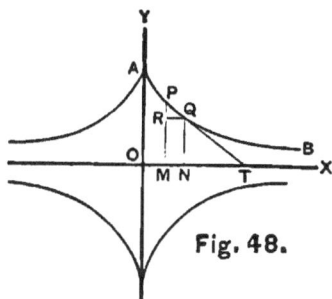

Fig. 48.

Let P and Q be two consecutive points on the curve; let (x, y) be the point P, and $OA = PT = a$.   Then

$$\frac{PQ}{PR} = \frac{a}{y};$$

$$\therefore \frac{ds}{dy} = -\frac{a}{y}$$

(the minus sign being taken since y is a decreasing function of s or x).

Hence, $s = -a \int_a^y \frac{dy}{y} = -a \left[\log y \right]_a^y$

$$= a \log \left(\frac{a}{y}\right).$$

This example furnishes an instance of our being able to determine the length of a curve from a geometric property of the curve, without previously finding its equation.

The *equation* of the tractrix may be found as follows :

$$\frac{PR}{RQ} = \frac{PM}{MT};$$

hence

$$\frac{dy}{dx} = -\frac{y}{\sqrt{a^2 - y^2}};$$

$$\therefore \quad x = -\int_a^y \frac{\sqrt{a^2 - y^2}}{y} dy$$

$$= -\int_a^y \frac{a^2 dy}{y\sqrt{a^2 - y^2}} + \int_0^y \frac{y dy}{\sqrt{a^2 - y^2}}$$

$$= a \log \frac{a + \sqrt{a^2 - y^2}}{y} - (a^2 - y^2)^{\frac{1}{2}}.$$

(See Ex. 17, Art. 146.)

This curve is sometimes considered as generated by attaching one end of a string of constant length $(= a)$ to a weight at A, and by moving the other end of the string along OX ; the weight is supposed to trace out the curve, and hence arises the name *Tractrix* or *Tractory*. This mode of generation is incorrect, unless we also suppose the friction produced by traction to be infinitely great, so that the weight momentum which is caused by its motion may be instantly destroyed. Price's Calculus, Vol. I, p. 315.

10. A fox started from a certain point and ran due east 300 yards, when it was overtaken by a hound that started from a point 100 yards due north of the fox's starting-point, and ran directly towards the fox throughout the race. Find the length of the curve described by the hound, both having started at the same instant, and running with a uniform velocity. *Ans.* 354.1381 yards.

This example, like the preceding, may be solved without finding the equation of the curve.

11. Find the length of the helix, estimating it from the plane xy, its equations being

$$x = a \cos \phi, \qquad y = a \sin \phi, \qquad z = c\phi.$$

$$Ans. \ s = (a^2 + c^2)^{\frac{1}{2}} \phi.$$

12. Find the length, measured from $\phi = 0$, of the curve which is represented by the equations

$$x = (2a - b) \sin \phi - (a - b) \sin^3 \phi,$$

$$y = (2b - a) \cos \phi - (b - a) \cos^3 \phi.$$

$$Ans. \ s = \tfrac{1}{2}(a + b) \phi + \tfrac{3}{2}(a - b) \sin \phi \cos \phi.$$

13. Find the length of the curve of intersection of the elliptic cylinder $a^2y^2 + b^2x^2 = a^2b^2$, with the sphere

$$x^2 + y^2 + z^2 = a^2.$$

$$Ans. \ 2\pi a.$$

CHAPTER VII.

AREAS OF PLANE CURVES.

184. Areas of Curves.—Let PM and QN be two con-secutive ordinates of the curve AB, and let (x, y) be the point P; let A denote the area included between the curve, the axis of x, and two ordinates at a finite distance apart. Then the area of the trapezoid MPQN is an infinitesimal element whose breadth is dx and whose parallel sides are y and $y + dy$; therefore we have

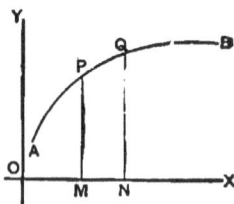

Fig. 49.

$$dA = \frac{y + (y + dy)}{2} \, dx = y\,dx,$$

since the last term, being a differential of the second order, must be dropped.

$$\therefore \quad A = \int y \, dx,$$

the integration being taken within proper limits. If, for example, we want the area between the two ordinates whose abscissas are a and b, where $a > b$, we have

$$A = \int_b^a y \, dx. \tag{1}$$

In like manner, if the area were included between the curve, the axis of y, and two abscissas at a finite distance apart, we would have

$$A = \int_d^c x \, dy, \tag{2}$$

where c and d are the y-limits.

185. Area between Two Curves.—If the area were included between the two curves AB and ab, whose equations are respectively $y = f(x)$ and $y = \phi(x)$, and two ordinates CD and EH, where $OD = b$ and $OH = a$, we should find by a similar course of reasoning,

$$A = \int_b^a [f(x) - \phi(x)]\, dx.$$

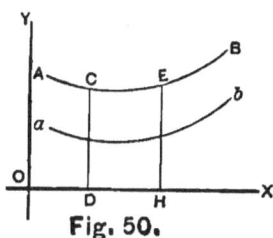
Fig. 50.

The determination of the area of a curve is called its *Quadrature.*

186. The Circle.—The equation of the circle referred to its centre as origin, is $y^2 = a^2 - x^2$; therefore the area of a quadrant is represented by

$$A = \int_0^a (a^2 - x^2)^{\frac{1}{2}}\, dx$$

$$= \left[\frac{x(a^2 - x^2)^{\frac{1}{2}}}{2} + \frac{a^2}{2} \sin^{-1} \frac{x}{a} \right]_0^a \quad \text{(See Ex. 4, Art. 151.)}$$

$$= \frac{a^2\pi}{4};$$

therefore the area of the circle $= \pi a^2$.

Also, if $OM = x$, the area of OBDM becomes

$$A = \int_0^x (a^2 - x^2)^{\frac{1}{2}}\, dx$$

$$= \left[\frac{x(a^2 - x^2)^{\frac{1}{2}}}{2} + \frac{a^2}{2} \sin^{-1} \frac{x}{a} \right]_0^x$$

$$= \frac{x(a^2 - x^2)^{\frac{1}{2}}}{2} + \frac{a^2}{2} \sin^{-1} \frac{x}{a}.$$

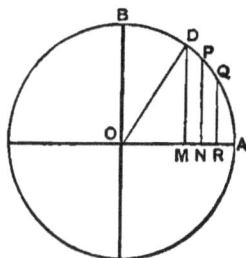
Fig. 51.

This result is also evident from geometric considerations

16

for the area of the triangle OMD $= \dfrac{x}{2}(a^2 - x^2)^{\frac{1}{2}}$, and the

area of the sector ODB $= \dfrac{a^2}{2}\sin^{-1}\dfrac{x}{a}$.

REMARK.—The student will perceive that in integrating between the limits $x = 0$ and $x = a$, we take in every elementary slice PQRN in the quadrant ADBO; also integrating between the limits $x = 0$ and $x = x = $ OM, we take in every elementary slice between OB and MD.*

187. The Parabola.—From $y^2 = 2px$, we have

$$y = \sqrt{2px}.$$

Hence, for the area of the part OPM, we have

$$A = \sqrt{2p}\int_0^x x^{\frac{1}{2}}dx = \tfrac{2}{3}\sqrt{2p}\,x^{\frac{3}{2}};\ i.\ e.,\ \tfrac{2}{3}xy.$$

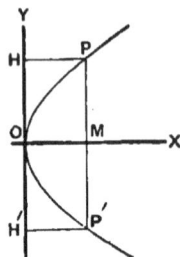

Fig. 52.

Therefore the area of the segment POP', cut off by a chord perpendicular to the axis, is $\tfrac{2}{3}$ of the rectangle PHH'P'.

188. The Cycloid.—From the equation

$$x = r\,\text{vers}^{-1}\dfrac{y}{r} - \sqrt{2ry - y^2},$$

we have

$$dx = \dfrac{y\,dy}{\sqrt{2ry - y^2}}.$$

$$\therefore\ A = \int_0^{2r}\dfrac{y^2 dy}{\sqrt{2ry - y^2}}$$

$$= \tfrac{3}{2}\pi r^2.$$

(See Ex. 6, Art. 151) $= \tfrac{1}{2}$ the area of the cycloid. Since integrating between the limits includes half the area of the figure.

---

* The student should pay close attention in every case to the limits of the integration.

Therefore the whole area $= 3\pi r^2$, or *three times the area of the generating circle.**

189. The Ellipse.—The equation of the ellipse referred to its centre as origin, is

$$a^2 y^2 + b^2 x^2 = a^2 b^2 ;$$

therefore the area of a quadrant is represented by

$$A = \frac{b}{a} \int_0^a (a^2 - x^2)^{\frac{1}{2}} \, dx$$

$$= \frac{b}{a} \frac{a^2 \pi}{4} \quad \text{(See Art. 186)} = \tfrac{1}{4} a b \pi .$$

Therefore the area of the entire ellipse is $\pi a b$.

190. The Area between the Parabola $y^2 = ax$ and the Circle $y^2 = 2ax - x^2$.—These curves pass through the origin, and also intersect at the points A and B, whose common abscissa is a. Hence, to find the area included between the two curves on the positive side of the axis of x, we must integrate between the limits $x = 0$ and $x = a$. Therefore, by Art. 185, we have

$$A = \int_0^a [(2ax - x^2)^{\frac{1}{2}} - (ax)^{\frac{1}{2}}] \, dx$$

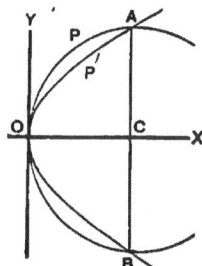

Fig. 53.

$$= \frac{\pi a^2}{4} - \tfrac{2}{3} a^2 ; \quad \text{(See Ex. 6, Art. 151.)}$$

which is the area of OPAP′.

---

* This quadrature was first discovered by Roberval, one of the most distinguished geometers of his day. Galileo, having failed in obtaining the quadrature by geometric methods, attempted to solve the problem by weighing the area of the curve against that of the generating circle, and arrived at the conclusion that the former area was nearly, but not exactly, three times the latter. About 1628, Roberval attacked it, but failed to solve it. After studying the ancient Geometry for six years, he renewed the attack and effected a solution in 1634. (See Salmon's Higher Plane Curves, p. 266.)

191. Area in Polar Co-ordinates.—Let the curve be referred to polar co-ordinates, O being the pole, and let OP and OQ be consecutive radii-vectores, and PR an arc of a circle described with O as centre; let (r, θ) be the point P. Then the area of the infinitesimal element OPQ = OPR + PRQ; but PRQ is

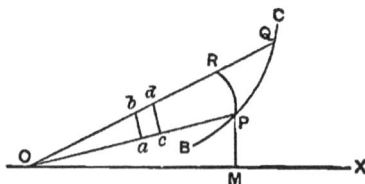

Fig. 54.

an infinitesimal of the second order in comparison with OPR, when P and Q are infinitely near points; consequently the elementary area $\text{OPQ} = \text{area OPR} = \dfrac{r^2 d\theta}{2}.$

Hence if A represents the area included between the curve, the radius-vector OP, and the radius-vector OB drawn to some fixed point B, we have

$$A = \tfrac{1}{2} \int r^2 d\theta.$$

If β and α are the values of θ corresponding to the points B and C respectively, we have

$$A = \tfrac{1}{2} \int_{\beta}^{a} r^2 d\theta.$$

192. The Spiral of Archimedes.—Let $r = \dfrac{\theta}{2\pi}$ be its equation. Then

$$A = \pi \int r^2 dr = \tfrac{1}{3}\pi r^3 + C.$$

If we estimate the area from the pole, we have $A = 0$ when $r = 0$, and $\therefore C = 0$; hence,

$$A = \tfrac{1}{3}\pi r^3,$$

which is the value of the area passed over by the radius-vector in its revolution from its starting at c to any value, as r.

If we made $\theta = 2\pi$, we have $r = 1$; therefore

$$A = \tfrac{1}{3}\pi,$$

which is the area described by one revolution of the radius-vector. Hence the area of the first spire is equal to one-third the area of the measuring* circle.

If we make $\theta = 2(2\pi)$, $r = 2$; therefore

$$A = \tfrac{8}{3}\pi,$$

which is the whole area described by the radius-vector during two revolutions, and evidently includes twice the first spire + the second. Hence the area of the first two spires $= \tfrac{8}{3}\pi - \tfrac{1}{3}\pi = \tfrac{7}{3}\pi$, and so on.

E X A M P L E S.

1. Find the area of $y = x - x^3$ between the curve and the axis of x. *Ans.* $\tfrac{1}{2}$.

The limits will be found to be $x = 0$, $x = +1$; also $x = 0$, $x = -1$.†

2. Find the area of $y = x^3 - b^2x$ between the curve and the axis of x. *Ans.* $\tfrac{1}{2}b^4$.

3. Find the area of $y = x^3 - ax^2$ between the curve and the axis of x. *Ans.* $\tfrac{1}{12}a^4$.

4. Find the whole area of the two loops of $a^2y^2 = x^2(a^2 - x^2)$. *Ans.* $\tfrac{4}{3}a^2$.

5. Find the area of $xy^2 = a^3$ between the limits $y = b$ and $y = c$. *Ans.* $2a^3\dfrac{b-c}{bc}$.

6. Find the whole area of the two loops of $a^4y^2 = a^2b^2x^2 - b^2x^4$. *Ans.* $\tfrac{4}{3}ab$.

---

* See Anal. Geom., Art. 158.

† The student should draw the figure in every case, and determine the limits of the integrations.

7. Find the whole area of $a^2y^2 = x^3 (2a - x)$. (See Arts. 150 and 188.) *Ans.* πa^2.

8. Find the whole area between the Cissoid $y^2 = \dfrac{x^3}{2a - x}$ and its asymptote. (See Art. 103.)
Ans. $3\pi a^2$.

9. The equation of the hyperbola is $a^2y^2 - b^2x^2 = -a^2b^2$; find the area included between the curve, the axis of x, and an ordinate.
$$Ans. \quad \frac{xy}{2} - \frac{ab}{2} \log \left(\frac{x + \sqrt{x^2 - a^2}}{a} \right).$$

10. The equation of the Witch of Agnesi is
$$x^2y = 4a^2 (2a - y) ;$$
find the area included between the curve and its asymptote.
Ans. $4a^2\pi$.

11. Find the area of the catenary VPMO, Fig. 45.
$$Ans. \quad \frac{a^2}{2} \left(e^{\frac{x}{a}} - e^{-\frac{x}{a}} \right) = a (y^2 - a^2)^{\frac{1}{2}}.$$

12. Find the area of the oval of the parabola of the third degree whose equation is $cy^2 = (x - a) (x - b)^2$. (See Art. 142.)
$$Ans. \quad \frac{8}{15\sqrt{c}} (b - a)^{\frac{5}{2}}.$$

13. Find the area of one loop of the curve
$$ay^2 = x^2 (a^2 - x^2)^{\frac{1}{2}}.$$
Ans. $\frac{1}{6}a^2$.

14. Find the whole area between the curve
$$x^2y^2 + a^2b^2 = a^2y^2$$
and its asymptotes. *Ans.* $2\pi ab$.

15. Find the whole area of the curve
$$\left(\frac{x}{a} \right)^2 + \left(\frac{y}{b} \right)^{\frac{2}{3}} = 1.$$
Ans. $\frac{3}{4}\pi ab$.

16. Find the area included between the parabola $y^2 = 2px$ and the right line $y = ax$.

These two loci intersect at the origin and at the point whose abscissa is $\dfrac{2p}{a^2}$; hence the x-limits are 0 and $\dfrac{2p}{a^2}$; therefore, Art. 185,

$$A = \int_0^{\frac{2p}{a^2}} (\sqrt{2px} - ax)\, dx = \frac{2p^2}{3a^3}, \quad Ans.$$

17. Find (*1*) the area included between the parabola

$$y^2 = 2px,$$

the right line passing through the focus and inclined at 45° to the axis of x, and the left-hand double ordinate of intersection. (See Art. 185.)

Also find (*2*) the whole area between the line and parabola.

(*1.*) Here the x-limits are found to be $\dfrac{p}{2}\left(\sqrt{2}+1\right)^2$ and $\dfrac{p}{2}\left(\sqrt{2}-1\right)^2$; hence we have

$$A = \int_{\frac{p}{2}(\sqrt{2}-1)^2}^{\frac{p}{2}(\sqrt{2}+1)^2} \left[\sqrt{2p}\, x^{\frac{1}{2}}\, dx - (x - \tfrac{1}{2}p)\, dx \right]$$

$$= \left[\tfrac{2}{3}\sqrt{2p}\, x^{\frac{3}{2}} - \tfrac{1}{2}x^2 + \tfrac{1}{2}px \right]_{\frac{p}{2}(\sqrt{2}-1)^2}^{\frac{p}{2}(\sqrt{2}+1)^2}$$

$$= \tfrac{2}{3}\sqrt{2p}\left(\tfrac{p}{2}\right)^{\frac{3}{2}}(12+2) - \tfrac{1}{2}(24\sqrt{2})\frac{p^2}{4} + \tfrac{1}{2}p(4\sqrt{2})\frac{p}{2}$$

$$= \tfrac{14}{3}p^2 - 3\sqrt{2}p^2 + \sqrt{2}p^2 = p^2\left(\tfrac{14}{3} - 2\sqrt{2}\right), \quad Ans.$$

(*2.*) Ans. $\tfrac{4}{3}p^2\sqrt{2}$.

18. Find the whole area included between the four infinite branches of the tractrix. *Ans.* πa^2.

19. Find the area of the Naperian logarithmic spiral.

Ans. $\tfrac{1}{4}r^2$.

20. Find the whole area of the Lemniscate $r^2 = a^2 \cos 2\theta$.

Ans. a^2.

21. Find the whole area of the curve

$$r = a \,(\cos 2\theta + \sin 2\theta).$$

Ans. πa^2.

22. Find the area of the Cardioide.   (See Art. 181.)

Ans. $\frac{3}{2}\pi a^2$.

23. Find the area of a loop of the curve $r = a \cos n\theta$.

Ans. $\dfrac{\pi a^2}{4n}$.

24. Find the area of a loop of the curve

$$r = a \cos n\theta + b \sin n\theta.$$

Ans. $\dfrac{a^2 + b^2}{4} \cdot \dfrac{\pi}{n}$.

25. Find the area of the three loops of the curve

$$r = a \sin 3\theta.   \text{(See Fig. 33.)}$$

Ans. $\dfrac{\pi a^2}{4}$.

26. Find the area included between the involute and the evolute in Fig. 46, when the string has made one revolution.

Ans. $\frac{4}{3}r^2\pi^3$.

CHAPTER VIII

AREAS OF CURVED SURFACES.

193. Surfaces of Revolution.—If any plane be supposed to revolve around a fixed line in it, every point in the plane will describe a circle, and any curve lying in the plane will generate a surface. Such a surface is called a *surface of revolution;* and the fixed line, round which the revolution takes place, is called the *axis* of revolution.

Let P and Q be two consecutive points on the curve AB; let (x, y) be the point P, and s the length of the curve AP measured from a fixed point A to any point P. Then $MP = y$, $NQ = y + dy$, and $PQ = ds$.

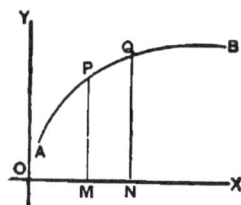

Fig. 55.

Denote by S the area of the surface generated by the revolution of AP around the axis OX; then the surface generated by the revolution of PQ around the axis of x is an infinitesimal element of the whole surface, and is the convex surface of the frustum of a cone, the circumferences of whose bases are $2\pi y$ and $2\pi (y + dy)$, and whose slant height is $PQ = ds$; therefore we have

$$dS = \frac{2\pi y + 2\pi (y + dy)}{2} PQ = 2\pi y ds,$$

since the last term, being an infinitesimal of the second order, must be dropped. Therefore, for the whole surface, we have

$$S = 2\pi \int y ds = 2\pi \int y \sqrt{dx^2 + dy^2},$$

the integral being taken between proper limits. If for example, we want the surface generated by the curve between the two ordinates whose abscissas are a and b, where $a > b$, we have

$$S = 2\pi \int_b^{,a} y \left(1 + \frac{dy^2}{dx^2}\right)^{\frac{1}{2}} dx.$$

In like manner it may be shown that to find the surface generated by revolving the curve round the axis of y, we have

$$S = 2\pi \int x ds.$$

194. The Sphere.—From the equation of the generating curve, $x^2 + y^2 = r^2$, we have

$$y = (r^2 - x^2)^{\frac{1}{2}} \quad \text{and} \quad \frac{dy}{dx} = -\frac{x}{y};$$

$$\therefore \quad S = 2\pi \int y \left(1 + \frac{x^2}{y^2}\right)^{\frac{1}{2}} dx = 2\pi \int r dx = 2\pi r x + C.$$

Hence, the surface of the zone included between two planes corresponding to the abscissas a and b is

$$S = 2\pi \int_b^{,a} r dx = 2\pi r (a - b);$$

that is, the area of the zone is the product of the circumference of a great circle by the height of the zone.

To find the surface of the whole sphere, we integrate between $+ r$ and $- r$ for the x-limits; hence we have

$$S = 2\pi r \int_{-r}^{,r} dx = 2\pi r [r - (- r)] = 4\pi r^2;$$

that is, the whole surface of the sphere is four times the area of a great circle.

Remark.—If a cylinder be circumscribed about a sphere, its convex surface is equal to $2\pi r \times 2r = 4\pi r^2$, which is the same as the surface of the sphere. If we add $2\pi r^2$ to this, which is the sum of the areas of the two bases, we shall have for the whole surface of the cylinder

$6\pi r^2$. Hence the whole surface of the cylinder is to the surface of the sphere as 3 is to 2. This relation between the surfaces of these two bodies, and also the same relation between the volumes, was discovered by Archimedes, who thought so much of the discovery that he expressed a wish to have for the device on his tombstone, a sphere inscribed in a cylinder. Archimedes was killed by the soldiers of Marcellus, B. C. 212, though contrary to the orders of that general. The great geometer was buried with honors by Marcellus, and the device of the sphere and cylinder was executed upon the tomb. 140 years afterward, when Cicero was questor in Sicily, he found the monument of Archimedes, in the shape of a small pillar, and showed it to the Syracusans, who did not know it was in being, he says it was marked with the figure of a sphere inscribed in a cylinder. The sepulchre was almost overrun with thorns and briars. See article " Marcellus," in Plutarch's Lives, Vol. III, p. 120.

195. The Paraboloid of Revolution. — From the equation of the generating curve $y^2 = 2px$, we have

$$y = \sqrt{2px}, \quad \text{and} \quad \frac{dy}{dx} = \tfrac{1}{2}\sqrt{\frac{2p}{x}}.$$

$$\therefore \quad S = 2\pi \int_0^x \sqrt{2px}\left(1 + \frac{p}{2x}\right)^{\frac{1}{2}}dx = 2\pi\sqrt{p}\int_0^x (p+2x)^{\frac{1}{2}}dx$$

$$= \left[\tfrac{2}{3}\pi\sqrt{p}\,(p + 2x)^{\frac{3}{2}}\right]_0^x = \tfrac{2}{3}\pi\sqrt{p}\,[(p + 2x)^{\frac{3}{2}} - p^{\frac{3}{2}}], \quad (1)$$

which is the surface generated by the revolution of the part of the parabola between its vertex and the point (x, y).

We might have found the surface in terms of y instead of x, as follows:

$$\frac{dx}{dy} = \frac{y}{p}.$$

$$\therefore \quad S = 2\pi \int_0^y y\left(1 + \frac{dx^2}{dy^2}\right)^{\frac{1}{2}}dy$$

$$= \frac{2\pi}{p} \int_0^y y\,(p^2 + y^2)^{\frac{1}{2}}dy$$

$$= \frac{2\pi}{3p} \left[(p^2 + y^2)^{\frac{3}{2}} - p^3 \right],$$

which result agrees with (1), as the student can easily verify.

196. The Prolate Spheroid (See Anal. Geom., Art 191).—From the equation of the generating curve

$$y^2 = (1 - e^2)(a^2 - x^2),$$

we have

$$2\pi \, y ds = 2\pi \sqrt{1 - e^2} \sqrt{a^2 - x^2} \, ds$$

$$= 2\pi \sqrt{1 - e^2} \sqrt{a^2 - e^2 x^2} \, dx \quad \text{(Art. 175.)}$$

$$= 2\pi \frac{b}{a} e \left(\frac{a^2}{e^2} - x^2 \right)^{\frac{1}{2}} dx,$$

therefore for half the surface of the ellipsoid, since the x-limits are a and 0, we have

$$S = 2\pi \frac{b}{a} e \int_0^a \left(\frac{a^2}{e^2} - x^2 \right)^{\frac{1}{2}} dx$$

$$= \frac{2\pi \, be}{a} \left[\frac{x \left(\frac{a^2}{e^2} - x^2 \right)^{\frac{1}{2}}}{2} + \frac{a^2}{2e^2} \sin^{-1} \frac{ex}{a} \right]_0^a$$

(See Ex. 4, Art. 151.)

$$= \frac{\pi be}{a} \left[a \left(\frac{a^2 - a^2 e^2}{e^2} \right)^{\frac{1}{2}} + \frac{a^2}{e^2} \sin^{-1} e \right]$$

$$= \pi ab \left[(1 - e^2)^{\frac{1}{2}} + \frac{\sin^{-1} e}{e} \right]$$

$$= \pi b^2 + \frac{\pi ab}{e} \sin^{-1} e.$$

197. The Catenary.—From the equation of the generating curve,

$$y = \frac{a}{2} \left(e^{\frac{x}{a}} + e^{-\frac{x}{a}} \right),$$

we have for the surface of revolution around the axis of x between the limits x and 0,

$$S = 2\pi \int_0^x y\,ds = \pi a \int_0^x \left(e^{\frac{x}{a}} + e^{-\frac{x}{a}} \right) ds$$

$$= \tfrac{1}{2}\pi a \int_0^x \left(e^{\frac{x}{a}} + e^{-\frac{x}{a}} \right)^2 dx \quad \text{(by Art. 177)}$$

$$= \frac{\pi a}{2} \int_0^x \left(e^{\frac{2x}{a}} + 2 + e^{-\frac{2x}{a}} \right) dx$$

$$= \frac{\pi a}{2} \left[\frac{a}{2} \left(e^{\frac{2x}{a}} - e^{-\frac{2x}{a}} \right) + 2x \right]$$

$$= \pi \left[\frac{a^2}{4} \left(e^{\frac{2x}{a}} - e^{-\frac{2x}{a}} \right) + ax \right]$$

$$= \pi (ys + ax), \quad \text{(where } s = VP, \text{ Fig. 45.)}$$

198. The Surface generated by the Cycloid when it revolves around its axis.—From its equation

$$y = r \operatorname{vers}^{-1} \frac{x}{r} + \sqrt{2rx - x^2}, \tag{1}$$

we have

$$\frac{dy}{dx} = \sqrt{\frac{2r - x}{x}}, \tag{2}$$

$$ds = \left(1 + \frac{dy^2}{dx^2} \right)^{\frac{1}{2}} dx = \sqrt{\frac{2r}{x}}\, dx. \tag{3}$$

$$\therefore \quad S = 2\pi \int y\,ds = 2\pi \int y \sqrt{\frac{2r}{x}}\, dx. \tag{4}$$

Put $\quad u = y, \qquad dv = \sqrt{\frac{2r}{x}}\, dx;$

$$\therefore \quad du = dy, \quad \text{and} \quad v = 2\sqrt{2rx};$$

therefore (by Art. 147) we have

$$\int y \sqrt{\frac{2r}{x}}\, dx \doteq 2y \sqrt{2rx} - 2 \sqrt{2r} \int \sqrt{x}\, dy$$

$$= 2y \sqrt{2rx} - 2 \sqrt{2r} \int \sqrt{2r - x}\, dx \quad \text{[by (2)]}$$

$$= 2 \sqrt{2rx} \left(r \operatorname{vers}^{-1} \frac{x}{r} + \sqrt{2rx - x^2} \right) + \tfrac{4}{3} \sqrt{2r} \,(2r - x)^{\frac{3}{2}}$$

[by (1) and integrating.]

$$\therefore \int_0^{2r} y \sqrt{\frac{2r}{x}}\, dx = 4\pi r^2 - \tfrac{16}{3} r^2,$$

which in (4) gives

$$S = 8\pi^2 r^2 - \tfrac{32}{3}\pi r^2 = 8\pi r^2 \left(\pi - \tfrac{4}{3}\right).$$

199. Surfaces of Revolution in Polar Co-ordinates.—If the surface is generated by a curve referred to polar co-ordinates, its area may be determined as follows : Let the axis of revolution be the initial line OX, see Fig. 54, and from P (r, θ) draw PM perpendicular to OX. Then PM $= r \sin \theta$, and the infinitesimal element PQ $= ds$ will, in its revolution round OX, generate an infinitesimal element of the whole surface, whose breadth $= ds$ and whose circumference $= 2\pi r \sin \theta$.   Hence,

$$S = \int 2\pi r \sin \theta\, ds^* = 2\pi \int r \sin \theta \left(r^2 + \frac{dr^2}{d\theta^2} \right)^{\frac{1}{2}} d\theta,$$

(Art. 179)

the integral being taken between proper limits.

200. The Cardioide.—From Art. 181, we have

$$ds = a \,(2 + 2 \cos \theta)^{\frac{1}{2}}\, d\theta = 2a \cos \frac{\theta}{2}\, d\theta.$$

---

* This expression might have been obtained at once by substituting in Art. 193, for y, its value $r \sin \theta$.

For the surface of revolution of the whole curve about the initial line, we have π and 0 for the limits of θ, therefore we have

$$S = \int_0^\pi 2\pi r \sin\theta\, ds$$

$$= 4\pi a^2 \int_0^\pi (1 + \cos\theta) \cos\frac{\theta}{2} \sin\theta\, d\theta$$

$$= 16\pi a^2 \int_0^\pi \cos^4\frac{\theta}{2} \sin\frac{\theta}{2}\, d\theta$$

$$= \left[-\tfrac{32}{5}\pi a^2 \cos^5\frac{\theta}{2} \right]_0^\pi = \tfrac{32}{5}\pi a^2.$$

201. Any Curved Surfaces.—Double Integration.—

Let (x, y, z) and $(x + dx, y + dy, z + dz)$ be two consecutive points p and q on the surface. Through p let planes be drawn parallel to the two planes xz and yz; also through q let two other planes be drawn parallel respectively to the first. These planes will intercept an infinitesimal element pq of the curved surface, and the projection of this element on the plane of xy will be the infinitesimal rectangle PQ, which $= dx\, dy$.

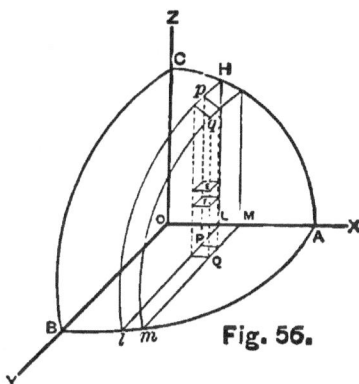

Fig. 56.

Let S represent the required area of the whole surface, and dS the area of the infinitesimal element pq, and denote by α, β, γ, the direction angles* of the normal at $p\,(x, y, z)$. Then, since the projection of dS on the plane of xy is the rectangle PQ $= dx\, dy$, we have by Anal. Geom., Art. 168,

$$dx\, dy = dS \cos\gamma. \tag{1}$$

---

* See Anal. Geom., Art. 170.

Similarly, if dS is projected on the planes yz and zx, we have

$$dy\, dz = dS \cos \alpha ; \qquad (2)$$

$$dz\, dx = dS \cos \beta. \qquad (3)$$

Squaring (1), (2) and (3), and adding, and extracting the square root, we have

$$dS = (dx^2 dy^2 + dy^2 dz^2 + dz^2 dx^2)^{\frac{1}{2}}$$

(since $\qquad \cos^2 \alpha + \cos^2 \beta + \cos^2 \gamma = 1,$

Anal. Geom., Art. 170).

$$\therefore\ S = \int\int (dx^2 dy^2 + dy^2 dz^2 + dz^2 dx^2)^{\frac{1}{2}}$$

$$= \int\int \left(1 + \frac{dz^2}{dx^2} + \frac{dz^2}{dy^2}\right)^{\frac{1}{2}} dx\, dy,$$

the limits of the integration depending upon the portion of the surface considered.

202. The Surface of the Eighth Part of a Sphere.—

Let the surface represented in Fig. 56 be that of the octant of a sphere ; then O being its centre, its equation is

$$x^2 + y^2 + z^2 = a^2.$$

Hence, $\qquad \dfrac{dz}{dx} = -\dfrac{x}{z}, \qquad \dfrac{dz}{dy} = -\dfrac{y}{z}.$

$$\therefore\ S = \int\int \left(1 + \frac{x^2}{z^2} + \frac{y^2}{z^2}\right)^{\frac{1}{2}} dx\, dy$$

$$= \int\int \frac{a\, dx\, dy}{\sqrt{a^2 - x^2 - y^2}}.$$

Now since pq is the element of the surface, the effect of a y-integration, x being constant, will be to sum up all the elements similar to pq from H to l; that is, from $y = 0$ to $y = Ll = y_1 = \sqrt{a^2 - x^2}$; and the aggregate of these elements is the strip Hpl. The effect of a subsequent x-integration will be to sum all these elemental strips that are comprised in the surface of which OAB is the projection, and the limits of this latter integration must be $x = 0$ and $x = $ OA $= a$. Therefore, we have

$$S = \int_0^a \int_0^{y_1} \frac{a\,dx\,dy}{\sqrt{a^2 - x^2 - y^2}}$$

$$= \int_0^a \int_0^{y_1} \frac{a\,dx\,dy}{\sqrt{y_1^2 - y^2}}$$

$$= \int_0^a \left[a\,dx\,\sin^{-1}\frac{y}{y_1} \right]_0^{y_1}.$$

$$= \int_0^a \tfrac{1}{2}\pi a\,dx = \frac{\pi a^2}{2}.$$

EXAMPLES.

1. Find the convex surface of a right circular cone, whose generating line is $ay - bx = 0$.

$$A\,ns.\ \ \pi b\,\sqrt{a^2 + b^2}.$$

REMARK.—It is evident that the projection of the convex surface of a right circular cone on the plane of its base, is equal to the base; hence it follows (Anal. Geom., Art. 168) that the convex surface of a right circular cone is equal to the area of its base multiplied by the secant of the angle between the slant height and the base. Thus, calling this angle α, we have in the above example,

$$S = \pi b^2 \sec \alpha = \pi b^2 \frac{\sqrt{a^2 + b^2}}{b} = \pi b\,\sqrt{a^2 + b^2},$$

which agrees with the answer.

2. Find the area of the surface generated by the revolution of a logarithmic curve, $y = e^x$, about the axis of x, between the y-limits 0 and y.

$$A\,ns.\ \ \pi \left\{ y\,(1 + y^2)^{\frac{1}{2}} + \log\left[y + (1 + y^2)^{\frac{1}{2}} \right] \right\}.$$

3. Find the area of the surface generated by the revolution of the cycloid (*1*) about its base, and (*2*) about the tangent at the highest point.

Ans. (*1*) $\frac{64}{3}\pi a^2$; (*2*) $\frac{32}{3}\pi a^2$.

4. Find the area of the surface generated by the revolution of the catenary about the axis of y, between the x-limits 0 and x. *Ans.* $2\pi [xs - a (y - a)]$.

By Art. 177, $$s = \frac{a}{2}\left(e^{\frac{x}{a}} - e^{-\frac{x}{a}}\right);$$

$$\therefore \ S = 2\pi \int_0^x x\,ds = 2\pi \left[xs - \int s\,dx\right]_0^x,$$

from which we soon obtain the answer.

5. Find the area of the surface of a spherical sector, the vertical angle being 2α and the radius of the sphere $= r$.

Ans. $4\pi r^2 \left(\sin \dfrac{\alpha}{2}\right)^2.$

6. Find the area of the surface generated by the revolution of a loop of the lemniscate about its axis, the equation being $r^2 = a^2 \cos 2\theta$. *Ans.* $\pi a^2 (2 - 2^{\frac{1}{2}})$.

Here find $r\,ds = a^2 d\theta$; $\therefore$ etc.

7. Find the area of the surface generated by the revolution of a loop of the lemniscate about its axis, the equation being $r^2 = a^2 \sin 2\theta$. *Ans.* $2\pi a^2$.

8. A sphere is cut by a right circular cylinder, the radius of whose base is half that of the sphere, and one of whose edges passes through the centre of the sphere. Find the area of the surface of the sphere intercepted by the cylinder.

Let the cylinder be perpendicular to the plane of xy; then the equations of the cylinder and the sphere are respectively $y^2 = ax - x^2$ and $x^2 + y^2 + z^2 = a^2$. It is easily seen that the y-limits are 0 and $\sqrt{ax - x^2} = y_1$, and the x-limits are 0 and a. Therefore, Art. 201, we have

$$S = \int_0^a \int_0^{y_1} \frac{a\, dx\, dy}{\sqrt{a^2 - x^2 - y^2}}$$

$$= a \int_0^a \sin^{-1} \frac{(ax - x^2)^{\frac{1}{2}}}{(a^2 - x^2)^{\frac{1}{2}}}\, dx$$

$$= a \int_0^a \sin^{-1} \left(\frac{x}{a + x}\right)^{\frac{1}{2}} d\,(a + x)$$

$$= a \left[(a + x) \sin^{-1} \left(\frac{x}{a + x}\right)^{\frac{1}{2}} - \sqrt{ax} \right]_0^a \text{ (Art. 147)}$$

$$= a^2 \left(\frac{\pi}{2} - 1\right).$$

Therefore, the whole surface $= 2a^2 (\pi - 2)$. (In Price's Calculus, Vol. II, p. 326, the answer to this example is $a^2 (\pi - 2)$, which is evidently only half of what it should be.)

9. In the last example, find the area of the surface of the cylinder intercepted by the sphere.

Eliminating y, we have $z = \sqrt{a^2 - ax}$ for the equation of the projection on the plane xz of the intersection of the sphere and the cylinder. Therefore the z-limits are 0 and $z_1 = \sqrt{a^2 - ax}$, and the x-limits are 0 and a; hence, Art. 201, we have

$$[dx^2\, dy^2 + dy^2\, dz^2 + dz^2\, dx^2]^{\frac{1}{2}} = \left[1 + \left(\frac{dy}{dx}\right)^2 + \left(\frac{dy}{dz}\right)^2\right]^{\frac{1}{2}} dx\, dz$$

$$= \frac{a\, dx\, dz}{2\sqrt{ax - x^2}} \text{ for an element of the surface of the cylinder.}$$

$$\therefore \quad S = \frac{a}{2} \int_0^a \int_0^{z_1} \frac{dx\, dz}{\sqrt{ax - x^2}} = \frac{a^{\frac{3}{2}}}{2} \int_0^a \frac{dx}{x^{\frac{1}{2}}} = a^2;$$

therefore the whole area of the intercepted surface of the cylinder is $4a^2$. (See Gregory's Examples, p. 436.)

10. The axes of two equal right circular cylinders intersect at right angles; find the area of the one which is intercepted by the other. *Ans.* $8a^2$.

Let the axes of the two cylinders be taken as the axes of y and z, and let $a =$ the radius of each cylinder. Then the equations are

$$x^2 + z^2 = a^2, \qquad x^2 + y^2 = a^2.$$

11. A sphere is pierced perpendicularly to the plane of one of its great circles by two right cylinders, of which the diameters are equal to the radius of the sphere and the axes pass through the middle points of two radii that compose a diameter of this great circle. Find the surface of that portion of the sphere not included within the cylinders.

Ans. Twice the square of the diameter of the sphere.

12. Find the area of the surface generated by the revolution of the tractrix round the axis of x. *Ans.* $4\pi a^2$.

13. If a right circular cone stand on an ellipse, show that the convex surface of the cone is

$$\frac{\pi}{2} (OA + OA') (OA \cdot OA')^{\frac{1}{2}} \sin \alpha,$$

where O is the vertex of the cone, A and A' the extremities of the major axis of the ellipse, and α is the semi-angle of the cone at the vertex. (See Remark to Ex. 1.)

CHAPTER IX.

VOLUMES OF SOLIDS.

203. Solids of Revolution.—Let the curve AB, Fig. 55, revolve round the axis of x, and let V denote the volume of the solid bounded by the surface generated by the curve and by two planes perpendicular to the axis of x, one through A and the other through P ; then as MP and NQ are consecutive ordinates, the volume generated by the revolution of MPQN round the axis of x is an infinitesimal element of the whole volume, and is the frustum of a cone, the circumferences of whose bases are $2\pi y$ and $2\pi (y + dy)$, and whose altitude is $MN = dx$; therefore we have

$$dV = \frac{\pi y^2 + \pi (y + dy)^2 + \pi y (y + dy)}{3} dx = \pi y^2 dx,$$

by omitting infinitesimals of the second order. Hence, for the whole volume generated by the area between the two ordinates whose abscissas are a and b, where $a > b$, we have

$$V = \int_b^a \pi y^2 dx.$$

In like manner, it may be shown that to find the volume generated by revolving the arc round the axis of y, we have

$$V = \pi \int x^2 dy.$$

204. The Sphere.—Taking the origin at the centre of the sphere, we have $y^2 = a^2 - x^2$; therefore we have

$$V = \pi \int_{-a}^a (a^2 - x^2) \, dx = \left[\pi \left(a^2 x - \tfrac{1}{3} x^3\right) \right]_{-a}^a = \tfrac{4}{3}\pi a^3,$$

for the whole volume of the sphere.

COR. 1.—To find the volume of a spherical segment between two parallel planes, let b and c represent the distances of these planes from the centre; then we have

$$V = \pi \int_c^b (a^2 - x^2)\, dx = \pi \left[a^2 (b - c) - \tfrac{1}{3} (b^3 - c^3) \right].$$

COR. 2.—To find the volume of a spherical segment with one base, let h be the altitude of the segment; then $b = a$ and $c = a - h$, and we have

$$V = \pi \int_{a-h}^a (a^2 - x^2)\, dx = \pi h^2 \left(a - \frac{h}{3} \right).$$

COR. 3. $\tfrac{4}{3}\pi a^3 = \tfrac{2}{3}$ of $\pi a^2 \times 2a = \tfrac{2}{3}$ of the circumscribed cylinder.   (See Art. 194, Remark.)

205. The Volume generated by the Revolution of the Cycloid about its Base.

Here $dx = \dfrac{y\,dy}{\sqrt{2ry - y^2}}$ (Art. 176);

and integrating between the limits $y = 0$ and $y = 2r$, we find for the whole volume

$$V = 2\pi \int_0^{2r} \frac{y^3 dy}{\sqrt{2ry - y^2}}$$

$$= 2\pi \tfrac{4}{3}r \int_0^{2r} \frac{y^2 dy}{\sqrt{2ry - y^2}} \quad \text{(by Ex. 6, Art. 151)}$$

$$= \tfrac{10}{3}r\pi \left(\tfrac{3}{2}r^2\pi \right) \quad \text{(by Ex. 6, Art. 151)}$$

$$= 5\pi^2 r^3.$$

We have $5\pi^2 r^3 = \tfrac{5}{8}\pi (2r)^2 \times 2\pi r$. .

Hence, *the volume generated by the revolution of the cycloid about its base is equal to five-eighths the circumscribing cylinder.*

206. The Cissoid when it revolves round its Asymptote.—Here $OM = x$, $MP = y$, $OA = 2a$, $MA = 2a - x$, $HD = dy$; hence an infinitesimal element of the whole volume is generated by the revolution of PQDH about AT, and is represented by $\pi (2a - x)^2\, dy$.

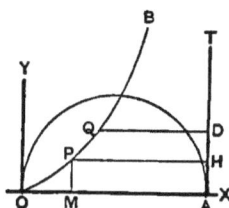
Fig. 57.

The equation of the Cissoid is

$$y^2 = \frac{x^3}{2a - x},$$

$$\therefore\ dy = \frac{(3a - x)\,(2ax - x^2)^{\frac{1}{2}}}{(2a - x)^2}\, dx;$$

hence, between the limits $x = 0$ and $x = 2a$, we have

$$V = 2\pi \int_0^{2a} (2a - x)^2\, dy = 2\pi \int_0^{2a} (3a - x)\,(2ax - x^2)^{\frac{1}{2}}\, dx$$

$$= 2\pi \int_0^{2a} \frac{6a^2x - 5ax^2 + x^3}{\sqrt{2ax - x^2}}\, dx = 2\pi^2 a^3$$

(by Ex. 6, Art. 151).

207. Volume of Solids bounded by any Curved Surface.—Let (x, y, z) and $(x + dx,\, y + dy,\, z + dz)$ be two consecutive points E and F within the space whose volume is to be found. Through E pass three planes parallel to the co-ordinate planes xy, yz, and zx; also through F pass three planes parallel to the first. The solid included by these six planes is an infinitesimal rectangular parallelopipe-don, of which E and F are two opposite angles, and the volume is $dx\, dy\, dz$; the aggregate of all these solids between

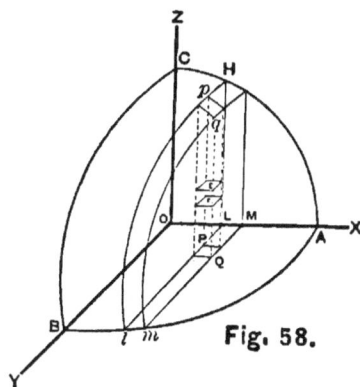
Fig. 58.

the limits assigned by the problem is the required volume. Hence, if V denote the required volume, we have

$$V = \int \int \int dx \, dy \, dz,$$

the integral being taken between proper limits.

In considering the effects of these successive integrations, let us suppose that we want the volume in Fig. 58 contained within the three co-ordinate planes.

The effect of the z-integration, x and y remaining constant, is the determination of the volume of an infinitesimal prismatic column, whose base is $dx dy$, and whose altitude is given by the equations of the bounding surfaces ; thus, in Fig. 58, if the equation of the surface is $z = f(x, y)$, the limits of the z-integration are $f(x, y)$ and 0, and the volume of the prismatic column whose height is Pp is $f(x, y) \, dx \, dy$; hence the integral expressing the volume is now a double integral and of the form

$$V = \int \int f(x, y) \, dx \, dy.$$

If we now integrate with respect to y, x remaining constant, we sum up the prismatic columns which form the elemental slice H$plmq$, contained between two planes perpendicular to the axis of x, and at an infinitesimal distance (dx) apart. The limits of y are Ll and 0, Ll being the y to the trace of the surface on the plane of xy, and which may therefore be found in terms of x by putting $z = 0$ in the equation of the surface ; or, if the volume is included between two planes parallel to that of xz, and at distances y_0 and y_1 from it, y_0 and y_1 being constants, they are in that case the limits of y ; in the same way we find the limits if the bounding surface is a cylinder whose generating lines are parallel to the axis of z. In each of these cases the result of the y-integration is the volume of a slice included between two planes at an infinitesimal distance apart, the length of which, measured parallel to the axis of y, is a function of its distance from the plane of yz; thus the limits of the y-integration may be functions of x, and we shall have

$$V = \int \int f(x, y)\, dx\, dy = \int F(x)\, dx,$$

where $F(x)\, dx$ is the infinitesimal slice perpendicular to the axis of x at a distance x from the origin, and the sum of all such infinitesimal slices taken between the assigned limits is the volume. Thus, if the volume in Fig. 58 between the three co-ordinate planes is required, and $OA = a$, then the x-limits are a and 0. If the volume contained between two planes at distances x_0 and x_1 from the plane of yz is required, then the x-limits are x_0 and x_1.

EXAMPLES.

1. The ellipsoid whose equation is

$$\frac{x^2}{a^2} + \frac{y^2}{b^2} + \frac{z^2}{c^2} = 1.$$

Here the limits of the z-integration are $c\left(1 - \dfrac{x^2}{a^2} - \dfrac{y^2}{b^2}\right)^{\frac{1}{2}}$ and 0, which call z_1 and 0; the limits of y are $Ll = b\left(1 - \dfrac{x^2}{a^2}\right)^{\frac{1}{2}}$ and 0, which call y_1 and 0; the x-limits are a and 0.

First integrate with respect to z, and we obtain the infinitesimal prismatic column whose base is PQ, Fig. 58, and whose height is Pp. Then we integrate with respect to y, and obtain the sum of all the columns which form the elemental slice H$plmq$. Then integrating with respect to x, we obtain the sum of all the slices included in the solid OABC.

$$\therefore \quad V = 8 \int_0^a \int_0^{y_1} \int_0^{z_1} dx\, dy\, dz$$

17

$$= 8c \int_0^a \int_0^{y_1} \left(1 - \frac{x^2}{a^2} - \frac{y^2}{b^2}\right)^{\frac{1}{2}} dx\, dy$$

$$= \frac{8c}{b} \int_0^a \int_0^{y_1} (y_1^2 - y^2)^{\frac{1}{2}} dx\, dy$$

$$= \frac{8c}{b} \int_0^a \left[\frac{y}{2}(y_1^2 - y^2)^{\frac{1}{2}} + \frac{y_1^2}{2} \sin^{-1}\frac{y}{y_1}\right]_0^{y_1} dx$$

$$= \frac{8c}{b} \int_0^a \frac{y_1^2}{2}\frac{\pi}{2}\, dx$$

$$= \frac{2\pi cb}{a^2} \int_0^a (a^2 - x^2)\, dx = \tfrac{4}{3}\pi abc.$$

2. The volume of a right elliptic cylinder whose axis coincides with the axis of x and whose altitude $= 2a$, the equation of the base being

$$c^2 y^2 + b^2 z^2 = b^2 c^2.$$

Here the z-limits are $\frac{c}{b}(b^2 - y^2)^{\frac{1}{2}}$ and 0, which call z_1 and 0; the y-limits are b and 0; the x-limits are a and 0.

$$\therefore \quad V = 8\int_0^a \int_0^b \int_0^{z_1} dx\, dy\, dz$$

$$= 8\frac{c}{b} \int_0^a \int_0^b (b^2 - y^2)^{\frac{1}{2}} dx\, dy$$

$$= 8\frac{c}{b} \int_0^a \left[\frac{y}{2}(b^2 - y^2)^{\frac{1}{2}} + \frac{b^2}{2} \sin^{-1}\frac{y}{b}\right]_0^b dx$$

$$= 8\frac{cb\pi}{4} \int_0^a dx = 2abc\pi.$$

(See Price's Cálculus, Vol. II, p. 356.)

3. The volume of the solid cut from the cylinder $x^2 + y^2 = a^2$ by the planes $z = 0$ and $z = x \tan \alpha$.

Here the z-limits are $x \tan \alpha$ and 0, or z_1 and 0; the y-limits are $(a^2 - x^2)^{\frac{1}{2}}$ and $-(a^2 - x^2)^{\frac{1}{2}}$, or y_1 and $-y_1$; the x-limits are a and 0.

$$\therefore \quad V = \int_0^a \int_{-y_1}^{y_1} \int_0^{z_1} dx\, dy\, dz$$

$$= \int_0^a \int_{-y_1}^{y_1} (x \tan \alpha)\, dx\, dy$$

$$= 2 \tan \alpha \int_0^a x\, (a^2 - x^2)^{\frac{1}{2}}\, dx = \tfrac{2}{3}\, a^3 \tan \alpha.$$

4. The volume of the solid common to the ellipsoid $\dfrac{x^2}{a^2} + \dfrac{y^2}{b^2} + \dfrac{z^2}{c^2} = 1$ and the cylinder $x^2 + y^2 = b^2$.

Here the limits of the z-integration are $c\left(1 - \dfrac{x^2}{a^2} - \dfrac{y^2}{b^2}\right)^{\frac{1}{2}}$ and 0, or z_1 and 0; the limits of the x-integration are $(b^2 - y^2)^{\frac{1}{2}}$ and 0, or x_1 and 0; the y-limits are 0 and b.*

$$\therefore \quad V = 8 \int_0^b \int_0^{x_1} \int_0^{z_1} dy\, dx\, dz$$

$$= 8c \int_0^b \int_0^{x_1} \left[1 - \frac{x^2}{a^2} - \frac{y^2}{b^2}\right]^{\frac{1}{2}} dy\, dx$$

$$= \frac{8c}{a} \int_0^b \left[\frac{x}{2}\left(a^2 - \frac{a^2 y^2}{b^2} - x^2\right)^{\frac{1}{2}} + \frac{a^2 - \dfrac{a^2}{b^2} y^2}{2} \sin^{-1} \frac{x}{a\left(1 - \dfrac{y^2}{b^2}\right)^{\frac{1}{2}}}\right]_0^{(b^2 - y^2)^{\frac{1}{2}}} dy$$

$$= \frac{4c}{a} \int_0^b \left\{ (b^2 - y^2)^{\frac{1}{2}} \left[a^2 - \frac{a^2}{b^2} y^2 - (b^2 - y^2)\right]^{\frac{1}{2}} \right.$$

$$\left. + a^2 \left(1 - \frac{y^2}{b^2}\right) \sin^{-1} \frac{b}{a} \right\} dy$$

$$= 4ac \int_0^b \left[\frac{(b^2 - y^2)\, (a^2 - b^2)^{\frac{1}{2}}}{a^2 b} + \left(1 - \frac{y^2}{b^2}\right) \sin^{-1} \frac{b}{a}\right] dy$$

$$= 4ac \left[\frac{b}{a^2} (a^2 - b^2)^{\frac{1}{2}} + \sin^{-1} \frac{b}{a}\right] \int_0^b \left(1 - \frac{y^2}{b^2}\right) dy$$

---

* In this example, this order of integration is simpler than it would be to take it with respect to y and then x.

$$= \tfrac{8}{3}abc\left[\frac{b}{a^2}(a^2 - b^2)^{\frac{1}{2}} + \sin^{-1}\frac{b}{a}\right].$$

(See Mathematical Visitor, 1878, p. 26.)

208. Mixed System of Co-ordinates.—Instead of dividing a solid into columns standing on *rectangular* bases, so that $z\,dx\,dy$ is the volume of the infinitesimal column, it is sometimes more convenient to divide it into infinitesimal columns standing on the *polar* element of area $abcd = r\,dr\,d\theta$, in which case

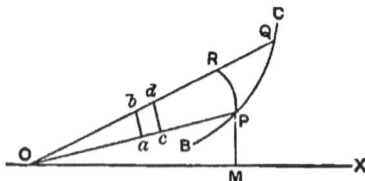

Fig. 59.

the corresponding parallelopipedon is represented by $zr\,dr\,d\theta$, and the expression for V becomes

$$V = \int\int zr\,dr\,d\theta,$$

taken between proper limits. From the equation of the surface, z must be expressed as a function of r and θ.

EXAMPLES.

1. Find the volume included between the plane $z = 0$, and the surfaces $x^2 + y^2 = 4az$ and $y^2 = 2cx - x^2$.

Here $z = \dfrac{x^2 + y^2}{4a} = \dfrac{r^2}{4a}$; hence the z-limits are $\dfrac{r^2}{4a}$ and 0. The equation of the circle $y^2 = 2cx - x^2$, in polar co-ordinates, is $r = 2c\cos\theta$; hence the r-limits are 0 and $2c\cos\theta$, or 0 and r_1; and the θ-limits are 0 and $\dfrac{\pi}{2}$.

$$\therefore \quad V = 2\int_0^{\frac{1}{2}\pi}\int_0^{r_1}\frac{r^3}{4a}\,d\theta\,dr$$

$$= \frac{2c^4}{a}\int_0^{\frac{1}{2}\pi}\cos^4\theta\,d\theta = \frac{2c^4}{a}\cdot\tfrac{3}{16}\pi. \quad \text{(Ex. 4, Art. 157.)}$$

$$= \frac{3}{8}\frac{\pi c^4}{a}.$$

2. The axis of a right circular cylinder of radius b, passes through the centre of a sphere of radius a, when $a > b$; find the volume of the solid common to both surfaces.*

Take the centre of the sphere as origin, and the axis of the cylinder as the axis of z; then the equations of the surfaces are $x^2 + y^2 + z^2 = a^2$ and $x^2 + y^2 = b^2$; or, in terms of polar co-ordinates, the equation of the cylinder is $r = b$.

Hence for the volume in the first octant, the z-limits are $\sqrt{a^2 - x^2 - y^2}$ or $\sqrt{a^2 - r^2}$ and 0; the r-limits are b and 0; the θ-limits are $\dfrac{\pi}{2}$ and 0.

$$\therefore\ V = 8 \int_0^{\frac{1}{2}\pi} \int_0^b z r\, dr\, d\theta$$

$$= 8 \int_0^{\frac{1}{2}\pi} \int_0^b r\, (a^2 - r^2)^{\frac{1}{2}}\, d\theta\, dr$$

$$= -\frac{8}{3} \int_0^{\frac{1}{2}\pi} \left[(a^2 - r^2)^{\frac{3}{2}} \right]_0^b d\theta$$

$$= \frac{4\pi}{3} \left[a^3 - (a^2 - b^2)^{\frac{3}{2}} \right].$$

(See Gregory's Examples, p. 428.)

209. The polar element of plane area is $r\, dr\, d\theta$ (Art. 208). Let this element revolve round the initial line through the angle 2π, it will generate a solid ring whose volume is $2\pi r \sin \theta r\, dr\, d\theta$, since $2\pi r \sin \theta$ is the circumference of the circle described by the point (r, θ). Let ϕ denote the angle which the plane of the element in any position makes with the initial position of the plane; then $d\phi$ is the angle which the plane in any position makes

---

* This example, as well as the preceding one, might be integrated directly in terms of x and y by the method of Art. 207, but the operation would be more complex than the one adopted.

with its consecutive position. The part of the solid ring which is intercepted between the revolving plane in these two consecutive positions, is to the whole ring in the same proportion as $d\phi$ is to 2π. Hence the volume of this intercepted part is

$$r^2 \sin\theta \, d\phi \, d\theta \, dr,$$

which is therefore an expression in polar co-ordinates for an infinitesimal element of any solid. Hence, for the volume of the whole solid we have

$$V = \int \int \int r^2 \sin\theta \, d\phi \, d\theta \, dr,$$

in which the limits of the integration must be so taken as to include all the elements of the proposed solid. In this formula r denotes the distance of any point from the origin, θ denotes the angle which this distance makes with some fixed right line through the origin (the initial line), and ϕ denotes the angle which the plane passing through this distance and the initial line makes with some fixed plane passing through the initial line. (See Lacroix Calcul Intégral, Vol. II, p. 209.)

The order in which the integrations are to be effected is theoretically arbitrary, but in most cases the form of the equations of surfaces makes it most convenient to integrate first with respect to r; but the order in which the θ- and ϕ-integrations are effected is arbitrary.

EXAMPLES.

1. The volume of the octant of a sphere. Let $a =$ the radius of the sphere; then the limits of r are 0 and a; hence,

$$V = \int \int \frac{a^3}{3} \sin\theta \, d\phi \, d\theta.$$

In thus integrating with respect to r, we collect all the elements like $r^2 \sin\theta \, d\phi \, d\theta \, dr$ which compose a pyramidal solid, having its vertex at the centre of the sphere, and for its base the curvilinear element of spherical surface which is denoted by $a^2 \sin \theta \, d\phi \, d\theta$.

Integrating next with respect to θ between the limits 0 and $\frac{\pi}{2}$, we have

$$V = \int \frac{a^3}{3} \Big[(-\cos \theta) \Big]_0^{\frac{1}{2}\pi} d\phi = \int \frac{a^3}{3} \, d\phi.$$

In thus integrating with respect to θ, we collect all the pyramids similar to $\frac{a^3}{3} \sin \theta \, d\phi \, d\theta$, which form a wedge-shaped slice of the solid contained between two consecutive planes through the initial line.

Lastly, integrating with respect to ϕ from 0 to $\frac{\pi}{2}$, we have

$$V = \frac{\pi a^3}{6}. \quad \text{(See Todhunter's Int. Cal., p. 183.)}$$

In this example the order of the integrations is immaterial.

2. The volume of the solid common to a sphere of radius a, and the right circular cone whose vertical angle is 2α and whose vertex is at the centre of the sphere.

Here the r-limits are 0 and a, the θ-limits are 0 and α, the ϕ-limits are 0 and 2π.

$$\therefore \quad V = \int_0^{2\pi} \int_0^{\alpha} \int_0^{a} r^2 \sin \theta \, d\phi \, d\theta \, dr$$

$$= \int_0^{2\pi} \int_0^{\alpha} \frac{a^3}{3} \sin \theta \, d\phi \, d\theta$$

$$= \int_0^{2\pi} \frac{a^3}{3} (1 - \cos \alpha) \, d\phi$$

$$= \tfrac{2}{3}\pi a^3 (1 - \cos \alpha).$$

1. Find the volume of a paraboloid of revolution whose altitude $= a$ and the radius of whose base $= b$.

$$Ans. \ \frac{\pi}{2} \, ab^2.$$

2. Find the volume of the prolate spheroid. Also of the oblate spheroid. *Ans.* The prolate spheroid $= \frac{4}{3}\pi ab^2$.
The oblate spheroid $= \frac{4}{3}\pi a^2 b$.

3. Find the volume of the solid generated by the revolution of $y = a^x$ about the axis of x, between the limits x and $-\infty$, where $a > 1$.

$$Ans. \ \frac{\pi}{2} \, a^{2x} (\log a)^{-1}.$$

4. Find the volume of the solid generated by the revolution of $y = a \log x$ about the axis of x, between the limits x and 0. *Ans.* $\pi a^2 x (\log^2 x - 2 \log x + 2)$.

5. Find the volume of the solid generated by the revolution of the tractrix round the axis of x. *Ans.* $\frac{2}{3}\pi a^3$.

6. Find the volume of the solid generated by the revolution of the catenary round the axis of x.

$$Ans. \ \frac{\pi}{2} \, a \, (ys + ax). \quad \text{(Compare with Art. 197.)}$$

7. Find the volume generated by the revolution of a parabola about its base $2b$, the height being h.[*] (See Art. 206.) *Ans.* $\frac{16}{15}\pi bh^2$.

8. The equation of the Witch of Agnesi being

$$x^2 = 4a^2 \, \frac{2a - y}{y},$$

find the volume of the solid generated by its revolution round the asymptote. *Ans.* $4\pi^2 a^3$.

---

[*] This solid is called a parabolic spindle.

9. Find the volume of a rectangular parallelopipedon, three of whose edges meeting at a point are a, b, c. (See Art. 207.)

Ans. abc.

10. Find the volume contained within the surface of an elliptic paraboloid * whose equation is

$$\frac{y^2}{a} + \frac{z^2}{b} = 2x,$$

and a plane parallel to that of yz, and at a distance c from it.

Ans. $\pi c^2 (ab)^{\frac{1}{2}}$.

11. The axes of two equal right circular cylinders intersect at right angles, their equations being $x^2 + z^2 = a^2$ and $x^2 + y^2 = a^2$; find the volume of the solid common to both.

Ans. $\frac{16}{3}a^3$.

12. A paraboloid of revolution is pierced by a right circular cylinder, the axis of which passes through the focus and cuts the axis of the paraboloid at right angles, the radius of the cylinder being one-fourth the latus-rectum of the generating parabola; find the volume of the solid common to the two surfaces.

Ans. $p^3 \left(\frac{2}{3} + \frac{\pi}{4} \right)$.

Here the equations of the surfaces are

$$y^2 + z^2 = 2px \qquad \text{and} \qquad x^2 + y^2 = px.$$

13. Find the volume of the solid cut from the cylinder $x^2 + y^2 = 2ax$ by the planes $z = x \tan \alpha$ and $z = x \tan \beta$.

Ans. $2 (\tan \beta - \tan \alpha) \dfrac{\pi a^3}{2}$.

14. Find the volume of the solid common to both surfaces in Ex. 8 of Art. 202. (See Art. 208.)

Ans. $\frac{2}{3} (3\pi - 4) a^3$.

15. Find the volume of the part of the hemisphere in the last example, which is not comprised in the cylinder.

Ans. $\frac{2}{3}a^3$.

---

* Called *elliptic paraboloid* because the sections made by planes parallel to the planes of xy and xz are parabolas, while those parallel to the plane of yz are ellipses. (Salmon's Anal. Geom. of Three Dimensions, p. 58.)

16. Find the volume of the solid intercepted between the concave surface of the sphere and the convex surface of the cylinder in Art. 208, Ex. 2. $\qquad$ *Ans.* $\frac{4}{3}\pi (a^2 - b^2)^{\frac{3}{2}}$.

17. Find the volume of the solid comprised between the surface $z = ae^{-\frac{x^2+y^2}{c^2}}$ and the plane of xy. $\qquad$ *Ans.* πac^2.

Here the r-limits are 0 and ∞; and the θ-limits are 0 and 2π.

18. Find the volume of the solid generated by the revolution of the cardioide $r = a (1 + \cos \theta)$ about the initial line.

Here $V = \int_{0}^{\pi} \int_{0}^{2\pi} \int_{0}^{a (1+\cos \theta)} r^2 \sin \theta \, d\theta \, d\phi \, dr = $ etc.

(See Art. 209.) $\qquad$ *Ans.* $\dfrac{8\pi a^3}{3}$.

19. Find the volume of the solid generated by the revolution of the Spiral of Archimedes, $r = a\theta$, about the initial line between the limits $\theta = \pi$ and $\theta = 0$.

Ans. $\frac{2}{3}\pi^2 a^3 (\pi^2 - 6)$.

20. A right circular cone whose vertical angle $= 2\alpha$, has its vertex on the surface of a sphere of radius a, and its axis coincident with the diameter of the sphere; find the volume common to the cone and the sphere.

Ans. $\dfrac{4\pi a^3}{3} (1 - \cos^4 \alpha)$.

21. Find the volume of a chip cut at an angle of $45°$ to the centre of a round log with radius r. (Mathematical Visitor, 1880, p. 100.) $\qquad$ *Ans.* $\frac{2}{3}r^3$.

22. Find the volume bounded by the surface

$$\left(\frac{x}{a}\right)^{\frac{1}{2}} + \left(\frac{y}{b}\right)^{\frac{1}{2}} + \left(\frac{z}{c}\right)^{\frac{1}{2}} = 1$$

and the positive sides of the three co-ordinate planes.

Ans. $\dfrac{abc}{90}$.

23. Find the volume of the solid bounded by the three surfaces $x^2 + y^2 = cz$, $x^2 + y^2 = ax$, and $z = 0$.

$$Ans. \; \frac{3\pi a^4}{32c}.$$

24. A paraboloid of revolution and a right circular cone have the same base, axis, and vertex, and a sphere is described upon this axis as diameter. Show that the volume intercepted between the paraboloid and cone bears the same ratio to the volume of the sphere that the latus-rectum of the parabola bears to the diameter of the sphere.

25. Find the volume included between a right circular cone whose vertical angle is 60° and a sphere of radius r touching it along a circle, by the formula

$$V = \int\int\int dx \, dy \, dz.$$

$$Ans. \; \frac{\pi r^3}{6}.$$

26. In the right circular cone given in Ex. 13 of Art. 202, prove that its volume is represented by

$$\frac{\pi}{3} (OA \cdot OA')^{\frac{3}{2}} \sin^2 \alpha \cos \alpha.$$

www.ingramcontent.com/pod-product-compliance
Lightning Source LLC
Chambersburg PA
CBHW021349210326
41599CB00011B/815